# A TREATISE ON THE PRINCIPLES AND PRACTICE OF HARBOUR ENGINEERING

**Brysson Cunningham**

Author of "Dock Engineering," ETC



MAXWELL PRESS

Chennai Trichy New Delhi

MAXWELL PRESS

This edition has been published in india by arrangement with Carson Books, UK

ISBN 978-93-5528-085-5

**Maxwell Press**
No. 44, Nallathambi Street,
Triplicane, Chennai 600 005

MJP 1295

Publisher : C. Janarthanan

# Publisher's Note

The legacy of a country is in its varied cultural heritage, historical literature, developments in the field of economy and science. The top nations in the world are competing in the field of science, economy and literature. This vast legacy has to be conserved and documented so that it can be bestowed to the future generation. The knowledge of this legacy is slowly getting perished in the present generation due to lack of documentation.

Keeping this in mind, the concern with retrospective acquiring of rare books has been accented recently by the burgeoning reprint industry. Maxwell Press is gratified to retrieve the rare collections with a view to bring back those books that were landmarks in their time.

In this effort, a series of rare books would be republished under the banner, "Maxwell Press". The books in the reprint series have been carefully selected for their contemporary usefulness as well as their historical importance within the intellectual. We reconstruct the book with slight enhancements made for better presentation, without affecting the contents of the original edition.

Most of the works selected for republishing covers a huge range of subjects, from history to anthropology. We believe this reprint edition will be a service to the numerous researchers and practitioners active in this fascinating field. We allow readers to experience the wonder of peering into a scholarly work of the highest order and seminal significance.

**MAXWELL PRESS**

# PREFACE.

The favourable reception accorded to his treatise on *Dock Engineering* led the author to consider the possibility of supplementing that work by a companion volume on the kindred subject of *Harbour Engineering*.

The principal difficulty lay in making such a treatise complete and self-contained without recapitulating a good deal of material which had appeared in the earlier book, and which was equally essential in the present instance, the two departments of maritime engineering having so much in common as to be generally practised together.

The most acceptable solution which suggested itself consisted in treating this common material in a somewhat different way from that previously adopted, by presenting fresh points of view, additional features of interest, and new illustrations. This plan has been followed, and it is hoped that it will meet with approval.

Once more the author has to express his indebtedness to a number of personal and professional friends who have rendered him kind and valued assistance in the execution of his task; to the writers of papers and to the societies who have courteously allowed him to make extracts from the various minutes of proceedings alluded to in the text; and to the editors of *Engineering* and *The Engineer* for much useful information gleaned from the columns of their respective journals.

As in the case of *Dock Engineering*, every care has been taken in regard to the accuracy of data and statistics, but mistakes are always possible, and any intimation of their presence will be gratefully appreciated.

BRYSSON CUNNINGHAM.

London,
*January* 1908.

# CONTENTS.

## CHAPTER I.

### INTRODUCTORY.



## CHAPTER II.

### HARBOUR DESIGN.



## CHAPTER III.

### SURVEYING, MARINE AND SUBMARINE.



## CHAPTER IV.

### PILING.



## CHAPTER V.

### STONE: NATURAL AND ARTIFICIAL.

## CHAPTER VI.

### BREAKWATER DESIGN.



## CHAPTER VII.

### BREAKWATER CONSTRUCTION.



## CHAPTER VIII.

### PIERHEADS, QUAYS, AND LANDING-STAGES.



## CHAPTER IX.

### ENTRANCE CHANNELS.



## CHAPTER X.

### CHANNEL DEMARCATION.

# LIST OF ILLUSTRATIONS.

---

Manufacturing Firms who have kindly furnished Illustrations of Plant and Apparatus:—

Siebe, Gorman & Co., London: Diving Apparatus.
The Ingersoll Sergeant Drill Co., London: Drills.
Nobel's Explosives Co., Glasgow: Explosives.
W. H. Bailey & Co., Manchester: Dynamometers.
Stothert & Pitt, Bath: Block-setting Plant.
Chance & Co., Birmingham: Lighthouses and Illuminating Apparatus.
Brown, Lenox & Co., London: Mooring Buoys.
Pintsch & Co., London: Channel Buoys and Lightships.

---

# LIST OF TABLES.

# HARBOUR ENGINEERING.

## CHAPTER I.

### INTRODUCTORY.

Harbour Engineering and Navigation—Natural and Artificial Harbours—Ancient Sea Routes—Phœnician, Egyptian, Grecian, Carthaginian, and Roman Harbours—Mediæval Navigation—The Cinque Ports—The Hanseatic League—National Interest in Harbours—State Subvention.

**Harbour Engineering and Navigation.**—The history of harbour engineering runs concurrently, through corresponding stages from origin to development, with the history of navigation. Nor is the fact at all surprising. From the very nature of the case little else could be expected, since the two sciences stand to one another in the closest inter-relationship of cause and effect. With the appearance on the seas of the first craft calling for the exercise of expert seamanship, there arose a need of havens in which it might not only find shelter during stress of weather, but also take in and discharge its cargoes under suitable conditions. And as vessels gradually increased in number, size, and importance, so the need for more spacious accommodation became the more pressing, and the demand for larger and better harbours the more imperative.

**Natural Harbours.**—Of natural creeks and basins, possessing intrinsically all the advantages which a haven of safe anchorage requires, there are in the world not a few, and, no doubt, at the outset they abundantly sufficed for the rudimentary necessities of the early mariner. But the accommodation afforded was in many cases limited, and, as time elapsed, it became less and less compatible with the exigencies of rapidly expanding navies, whether engaged in commerce or war. Neither did the situation of these inlets always prove convenient, more especially to trading vessels. Some of the most commodious of them are to be found far out of the track of well-established lines of communication, and away from the principal routes of over-sea trade. And of those which are conveniently accessible, few, if any, have realised the ideal of a completely sheltered haven. There has been almost invariably some inherent defect to be remedied, some deficiency to be made good. Accordingly, even the best of natural harbours have called for improvement and development, while, in the absence of natural advantages, steps have had to be taken to provide by artificial means the requisite degree of shelter and protection.

**Artificial Harbours.**—In every country, therefore, finding itself in the possession of a seaboard, and with any pretensions to maritime enterprise, there has been formed by degrees a series of artificial or semi-artificial enclosures, constructed at first somewhat crudely and informally, but later with full application of scientific method and technical skill.

There exists, indeed, no record on parchment, bronze, or stone to attest the date of the inception of the first artificial harbour. In the absence of any controverting evidence, the honour of creating the prototypes of modern maritime engineering undertakings of this kind is usually assigned to the public-spirited policy of monarchs of early Egyptian and Phœnician dynasties. Yet this ascription of priority is, after all, one more of inference and assumption than of definite knowledge, and there is reason to suspect that artificial harbours, in an embryonic form at any rate, are of much more ancient origin, dating back through the earlier civilisations of the remoter East to a period of time of which all historical traces have been lost, and concerning which it would, for that reason, be useless to inquire and idle to speculate.

**Ancient Sea Routes.**—We have no alternative but to confine our animadversions within the limits of fruitful historical research. It is known for a fact that both Egypt and Phœnicia possessed commercial navies, and that they carried on an elaborate system of trading operations. Their maritime traffic was not only characterised by regularity and importance, but it was also of a mutual nature, and the two countries were linked together by ties of common interest and advantage.

The sea trade of Western Asia and the contiguous portion of Africa was conducted in ancient times principally along two routes. The first of these led from the Phœnician ports, *via* Cyprus to Sicily and Malta, with an extension along the northern coast of Africa, finally reaching Tartessus in Spain, the site of which was probably near where Gibraltar now stands. The second sea route was from Ezion Geber at the head of the Gulf of Akabah, along the Red Sea, skirting the Southern coast of Arabia to Ophir at the mouth of the Indus, the "land of gold and precious stones."

Besides these two main routes, however, there were a number of subsidiary tracks intersecting one another in various directions. It has already been mentioned that close commercial relations existed between Tyre and Sidon and the deltaic ports of the Nile. Apart from these, there was considerable traffic which passed into the mouth of the Euphrates, both from the coasts of India and of Africa. Nor were the more adventurous spirits of the age restricted to beaten tracks. Right up to the Cassiterides (Scilly Isles) in the far West sailed the Cabots and Frobishers of that age, as also they may have circumnavigated India and penetrated to Burmah and the confines of Siam in the East.

**Phœnician Harbours.**—Artificial works were indubitably in existence at both Tyre and Sidon. The former town stood on a peninsula flanked on each side by a harbour formed of moles of loose or random rubble. Sidon possessed similar works of perhaps a little less extensive character. On the testimony of ancient historians, Tyre was a magnificent city and a flourishing port, with properly constructed quays, equipped with substantial warehouses, dating back between two and three thousand years prior to the commencement of the Christian era. The town underwent several vicissitudes in the course of its history, even to the extent of being destroyed by the princes of Assyria, and afterwards rebuilt. It fell finally at the hands of Alexander the Great, B.C. 332, and although the town of Sur still marks the site at the present day, scarcely a vestige of the glory of the ancient city remains, and the world-renowned harbours have sunk through successive stages of disrepair and decay to ruin.

**The Harbour of Alexandria.**—Remarkable as were the harbour works of Phœnicia, they were far outshone by the more elaborate undertakings at Alexandria, originated by the Conqueror of Tyre and brought to a successful conclusion under the first two Ptolemies about 200 B.C. In this case there was the customary mole inclosing a floating basin, and in addition thereto, the celebrated tower, or lighthouse, Pharos,[1] built on the island of that name. It passed for one of the seven wonders of the world, being constructed of white marble and visible at a distance of a hundred miles The cost of it has been variously estimated at £165,000, and at double that amount. Fires were kept constantly alight on the summit as an aid to the navigation of the bay.

**Grecian Harbours.**—The Greeks and cognate races were notable harbour engineers, and their handiwork was made manifest at Rhodes, Salamis, Corinth, Syracuse, and many other places. Perhaps the most noteworthy instance was Piræus, the harbour of Athens, situated at the mouth of the Cephisus, about three miles distant from the capital city. It was a most capacious harbour, inclosing three large basins called Cantharos, Aphrodisium, and Zea, and sufficiently commodious for the reception of a fleet of 400 ships.

**Harbour at Carthage.**—The Carthaginians, as might be expected from their blood relationship to the parent stock of the Phœnicians, also developed a talent for harbour construction, and they made Carthage a model port. It comprised two compartments inclosed by breakwaters and connected by a channel 70 feet in width. Around the inner basin, which afforded space for over 200 ships of war, were located the arsenals and stores. When Scipio blockaded the place in B.C. 146, cutting off communication with the sea by means of a dam across the entrance to the outer harbour, it is recorded that the Carthaginians, with characteristic energy, excavated for themselves a new outlet. Their exertions and strenuous defence were, however, without avail, and the downfall of the city took place shortly afterwards.

[1] The title *Pharology*, applied to the science of lighthouse construction, is derived from the name of this tower.

**Roman Harbours.**—At the close of the second Punic War and considerably prior to the event just narrated, the world sea-power had passed into the hands of the Romans, and with it the genius and capacity for harbour construction. Ostia, Ancona, Antium and Civita Vecchia, amongst hundreds of other instances, may be cited as evidences of this fact. Furthermore, the works carried out by the Latin race were of a solid and enduring character, which in many cases have defied alike the ravages of time, storm, and devastation. Civita Vecchia still possesses a serviceable harbour capable of receiving vessels of 20 feet draught. The works at Ostia also exist, although the town, formerly at the mouth of the Tiber, is now twenty miles or so inland.

**Mediæval Navigation.**—Passing to mediæval times, we find a vast expansion in maritime trade and a corresponding increase in the number and size of harbours. The whole of Europe was now engaged in avocations connected with the sea and embarking on nautical enterprises as adventurous as they were remunerative.

This is not the place, however, in which to attempt anything of the nature of a historical and analytical disquisition on the growth and expansion of seaborne commerce, nor even is there space to describe the provision made for its reception and accommodation at the various ports with which it was associated. We do not propose, therefore, to dwell further upon this part of the subject beyond making a brief allusion to two features of outstanding interest and importance, showing how closely the commercial and political welfare of a maritime country is bound up in the maintenance and development of its seaports and harbours.

**The Cinque Ports.**—The first of these features is the formation of a confederation, of which only a name and an office, and that a pure sinecure, remain at the present day. The Cinque Ports were, and for that matter are, as the name implies, a group of five ports in this country fronting the English Channel. The towns were originally Hastings, Romney, Hythe, Dover, and Sandwich, and subsequently there were added—Winchelsea, Rye, and Seaford. They represented the naval activity of this country, and they were responsible for the protection of the Kentish coast against the incursions of foreign foes. To this end they held certain levies of shipping constantly at the disposal of the crown, and, in return, they had conferred upon them several special distinctions and privileges.

At this distance of time, it is difficult (with, perhaps, one notable exception) to think of these insignificant villages as forming the forefront and backbone of England's naval power. Yet from the modest moorings and lowly quays of these Kentish harbours slipped away many a valiant little cog to confront the caravels of France and the galleons of Spain. But more than this, they were, in sooth, the very seed and nucleus of England's foreign trade; inferior, certainly, to London in importance, but, during their palmier days, vying with Bristol and Plymouth in the west in the honour and distinction of seaward enterprise, and forming the principal links in the chain connecting England with the Continent and with all the commercial products of the civilised world known at that epoch.

**The Hanseatic League.**—The other noteworthy feature was the Hanseatic League. This was an association of German cities inaugurated about the twelfth century, or perhaps earlier (for the real origin of the association is involved in some obscurity), for the protection and advancement of seaborne commerce generally, and more particularly to foster their own interests therein. The combination grew in importance and became ultimately exceedingly influential, embracing a number of ports in the Netherlands, France, Spain, and Italy, and also London in this country. For a considerable time the League enjoyed such power as to render it well-nigh independent of national jurisdiction, but gradually, by absorption and suppression, its privileges were curtailed, until they practically disappeared towards the close of the seventeenth century. What now remains of the confederacy is strictly limited to the three German ports of Hamburg, Bremen, and Lubeck. But during the period of its greatest glory and power, it exercised a far-reaching influence in the encouragement and development of trade both by land and sea, and especially in regard to the administration of port dues and charges.

These two historical episodes illustrate in a very marked degree the close inter-relationship of national policy and commercial enterprise, and they demonstrate how essential to the prosperity of maritime nations is the maintenance and protection of their seaports. There are few countries in the world which are so unfortunate as to possess no seaboard. What few there are, are insignificant in size and in political importance. It is the definite aim and object of most countries, where possible, to increase the extent of their sea frontage. More than one modern war has been really, if not ostensibly, due to the endeavour of a nationality handicapped by a restricted littoral to attain improved communication with the open sea, or, in some cases, even to gain simply direct access to it. The sea is the great highway of the world, a spacious and practically limitless expanse whereon transport is a process at once simple, economical, and direct.

**National Interest in Harbours.**—Such being the case, the inquiry can scarcely fail to arise: How far is the state responsible for the upkeep and development of its ports? Ought harbours to be under the control and tutelage of the nation, and, if so, what kind of patronage and protection, and how much of it, should the latter accord? Stated in concrete terms, should harbours be kept in a state of efficiency, not merely by means of local resources, but by direct governmental assistance, involving the contributions of inland towns? The question is a complex one, and admits of more than one answer.

In so far as a state is a naval power, it has absolute need for shelter and coaling places for its vessels of war. It is, therefore, without any question, entirely concerned in the provision and management of such depôts as are necessary for the purpose. Moreover, in states possessing a littoral frontier swept by fierce gales, it is also a matter of national expediency to produce at certain points works of a protective nature, which will enable imperilled shipping to survive the disastrous effects of sudden tempests.

So far the matter incontrovertibly affects the national welfare. Next, as regards interests which are open to the charge of being purely, or mainly, local.

In regard to ports which have grown up entirely on a commercial basis associated with markets and industries in the immediate vicinity, the same requisition for state interference is not so apparent, and while conditions are favourable and trade prosperous, there is little desire or need to raise the point for discussion. So long as local rates and charges are sufficient to meet all demands entailed in the upkeep of such commercial ports, it is difficult to see why they should not be allowed and encouraged to retain their own independence and work out their own programmes of development. It occasionally happens, however, that a commercial port falls behind the times; it may be from various causes—possibly from indifferent administration, mismanagement, culpable malversation, and so on, but generally and ultimately from lack of funds to carry out improvements which have become necessary by reason of the continually rising standards of ship-building. In such cases more than temporary stagnation is threatened. Maritime engineering works, having once become obsolete, cease to be utilisable at all in any practical sense, and there is no prospect before them of anything but a speedy decline of their trade. Now the state can hardly regard with equanimity the extinction of any one of its centres of commercial activity. Therefore it becomes a question—and the plea has been urged in at least one prominent instance of late—whether the state is not bound to step in with the necessary financial assistance or guarantorship, on the ground that by so doing she is favouring the interests of the community at large. In a general sense the contention is legitimate, but the application of such a principle must necessarily be governed very largely by the special conditions of each particular case, since there may be circumstances under which a grant would be injudicious as well as unjustifiable.[1]

Lastly, in regard to the great majority of harbours—small, almost insignificant havens, many of them—fringing the coasts of every civilised country, where a hardy race of fishermen wring a strenuous and ofttimes scanty harvest from the sea. The districts in many cases are poor, and the calling could hardly be classified as lucrative. Yet there are many thousands of people in this country alone dependent on it for sustenance, either directly or indirectly. Where local resources are so utterly inadequate to cope with

[1] "It may be urged that the expenditure involved in keeping ahead of the developments of shipping is greater than port authorities should be called upon to incur from their own resources, and there is doubtless, in some cases, something to be said in favour of that view, although I hold that such expenditure is reproductive in a variety of ways beyond the mere income arising from the exaction of dues. In my opinion, however, when port authorities, who have striven to provide and have provided certain facilities, are unable to incur the necessary expenditure for further development, it is desirable that the state should come to their assistance and thereby aid these authorities in developing ports on national rather than on local lines."—The Rt. Hon. Lord Pirrie on Harbour and Dock Requirements, Engineering Conference, 1907.

the difficulties of the situation, it is obviously impossible to look to them for the requisite outlay, and the principle of state subvention is now, indeed, fully recognised by the Board of Trade in all instances where little local communities are desirous of extending their accommodation within approved limits.

In making these grants[1] it has been stated that "it is regarded as essential that of the total cost required for construction, at least two-thirds should be provided from local or outside sources, and that the contribution from the exchequer should in no case exceed the remaining third." The action of the Committee has been limited "to the case of harbours serving, or likely to develop, a large fishing district either as points of departure and landing for the fleet, or as providing refuge on parts of the coast, where the nearest existing harbour is so distant as to destroy the value of fishing grounds which produce a good harvest of fish."

Taking the British Isles as a whole, there are something like 130 to 140 of these fishing centres distributed among Ireland, England (including Wales) and Scotland respectively, very closely in the proportion of 1, 2, 3. These harbours provide sheltered areas of water ranging from 2 or 3 acres upwards, though a very large proportion of them are under 10 acres. They are, therefore, individually small, but as already stated, since no inconsiderable proportion of the population derive their livelihood from connection with them, their importance is not to be gauged by size alone.

The commercial ports of this country are less numerous. They number about a score, and the accommodation they provide is largely in the form of docks and inclosed basins of considerable area, both individually and collectively.

National harbours of refuge and for naval purposes are still fewer in number. The areas inclosed, however, are correspondingly larger and attain to as much as 500 and 600 acres a-piece, and even more.

Here our subject leads us on from general observations to an organised investigation of the principles of harbour design, which we can deal with to better effect in another chapter.

[1] The following is a list of grants which have already been sanctioned up to the time of writing:—

| | | |
|---|---|---|
| Port Knockie, Banffshire | 1899 | £15,000 |
| Craigenroon, do | 1899 | 15,000 |
| Pwllheli, Carnarvonshire | 1901 | 22,500 |
| Fraserburgh, Aberdeenshire | 1902 | 15,000 |
| Macduff, Banffshire | 1904 | 2,000 |
| Peterhead, Aberdeenshire | 1904 | 28,000 |
| Lerwick, Shetland Islands | 1904 | 4,500 |
| Mevagissey, Cornwall | 1904 | 2,000 |
| Wick, Caithness | 1905 | 20,000 |
| Southwold, Suffolk | 1905 | 15,000 |
| Whitby, Yorkshire | 1907 | 24,400 |
| Scarborough, do. | 1907 | 6,000 |

# CHAPTER II.

## HARBOUR DESIGN.

Difficulties of the Subject—Classification—Definitions—Roadstead—Harbour—Basin—Dock Harbours of Refuge—Commercial Harbours—Fishery Harbours—Localisation—Coastal and Inland Ports—Procedure in Design of Harbours—Preliminary Considerations—Natural Phenomena—Prevalence and Intensity of Storms—Coastal Change—Accretion and Denudation—Effect of Artificial Interference—Influence of Effluents—Island Harbours—Harbour Areas and Entrance Widths—Illustrations of Harbours at Zeebrugge, Queenstown, Sandy Bay, Sunderland, Peterhead, Libau, Madras, Whitby, and elsewhere.

**Difficulties of Systematic Treatment.**—That maritime engineering is a science of much complexity and no little incertitude, is but a trite remark to make. It will be admitted, without any controversy, that its operations are of necessity founded largely upon assumption and carried out by tentative rather than confident measures. Hypothesis, analogy, and experiment constitute its working basis, alike in regard to theory as to practice, to design as to execution. The whole field of it is beset by many and peculiar difficulties, and scarcely any other department of constructive work finds so many hazards and obstacles in the way of satisfactory accomplishment. The task of the engineer who sets himself to contend with the almost bewildering array of antagonistic forces incidental to maritime operations, is exacting in the extreme. The data upon which his calculations must perforce be based are often defective and their origin obscure. He has to deal with agencies not only conflicting but frequently also co-operative, and as destructive as they are capricious. His work is subjected to the most trying of all ordeals, in that it is constantly exposed to the risk of unascertainable possibilities. Occasions arise when the profoundest sagacity and the ripest experience may well prove to be at fault. Laws which hold good in one locality seemingly reverse themselves in another. The success of certain dispositions in one case is no guarantee of their efficacy elsewhere, still less justification for their general application. Each place has its own definite characteristics, its peculiar defects, and its special advantages, differentiating it from all other places. There is no uniformity, and very little similarity. Generalisation, therefore, is impossible, and classification becomes difficult.

Yet, in spite of these deterrent considerations, it is manifest that some system of treatment must be adopted, unless the principles of harbour engineering are to rest on a haphazard, heterogeneous basis, contrary to the spirit of all scientific procedure. Our endeavour, therefore, in the following pages will be to collate such data as are definitely acceptable, to elucidate as far as possible those problems which present themselves within the range of ordinary experience, and to lay down certain rules which may serve for general guidance to those engaged in adapting some of the most variable forces in nature to the use and service of man.

A clear conception of our purview is essential, so we must commence with one or two definitions.

**Classification.**—A **harbour** is primarily a place of rest and refuge—a place where safety and hospitality are to be found. But round this central

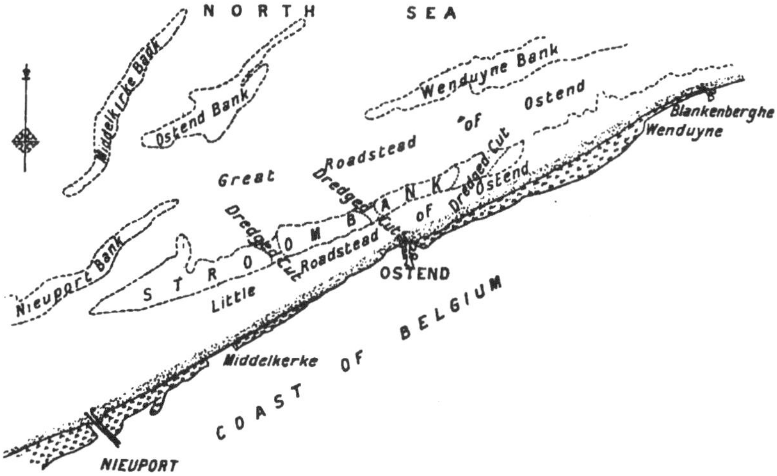


FIG. 1.—Roadsteads of Ostend.

idea have grown several accretions of meaning, the gradations of which it is desirable to point out.

Limiting our references, as is natural and proper, to the domain of navigation, we may observe that a vessel seeking shelter under stress of wind and weather may possibly obtain it as follows :—

(1) Within a tract of water, not necessarily inclosed in any way, even partially, but adjacent to or not far distant from the coast-line, where there is good holding-ground for anchors and some protection from the onset of heavy seas. Such conditions constitute a **Roadstead**. Roadsteads may be either natural or artificial. In the case of a natural roadstead, a deep channel, with an intervening bank or shoal to seaward, possesses the necessary characteristics, as exemplified in the offing of the Port of Ostend (fig. 1). An artificial roadstead may be created on similar lines by a breakwater, either parallel to the coast, or curvilinear, such as that at Zeebrugge (fig. 2).

(2) Within a definitely circumscribed area, almost completely inclosed, either naturally, as in the case of a creek or estuary, or artificially, by

projecting piers, moles, and jetties. The harbour of Queenstown (fig. 3) is an exemplification on a large scale of the former class, while the outstanding

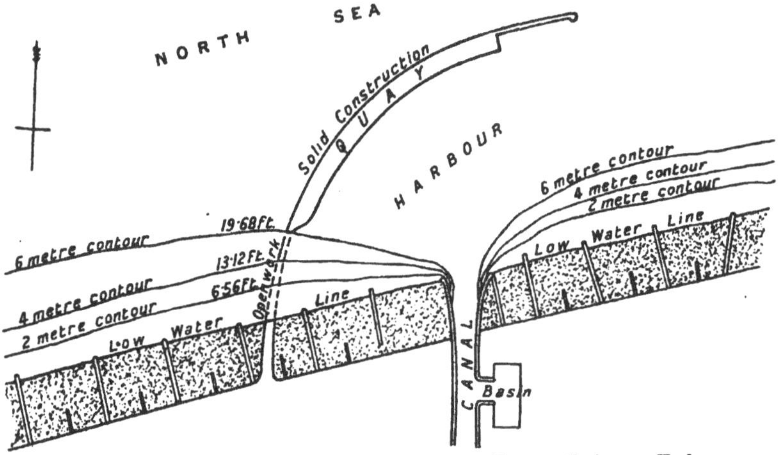


FIG. 2.—Zeebrugge Harbour.

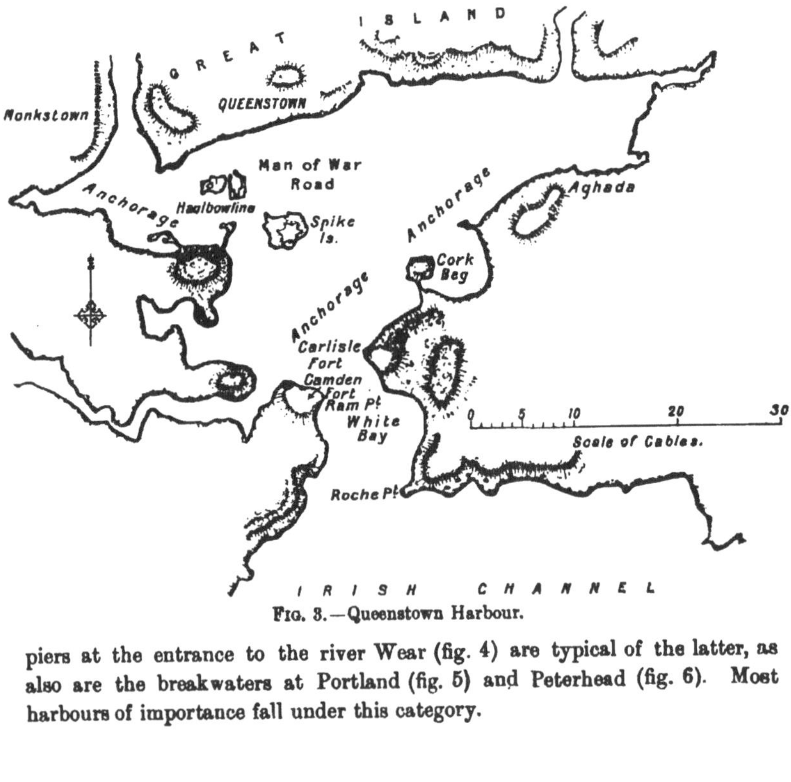


FIG. 3.—Queenstown Harbour.

piers at the entrance to the river Wear (fig. 4) are typical of the latter, as also are the breakwaters at Portland (fig. 5) and Peterhead (fig. 6). Most harbours of importance fall under this category.

(3) Within a confined **Basin** of comparatively small extent, having a narrow aperture only for the ingress and egress of vessels. There is little to distinguish this from what is termed a **Dock**, though the latter expression is commonly restricted to basins provided with entrance gates. An illustration of this class of harbour is to be found at Peterhead in the South Harbour (fig. 6), while a dock at Sunderland (North Dock) is shown in fig. 4. Fishery harbours generally belong to this class, and it is no uncommon feature to find a smaller inner harbour, or basin, constructed in conjunction with a large outer harbour, or a roadstead. These basins are provided with quays for the reception of cargoes.

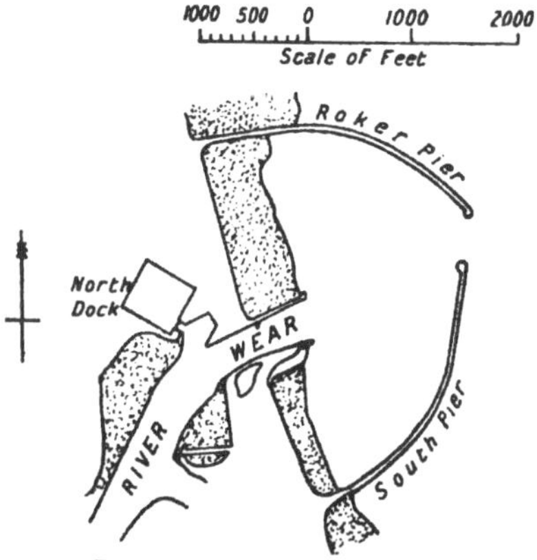


Fig. 4.—Sunderland Harbour.

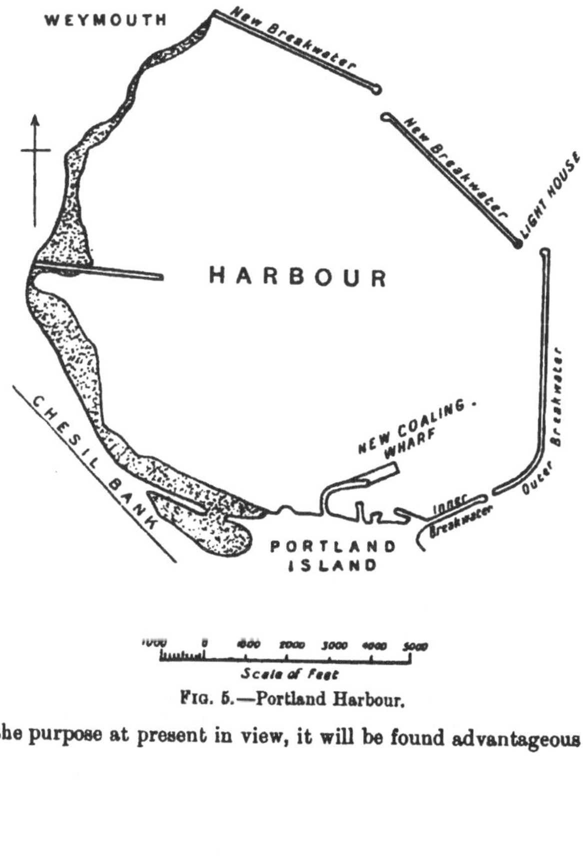


Fig. 5.—Portland Harbour.

For the purpose at present in view, it will be found advantageous to adopt a slightly different classification based on the object to be attained. From this standpoint there are three important divisions, as follows :—

(1) Harbours of Refuge.

(2) Commercial Harbours.

(3) Fishery Harbours.

Though fundamentally in unison, and oftentimes found in combination, the designs of these three classes are sufficiently distinct to justify us in

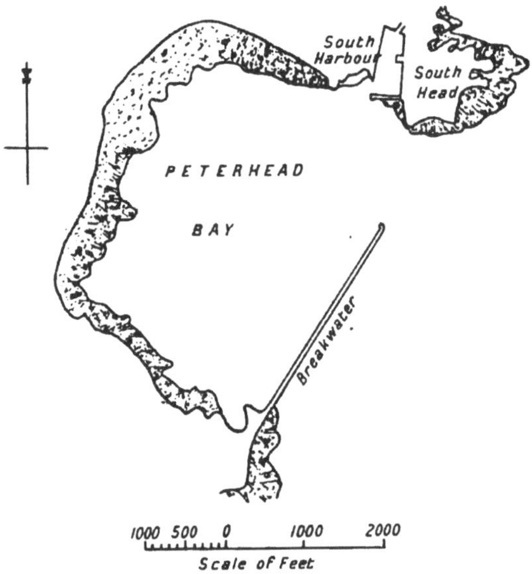


FIG. 6.—Peterhead Harbour of Refuge.

treating them separately, describing their particular functions and enumerating their special requirements.

**Harbours of Refuge.**—The principal duty of a Harbour of Refuge is, as the name implies, to provide a refuge for vessels overtaken by sudden stress of weather, or otherwise hard pressed or disabled. The proper *locale* for the construction of such harbours is obviously at conveniently accessible stations upon coasts which are inhospitable and dangerous. Yet, manifest as is the desideratum, the means of its accomplishment is not so obvious, and the subject has given rise to some conflict of opinion. Is the proper position for a harbour of refuge upon an outstanding frontage or within a bay? Ought it to be projected into the open, or recessed within the coast-line? In the former case, the goal is more easily reached and less delay is incurred in putting out again to sea; on the other hand, there is greater exposure, and this endangers the ingress of vessels, rendering them more liable to miss the entrance, in which case they will probably be driven on to the shore. Yet the risk of ultimate catastrophe must necessarily be greater in the case of a vessel missing the entrance of an embayed harbour. So it is scarcely safe to dogmatise upon the point. There may be advanced positions to which a lofty headland imparts all the advantages of a sheltered recess, and there are likewise cases in which deep coastal indentations afford very meagre protection from tempestuous seas. Where it can be assured without serious risk, the nearer the haven to the distressed ship, the better her chances of reaching it. Many seamen, however, prefer, where practicable, to ride at anchor in the open rather than make for the uncertainties of the shore.

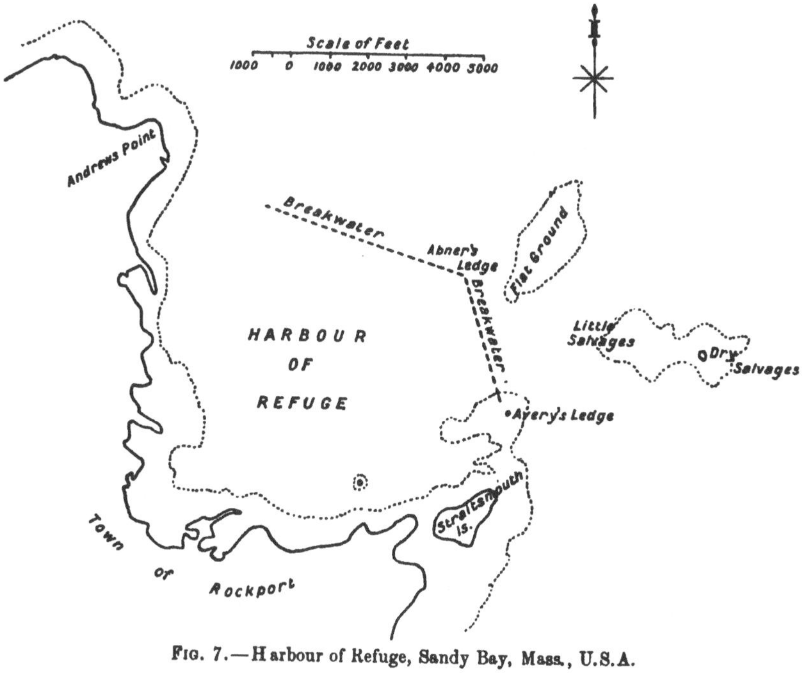


Fig. 7.—Harbour of Refuge, Sandy Bay, Mass., U.S.A.

The requirements of a harbour of refuge may be summed up as three:—

(a) Ready accessibility.

(b) Safe and commodious anchorage.

(c) Facilities for obtaining supplies and for executing minor repairs.

Upon the first point we have already dwelt a little. But accessibility depends not only upon the site of the harbour; it depends also upon its disposition. The entrance must be conveniently placed and designed, so as to allow of its being easily taken by ships driving before a storm. A narrow entrance is difficult to negotiate, but, on the other hand, a wide entrance exposes the interior to the effects of rolling seas. Local circumstances will largely influence the determination of the dimension to be accorded thereto; at the same time, it may be said that from 600 to 800 feet approximately represents the expression of modern British practice. It is not unusual to provide more than one entrance; sometimes two in opposite directions, so as to afford a choice of approaches to vessels under varying conditions of wind. An entrance may also be deflected in order to afford cover to the interior. Provided a vessel is able to reach the shelter of an outer breakwater, it becomes more amenable to control, and it may be navigated into an inner berth with much less risk and trouble. An entrance may, accordingly, be placed to receive ships direct from the most exposed quarter, while at the same time the channel, or passage, may be diverted towards the interior, in such a way as to mitigate the influx of rough seas. A roadstead forms a useful vestibule to a harbour in this respect.

For the efficiency of its anchorage, a harbour depends upon the nature of the ground and the depth of water obtainable. Light, sandy bottoms do not, as a rule, afford a good hold; but firm, tenacious material of any kind is most suitable. The depths should be sufficient to allow adequate flotation for ships of the largest class, and the area ample enough for them to ride at anchor without inconvenience or danger of fouling one another. In the event of a change of wind, and inevitably as regards a tidal harbour, there must be sufficient space allocated to each vessel to allow of its swinging round upon its anchor. For merchant vessels a radius of one cable's length (120 fathoms or 720 linear feet) will suffice; but in the case of battleships, an allowance of one and a half times to twice this dimension should be made. The draught of the most modern vessels in the mercantile marine now attains to nearly 40 feet, but by far the greater number draw less than 30 feet of water, and from 5 to 6 fathoms will prove ample for all present requirements of anchorage, especially in regard to vessels which are likely to seek a harbour for refuge purposes only. A battleship of the highest class has a draught of about 27 or 28 feet.[1]

Ships once within the shelter of a harbour should certainly find facilities for obtaining supplies of stores and coal, and for executing any repairs which may be necessary to render them seaworthy, or to enable them at least to proceed to some neighbouring port where they may receive proper attention. The extent to which this accommodation is possible, or desirable, will vary with the locality. It is not often that a ship, sufficiently large to require a graving dock for its overhaul, will be driven to seek refuge, but the contingency is possible, and in case of war is not unlikely. Smaller boats may be beached or laid upon a gridiron or slipway.

The circumstances attending an outbreak of hostilities are such as to render it eminently desirable for harbours of refuge to be equipped with some means of defence. Under modern conditions of torpedo attack, it has become necessary to completely close the entrances of naval stations at night by booms

[1] Insufficient depth is fatal to the utility of all harbours, but more particularly in the case of commercial harbours, which have to meet the requirements of shipowners and shipbuilders. The dimensions of vessels in the mercantile marine are expanding continuously, and the greatest difficulty is being experienced in obtaining a proper augmentation of depth, owing to the shallowness of the entrances to many harbours and of main waterways, such as the Suez Canal.

and other obstructions. These precautions, however, appertain to the province of military engineering.

**Commercial Harbours.**—Passing on to special aspects of Commercial Harbours, we may describe them as a class forming most important appanages to ports. They are, in fact, the great termini of the highways of the sea. Their province is the accommodation of the mercantile marine during the operations of loading and discharging cargoes, and for the transaction of trade. Thus, in addition to the obviously fundamental needs of accessibility and accommodation already discussed, we meet with the more special requirements of Quays and Sheds, and also of Inner Basins and Repairing Docks.

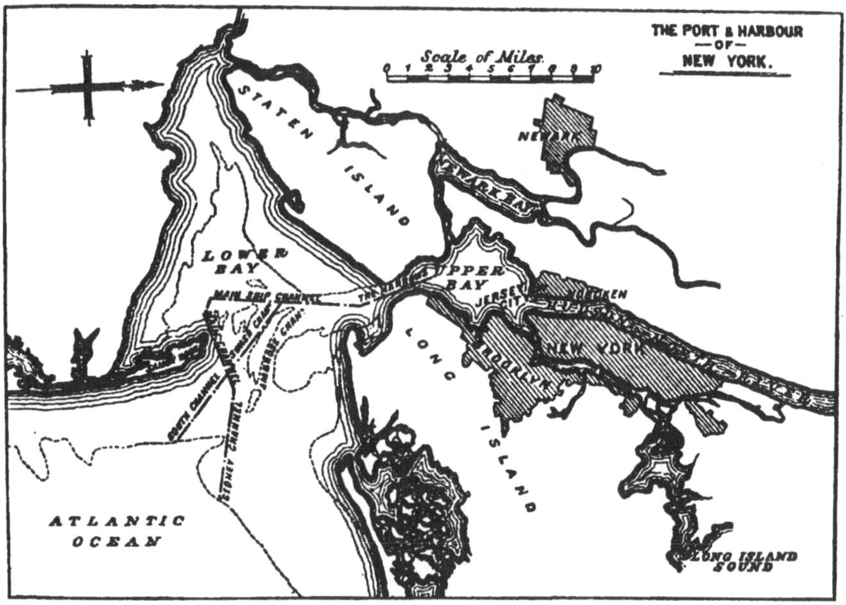


FIG. 8.—New York Harbour.

Commercial harbours are to be found in a variety of situations: upon the seacoast, at the mouths of rivers, inside sheltered estuaries, and even some considerable distance inland along the banks of rivers and canals. They require more shelter than that which suffices for simple purposes of refuge. It is indispensable to the conditions of modern trade that there should be the least possible delay in the reception and despatch of vessels; hence everything must be done to ensure continuity of operations, and for this purpose protected quays are a first consideration.

Coastal Harbours present most difficulty in regard to this point. The mere protection afforded by a breakwater is not sufficient to impart that tranquillity which is essential to the loading and unloading of ships. There are, of course, cases like that of Zeebrugge (fig. 2) where a single mole built out into the sea is made to suffice, being provided with a level quay and covered sheds for the reception of merchandise; but such cases are rare, and, generally speaking, it will be found necessary to provide an enclosure, practically complete, with an inner harbour, or docks, for commercial purposes. Indeed, in the case of a coast exposed to heavy seas, the adjunct becomes imperative.

Estuarine Harbours find the requisite shelter already provided, in many cases, by rising ground flanking their entrances, and, indeed, many harbours situated in creeks and inlets are also admirably protected by adjacent hills so as to require no further defence. In addition to Cork and Queenstown, the harbours of Sydney, San Francisco, and Rio de Janeiro may be cited as cases in point. Instances of estuarine harbours are afforded by Liverpool, at the mouth of the Mersey; Dublin, at the mouth of the Liffey; Havre, at the mouth of the Seine, and numerous other ports. The case of Liverpool is an admirable example, the form of the river at its mouth being excellently adapted for harbourage. The Mersey is broad and deep, expanding inwards from a narrow-necked entrance, while rising ground on both sides contributes the necessary shelter.

But while estuarine harbours possess many advantages, they have corresponding defects. The majority of rivers are afflicted with bars, that is to say ridges of material lying across their entrances in such a manner as to reduce the available depth of water, and so impede navigation. The point is only mentioned here in passing: it is of such grave importance as to call for detailed treatment, and this must be reserved for a later chapter.

River Ports, such as Bremen on the Weser, Hamburg on the Elbe, Glasgow on the Clyde, etc., and Canal Ports, such as Manchester, Bruges, etc., are comparatively free from many of the evils which affect harbours on the littoral, but they are attended by certain inconveniences of another kind. The navigation of inland waterways is a slow and tedious process for sea-going ships, and it involves considerable delay, which, in these days of rapid transit, counts for a great deal. Ports like Antwerp, Hamburg, and Rotterdam are very unfavourably situated for competing as regards expedition and despatch with ports on the seacoast. True, there are advantages attaching to water carriage throughout, which outweigh the alternative of unshipment and transport by rail from the nearest seaport; but this fact does not invalidate the contention that the nearer the port to the great ocean highways, the speedier and better the distribution of freight. Moreover, river harbours are subject to the adverse influences of strong currents, freshets, and floods; they have a continuous tendency to silt up, and they are even invaded at times by floating masses of weeds and mud. Under these particular circumstances, a river harbour should possess an entrance pointing downstream, making an acute angle with the bank, and the width of entrance should not be greater than is necessary for the manœuvring of the vessels which frequent the port.

**Fishery Harbours.**—Fishery Harbours, though a numerous class, are not, generally speaking, constructionally important, but they possess characteristics sufficiently notable to merit some attention. In the first place, the scarcity of capital available for artificial operations renders it necessary to take the utmost advantage of natural features. Then, since fishermen require a maximum amount of time for their expeditions, with a minimum of delay in despatching their hauls, on account of the perishable nature of fish, every facility should be afforded them for making the harbour at the last possible moment consistent with the state of the tide. This intermittent accessibility, however, characteristic as it is of all tidal harbours in

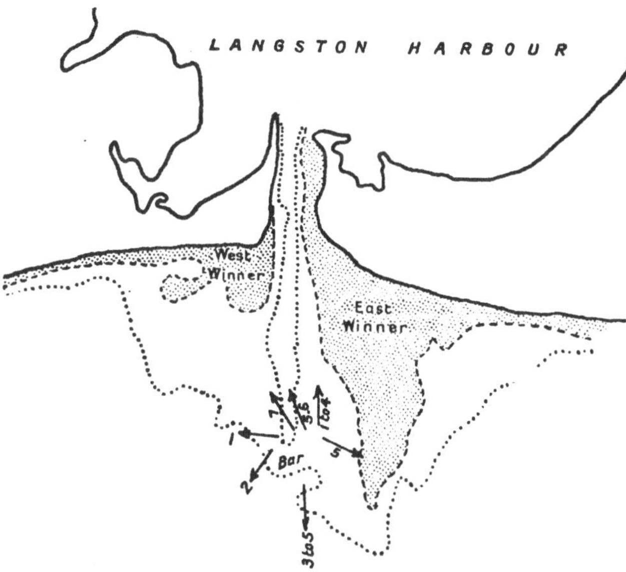


FIG. 9.—Entrance to Langston Harbour. Direction of currents. The numerals represent hours of ebb and flood tide.

which vessels take the ground at low water, or where there is a shallow bar, is inimical to the interests of sea fishermen. Their requirements demand a harbour constantly open for arrival and departure alike. A shoal of fish may be missed as easily from sheer inability to proceed to sea as from the deterrent effect of impending foul weather.[1]

The entrance of a fishery harbour, while not unduly wide, must not be made too narrow. It is liable to be thronged at times by craft anxious to enter in order to escape a rising gale, or to catch the early market. Cases

[1] Depths below low-water level of ordinary spring tides at some of the more important fishery harbours in this country are:—Hartlepool, 15′; Aberdeen, 14′; Dundee, Yarmouth, and Newlyn, 12′; Lowestoft, 10′. The majority of fishery harbours, however, have depths much less than these, and, in many cases, they become practically or actually dry at low water.

2

have occurred in which boats have become jammed in an entrance through excessive crowding, with, unfortunately, disastrous consequences. Fishing smacks of the present day have a beam of 20 feet or so, and allowance should be made for, at least, three or four, or even more, entering abreast, according to the size of the fleet.

The length of ordinary fishing-boats ranges from 50 to 60 feet, and their loaded draught lies generally between 6 feet and 10 feet; but steam trawlers draw as much as 14 or 15 feet, and their length is rarely less than 110 or 120 feet.

In a fishery harbour, broad open quays with a large covered hall, or market, are required for sorting the fish and conducting sales.

It has to be borne in mind that under modern conditions of trade, a fishing fleet of steam trawlers must needs be furnished with supplies of coal and ice, and for this purpose railway sidings at the quayside become a necessity. For large vessels of this class, the fishery harbour becomes of a more important character, trenching, in many respects, on the commercial harbour.

**Selection of Locality.**—Reviewing the subject as a whole, and postulating the choice of a situation for a harbour, the conditions which would govern that choice divide themselves into three heads.

First and foremost, there is the obvious advantage to be derived from a position adjacent to some existing means of internal communication, such as river, canal, or railway. In the absence of all these, it is still possible to consider the feasibility of the formation of two of them, and any obstacles likely to prove antagonistic should be carefully weighed and avoided.

Then the extent of adverse meteorological and climatic influences claims attention. Fog and frost are dual evils which infest very many harbours to the infinite detriment of their usefulness. The former jeopardises shipping, and both impede navigation. To be ice-locked for several months in a year, like Montreal or St Petersburg, is a serious outlook for any progressive port. The ice-breaker, however, has done much to relax the rigorous grasp of winter, and very few ports need now resign themselves to the entire cessation of their sea-trading operations. For the fog-fiend, there is unfortunately no practicable remedy as yet. Certain experiments conducted in scientific quarters seem to suggest a possible amelioration, but on too small a scale for general application. As regards protection from storms, the value of headlands and promontories has already been pointed out. Cover should, of course, be sought from the most tempestuous quarter.

Lastly, the facilities afforded for providing such artificial protection as is requisite should be taken into account. This constitutes the economical aspect of the question, and some of the numerous matters deserving attention are the length of breakwaters and jetties required, the nature of their construction, the source of suitable material, and the expense of carriage, the means of procuring labour and plant, and the resources of the district generally. In regard to all these, certain localities will be more favourably situated than others.

But, except in rare instances, the engineer can hardly expect to have the opportunity of allocating a harbour and of designing it *in toto*. Trade routes are sufficiently firmly established to preclude the diversion of much traffic to other lines. An occasional harbour of refuge, with a fishery station or so, marks the limits of entirely new construction at the present day. Yet, at the same time, there is great and increasing scope for the development of maritime works already in existence, and the enlargement and improvement of harbours forms one of the most important fields of civil engineering.

**Procedure in Design.**—Such being the case, the unrestricted choice of a site will rarely lie within the province of the engineer. The locality, at anyrate, will already have been determined and the preliminary dispositions established, before his services are requisitioned. It falls to his lot, therefore, to utilise existing conditions and to devise a *modus vivendi* out of circumstances beyond his control.

Assuming, momentarily, for the purpose of discussing the question in all its bearings, that the site is a virgin one, there are certain preliminaries to be carried out before any scheme can be laid down. We will deal with them in their natural order of procedure. Thus, the first point would be to make a survey of the neighbourhood, and to prepare a chart indicating the depths of water in the vicinity. Not only should a complete set of soundings be taken, but borings should also be made to ascertain the nature of the ground, its fitness for anchorage, and the extent to which it lends itself to an economical increase of depth, should this be or become necessary. The depths obviously must be sufficient to meet the requirements of the deepest draughted vessels which are likely to frequent the place,[1] and it should not be overlooked that some allowance is necessary for the pitch or surge of a vessel in rough water, whereby its keel descends below the normal level.

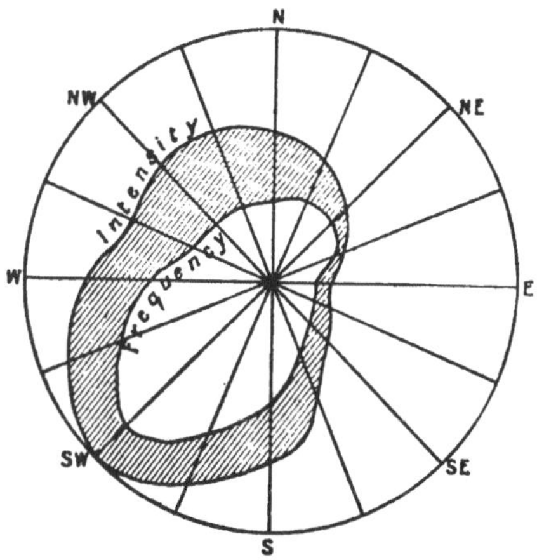


FIG. 10.—Wind Diagram. Frequency ordinates set off from centre; intensity ordinates from frequency curve.

**Natural Phenomena.**—After the preparation of the survey and the plotting of the contour lines (or lines of uniform level, as shown in fig. 2), the engineer will search local records for data, and also make

[1] See footnote, p. 14.

observations himself, in reference to various natural and meteorological phenomena, and the following will specially claim his attention:—

1. The direction and intensity of the winds and the frequency of storms.

2. The height and force of the waves.

3. The range of the tides.

4. The direction and velocity of the currents.

5. Evidences of silting, littoral drift, or coast erosion.

6. The extent of exposure and the maximum "fetch."

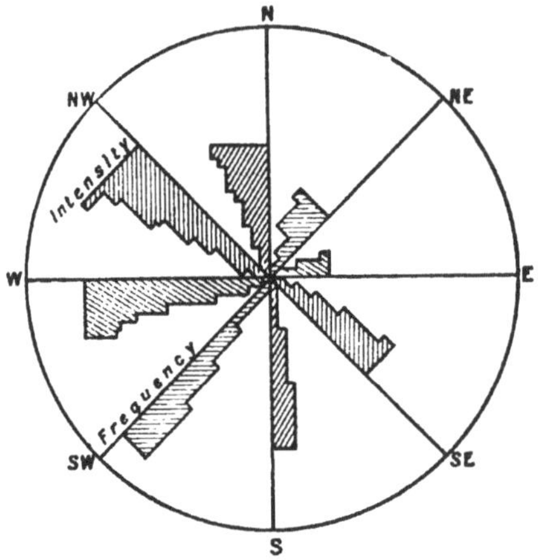


FIG. 11.—Wind Diagram. Frequency ordinates set off from centre; intensity ordinates from radial lines.

With reference to the first of these features, it may be pointed out that nearly every place is subject to what is called a **Prevailing Wind**, that is, a wind blowing with great constancy for a considerable portion of the year from a certain point of the compass. But while the prevailing wind is the most frequent, it must not, by any means, be concluded that it is the wind which is most to be feared. A single gale arising from a totally different quarter may cause more havoc and destruction than a whole twelve-month of the prevailing wind. The importance of the latter lies rather in the effect it has upon the coastal contour in its relationship with other agencies, the effects of which, though momentarily insignificant, are continuous and cumulative. Such agencies are the ebb and flow of the tide and the erosive and transportive power of currents.

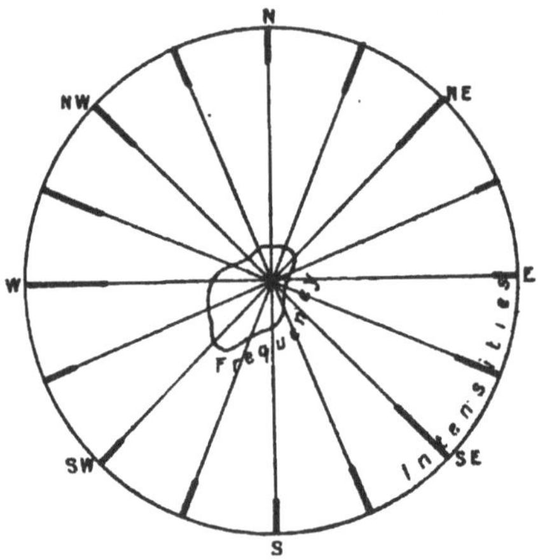


FIG. 12.—Wind Diagram. Frequency ordinates set off from centre; intensity ordinates from circumference.

Ways of recording wind frequency and intensity are numerous. Three examples are shown in figs. 10, 11, and 12, with explanatory notes. Time ordinates are not difficult to plot, possessing, as they do, a direct numerical value. It is a different matter with the intensity ordinates, which have to be to a certain extent conjectured. The velocity of a wind may be gauged more or less accurately by an anemometer, if one is available; in other cases it is estimated. But the velocity is fitful, and, as a measure of pressure, by no means an ideal standard. The range of intensity is usually divided into twelve sections, forming a scale, known as Beaufort's scale, which is given below.

BEAUFORT SCALE FOR WIND.

| | | | |
|---|---|---|---|
| 0 denotes | Calm | Velocity in miles per hour | =0 |
| 1 ,, | Light Air | ,, ,, | 7 |
| 2 ,, | Light Breeze | ,, | 14 |
| 3 ,, | Gentle Breeze | | 21 |
| 4 ,, | Moderate Breeze | ,, | 28 |
| 5 ,, | Fresh Breeze | | 35 |
| 6 ,, | Strong Breeze | | 42 |
| 7 ,, | Moderate Gale | | 49 |
| 8 ,, | Fresh Gale | | 56 |
| 9 ,, | Strong Gale | | 63 |
| 10 ,, | Whole Gale | | 70 |
| 11 ,, | Storm | ,, | 77 |
| 12 ,, | Hurricane | ,, | 84 |

**Coastal Change.**—The influence of the wind in relation to the magnitude of waves will be more fully considered in another chapter. At present, having regard to the general outlines of harbour design, we will simply notice its bearing upon the coastal contour in the vicinity of any artificial works.

That the seacoast is undergoing a gradual change must be evident to the most superficial observer. In certain districts, notably the borders of Yorkshire and East Anglia, there are manifest signs of sea encroachment. Every year witnesses the retrogression of some extent of shore frontage, and, in the course of a few centuries, whole tracts, such as the Goodwin Sands and districts including villages and townships, have disappeared. On the other hand, in other quarters there has been a gradual gain and accretion. Southport in Lancashire, formerly, as its name implies, situated at the water's edge, now lies at a perceptible distance inland. At Dungeness in Kent, a headland of shingle is accumulating at something like the rate of 200,000 tons per annum. Instances of both kinds might be multiplied indefinitely.

The essential point to consider is the probable effect of any artificial projection from the coast in accentuating or mitigating the natural process of mutation. This is not altogether an easy matter to determine, owing to the predominating influence of local circumstances, quite apart from the fact that the causes of coastal denudation and accretion are but imperfectly understood. The carriage of material from one point to another is assigned by one school of engineers entirely to wave-action, and by another school, mainly to current flow. It seems, on the whole, not improbable that both agencies are involved, in varying degree: the breaking of waves on a beach serves to stir up the sand and shingle, the former of which the water, in its troubled state, retains in suspension long enough for it to be projected some distance along the shore by the resolved component of wave force in that direction,

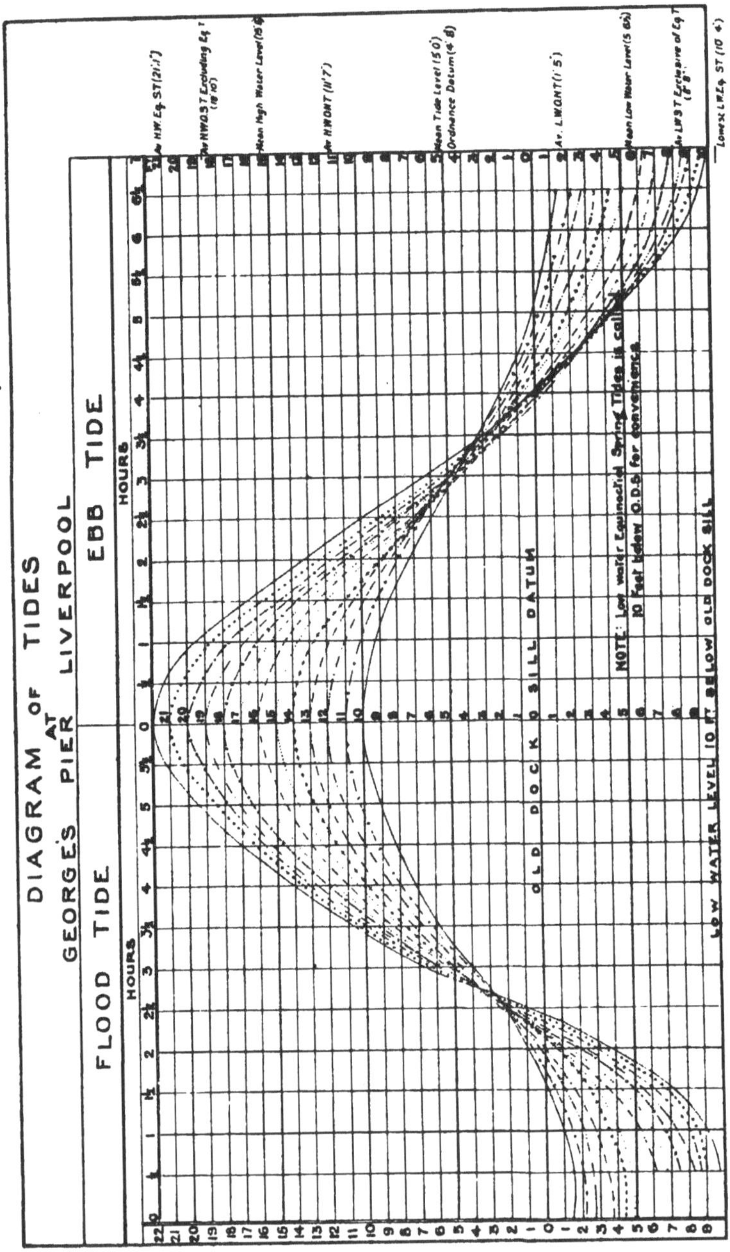


FIG. 18.—Tidal Diagram, showing range of tides at Liverpool.

being aided therein by the littoral current. The heavier particles are rolled along and partake of a zig-zag movement, as shown in fig. 14. It is generally agreed that the action is practically confined to the region between high and low water marks.

The trend of littoral drift is therefore attributable, in the first instance, to the wind which governs the predominant direction of the waves.

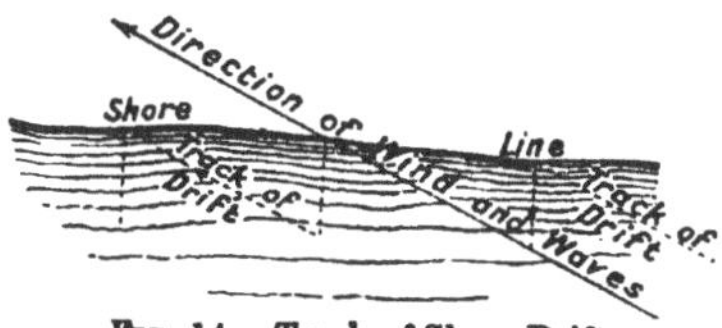


FIG. 14.—Track of Shore Drift.

To illustrate in some way, however imperfectly, the general effect of wind and flow upon a coast-line, with the modifications brought about by intrusive structures, figs. 15–22 have been prepared. A simple case only has been taken; the action, as can well be imagined, is often much more complex. The supposition made is that of a coast-line with the dominant current flowing parallel thereto (from right to left) coincident with the direction of the prevailing wind. Fig. 15 and those which follow serve to indicate the tendency towards shoaling in various parts, which is brought about by the construction of harbour works of typical kinds.

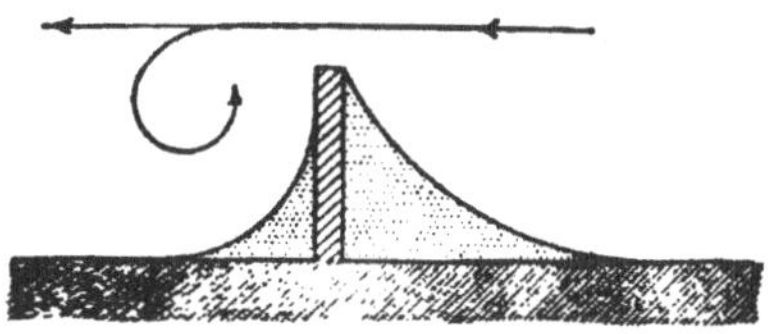

FIG. 15.

The straight pier or breakwater at right angles to the coast-line (fig. 15) induces an accretion of sand and shingle along each of its sides. The windward accumulation is the more pronounced, the leeward deposit being reduced by eddying round the outer end of the pier. The returned pier (fig. 16) serves to increase the leeward deposit, there being a circular motion of the water round the pier-head with a tendency to scour at that point, while the slacker water inside leads to settlement of suspended material. An example of this is to be found in the harbour of Salina Cruz on the Pacific Coast (fig. 18), where the initiation of a breakwater of this type brought the low-water line forward, temporarily, at anyrate, to the 26 feet contour of six years previously. Much the same effect is apparent with double piers (fig. 17), the accretion being emphasised by reason of the additional extent of quiescent area. Evidence in support of this is forthcoming from Madras, where the harbour entrance is slowly but surely silting up in this way. Apparently the only practicable means of remedying the evil due to solid structures is that of substituting openwork for the portion of the jetty which immediately joins the land. It has even been suggested that the most logical method is that of "Island Harbours," formed in deep water out

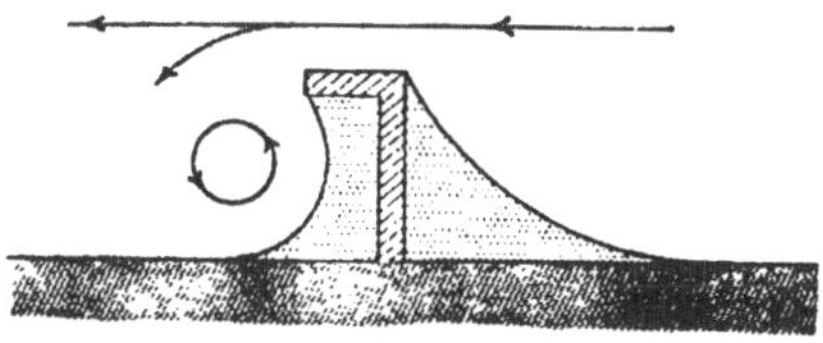

FIG. 16.

of the range of accretion, and connected with the shore by means of openwork jetties, as exemplified at Arnager, Snogebœk and Hundested (page 41). Columnar structures offer little, if any, perceptible obstruction to current flow, and consequently should not give rise to a deposit of any appreciable extent. Whatever shoaling takes place abreast of the outermost and solid

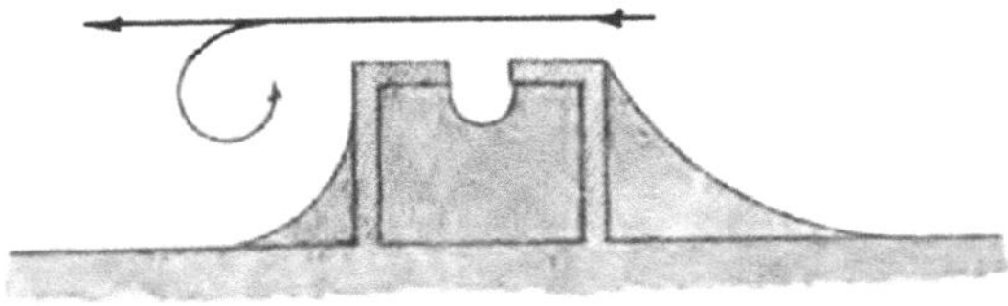

FIG. 17.

portion of the pier, should be, and generally is, comparatively insignificant. The breakwater at Zeebrugge has been designed on these lines, as also the pier at Rosslare on the east coast of Ireland. Unfortunately, for the general application of the principle, a disposition of the same kind adopted at Ceara harbour, in Brazil, has proved strikingly unsuccessful, and has entailed consequences disastrous to the port. The entire harbour is blocked with

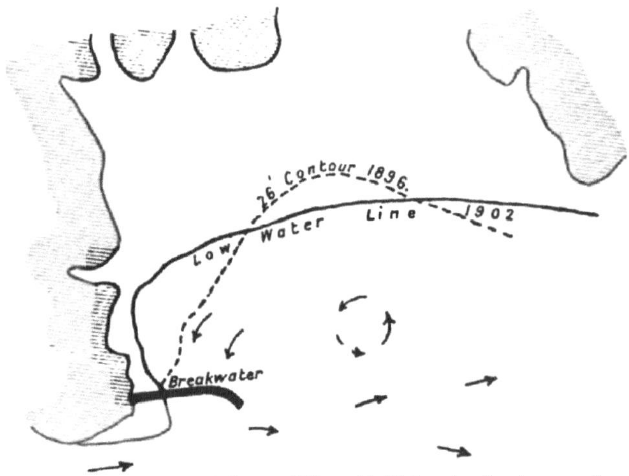


FIG. 18.—Salina Cruz Harbour. Effect of initiatory breakwater operations.
*Note.*—The design was subsequently modified.

sand (fig. 19), despite the fact that an openwork viaduct lies between the breakwater and the shore.

Land water entering the sea at any point is deflected by tidal currents, where they exist, to each side alternately, with the result that with the coastal sediment there is a tendency to shoal at some short distance outwards, forming a bar of various contours. With a strong wind and littoral current in one predominant direction, there will most likely be produced, in addition to a bar, a spit or horn (fig. 20), through which the river, in times of flood, may break, but which generally reforms. In either case the shoaling is

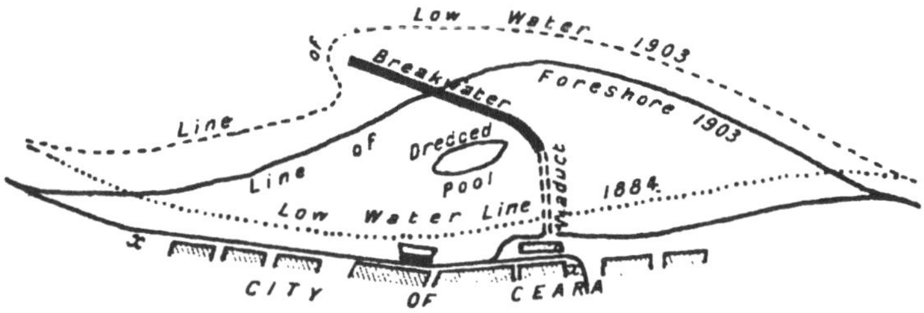


FIG. 19.—Harbour of Ceara. Effect of breakwater.

detrimental, and various means have been tried to remove it. The most generally accepted methods are by means of projecting jetties, or training walls, and dredging. Unless the former are extended into very deep water, they are not likely of themselves to prove completely efficacious. Accretion takes place as shown in fig. 22, and matters readjust themselves a little further out much to the same effect as before. The history of Dunkirk, Calais, and other French ports, proves this, the jetties constructed at those places having to be prolonged from time to time, as shoaling has accrued. The works at the mouth of the Mississippi (where there is, of course, some

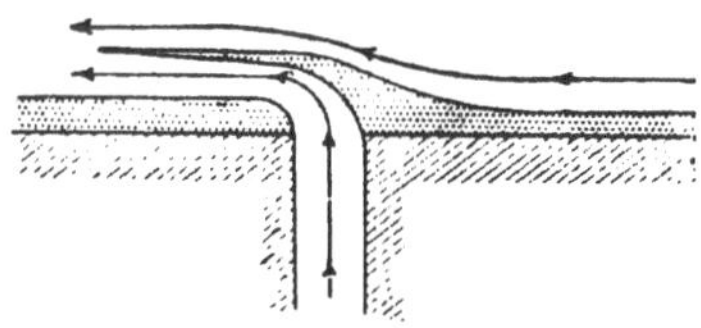

FIG. 20.

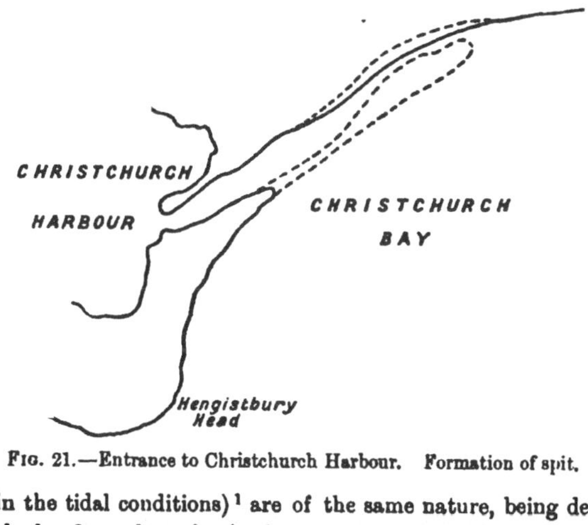


FIG. 21.—Entrance to Christchurch Harbour. Formation of spit.

difference in the tidal conditions)[1] are of the same nature, being designed for the removal of a bar; but the jetties are of considerable length, and, being

[1] The tidal range in the Gulf of Mexico is very small.

extended to a point sufficiently seaward to bring the fluvial deposits within range of powerful submarine currents, there is less fear of the deposit becoming permanently localised to any serious extent. At the same time, the jetties may need extending ultimately on account of the advancement of the coast-line.

With dredging employed as an auxiliary, the silting up of outlet channels may, of course, be held in check as long as and to whatever extent may be deemed expedient. We revert to this part of the subject later on (Chapter IX.).

In considering the diagrams employed in illustration of the foregoing principles and hypotheses, it is to be distinctly understood that the shoaling indicated does not necessarily appear above water level, even at lowest water, nor indeed does it, in a number of cases, manifest itself to any pronounced extent. Certain tendencies only are outlined, which become more or less marked, according to the absence or presence of counteracting influences. It is the province of the engineer to secure or provide these counteracting influences by natural or artificial means, so as to maintain a state of equilibrium in so far as it is possible to do so. Some of the means employed for the purpose will be discussed in a later chapter, but we cannot hope to deal exhaustively with a problem which admits of innumerable variations in accordance with local peculiarities. The subject of littoral drift and its complementary phenomenon, coast erosion, is so involved, and its ramifications so extensive, as to require special and individual treatment.

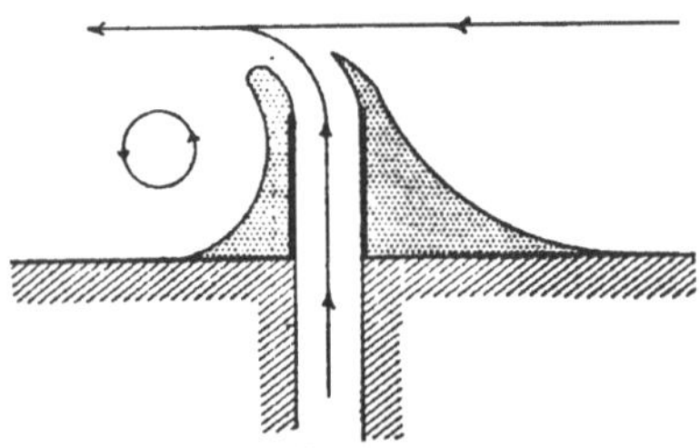

Fig. 22.

As a forcible illustration of the complex nature of currents set up by artificial works, a series of illustrations (figs. 23–28) are given, showing the vagaries in the tidal flow exhibited at Dover Harbour since the formation of the new breakwaters. Beforetime, the set of the tide was almost parallel to the shore line, running from N.E. to S.W. from 4½ hours after to 1¾ hours before high water, and from S.W. to N.E. from 2 hours before to 4 hours after high water. At high water of spring tides the rate of the east-going stream was about 4 knots, and at low water the west-going stream had a velocity of 2½ knots. Both direction and rate of flow are now completely altered in every respect.[1]

The currents at the entrance to Langston Harbour are indicated in fig. 9.

**Harbour Areas and Entrance Widths.**—We turn now to the subject of wave power, regarding it simply as it affects the general question

[1] When these diagrams (issued by the Admiralty) were prepared, the western portion of the Island breakwater was still incomplete and the site occupied by open constructional piling. No doubt further changes will manifest themselves with the progress of the solid work.

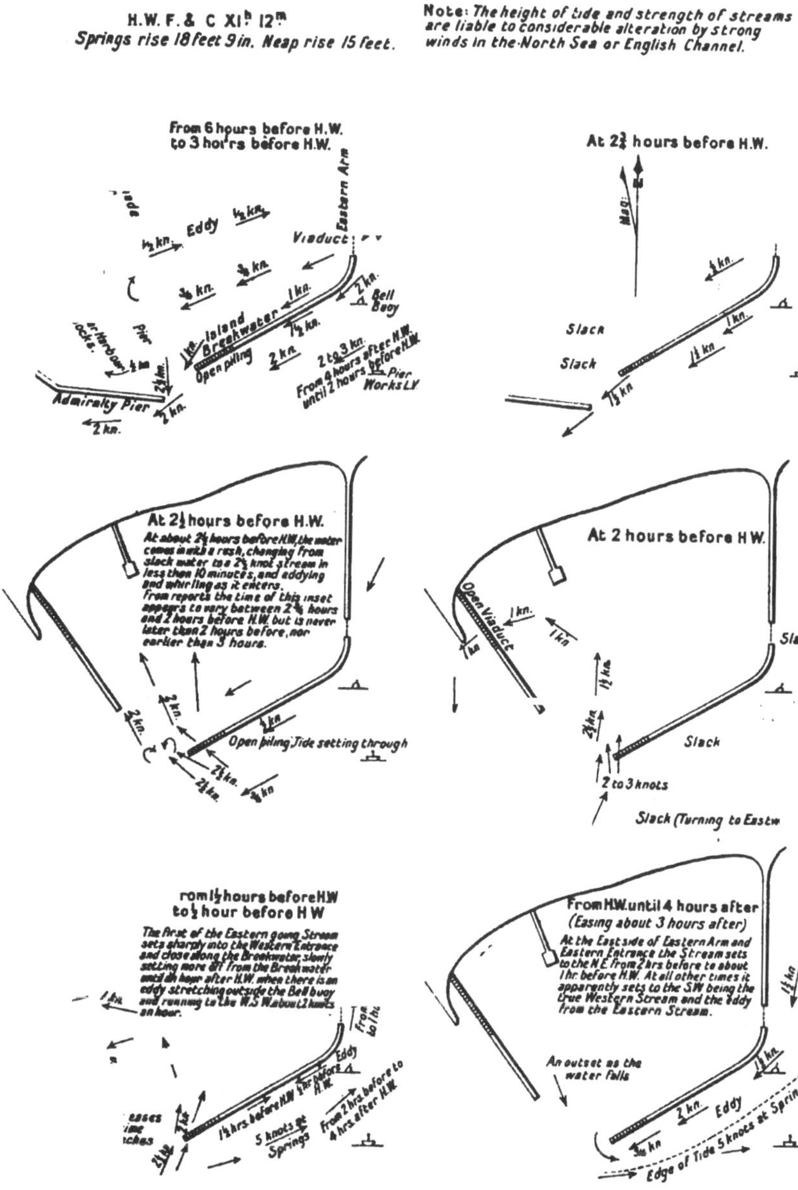


Figs. 23–28.—Tidal Currents in and out of Dover Harbour.

of harbour design. At a later stage it will be necessary to take it into detailed consideration from the point of view of its influence on structural features. For the moment, however, we are only concerned with it in its general aspect, that is, in so far as it affects the important relationship existing between the area of a harbour and the width of its entrance.

The determination of area is the primary consideration. Obviously, small harbours will be more appreciably affected by external commotion than large harbours, assuming the inclosed areas in each case to be equally well protected, for it is easier to transmit agitation to a small body of water than to a large one. But, on the other hand, large harbours, unless most effectively screened and sheltered, are themselves liable to act in some degree as wave generating areas. Hence some discrimination is necessary, and the question of area is more likely to be determined by other considerations than those immediately connected with exposure. The required accommodation, the dictates of convenience to navigation, and the adaptability of natural features, in fact, have foremost place in the determination of area.

As regards entrance width, there can be little doubt that the narrower the entrance, the more effectually is the interior secured from the ingress of disturbing elements. In addition thereto, a narrow inlet very materially reduces the power of those waves which do find an entrance. On the other hand, an entrance must have adequate width for vessels entering not only singly and in calm water, but also when driven in groups under stress of weather. Accordingly, the entrance bears a double relationship to the harbour, viz., (*a*) as regards shelter, and (*b*) as regards accommodation.

From the first of these points of view, Stevenson has evolved an empirical formula to connect the extent of the reduction in the height of waves with the width of the inlet and the width of the sheltered area.

Thus, calling H the height of the unrestricted wave at the mouth of the harbour having an entrance width $b$, the reduced height of the wave, $h$, within the harbours at a distance, D, from its mouth and at a point where the breadth of the harbour is B, is given by the expression

$$h = \mathrm{H}\left\{\sqrt{\frac{b}{\mathrm{B}}} - \cdot 02\sqrt[4]{\mathrm{D}}\left(1+\sqrt{\frac{b}{\mathrm{B}}}\right)\right\}.$$

For example, when D = 256′, $b$ 100′, B 400′, and H 10′, we obtain $h = 3{\cdot}8'$.

The entrance width, however, is subject to other and further considerations. In tidal harbours there is the outrun of the ebb tide with the cumulative effect of the discharge of any upland waters, all tending to produce a rapid current in a narrow waterway. And while the scour induced by this means is beneficial within certain limits in maintaining a deep channel, yet, carried to excess, it is likely to prove prejudicial to the stability of walls and piers by undermining their foundations, and, moreover, the rate of flow may be such as to interfere with and possibly prevent safe navigation. To be precise, a velocity of from 3½ to 4 knots per hour should be looked upon as the maximum current permissible.

Such, in brief compass, are some of the more important matters bearing on the general question of harbour design, from which it will be seen that there are many weighty considerations which contribute towards a determination of the proper form and arrangement of areas reserved for the reception of shipping. In the ensuing chapters, it will be our duty to investigate some of these features more closely and in greater detail.

Meanwhile, we conclude the present section with a brief description of three ports selected as furnishing fairly representative examples of the three principal types of harbours, viz., national harbours, commercial harbours, and fishery harbours, and also of a trio of harbours remarkable more for their form than for their size, and possessing interest out of proportion to their commercial importance.

**The Military Outport at Libau.**[1]—Libau, as a commercial harbour, dates from the thirteenth century, and various extensions have been made in its accommodation from time to time. In 1870, when the Libau-Romen railway was constructed, the port came into considerable prominence, and in 1887 the Russian Government determined to make also a military harbour the site of which was to be immediately to the north of and connected with, the commercial harbour.

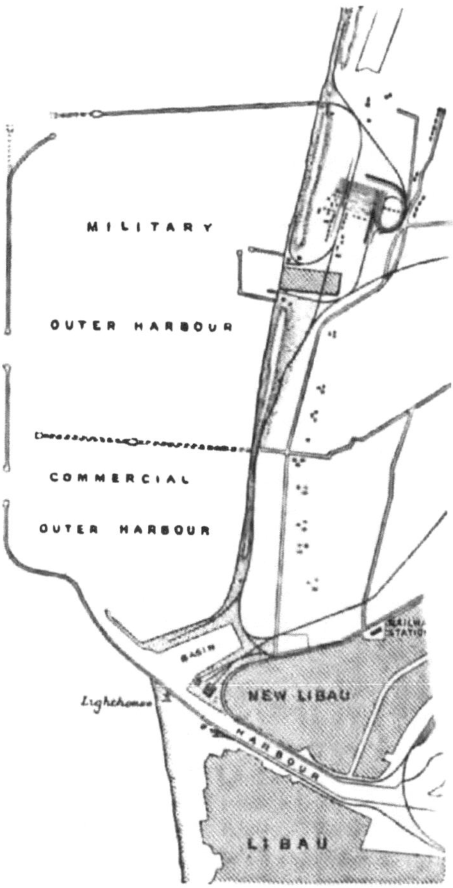


Scale, 1 inch = 5,000 feet.

Fig. 29.—The Port of Libau.

"In designing the general arrangement of the sheltering constructions of the outport, two questions had to be taken into consideration: (1) to lessen as much as possible the risk of the entrances and of the interior of the port being silted up by the coast drifts; (2) to prevent floating ice from accumulating in front of the walls and to assist the escape of the ice formed

[1] Jarintzoff on "The Military Outport of Libau," *Min. Proc. Inst. C.E.*, vol. cxxvi.

within the basin. For both purposes the design adopted may be considered as the most suitable. The movement of the sand along the coast is of a two-fold character. In shallow water the sand is carried by the waves along the shore and accumulates at each exposed point, which tends to prevent its further movement. For that reason the more the southern mole of the commercial port was extended into the sea, the more rapid was the growth of the coast in the angle between the mole and the shore; but, in the future, this growth will be slower, first, because the depth of the sea increases further from the shore, and secondly, because the mole was built out at once to a considerable distance and to a great depth, which obliged the waves from the west and south-west to glide along the mole and dash against the coast, thus scattering the sand collected. Certainly this does not prevent the harbour from silting up, but the sand is carried for a long distance along the coast, and therefore the danger of accumulations at the entrance of the harbour is considerably diminished. Beyond the breakwater the movement of the sand is produced by the coast current, in which the particles of sand are suspended. If the currents do not meet with any obstacles, the greater part of the sand is carried along the coast and is left in sheltered places, and this action is favoured by the circumstance that the breakwater and the point of the southern mole form a straight line. As regards the ice, which generally moves backwards and forwards from north to south, the arrangement of the walls in one line is very convenient. There is nothing to stop the ice and give isolated masses time to freeze together under the influence of the cold coast winds. Consequently there can be no accumulation of large ice masses, and a strong ice-breaker can at all times easily make a way out of the port into the open sea. The ice in the harbour, broken up by the ice-breaker, passes without difficulty through the three outlets; but this ice, owing to the mildness of the climate, is never so thick as to be a serious obstacle to the movement of the ships."

The military port, as formed, is 7,700 feet long, 7,000 feet wide, and occupies about 1,200 acres. Its natural depth is 14 feet at a distance of 1,400 feet from the coast, 22 feet at a distance of 3,500 feet, and it gradually increases to 29 and 30 feet as it nears the breakwater. The width of each of the three entrances is 700 feet, and the general depth seaward is 30 feet, though diminished in places to as little as 24 feet.

**Madras Harbour.**—The case of Madras Harbour is a striking instance, amongst many which might be cited, of the difficulties attending the effective realisation of benefits theoretically deducible from principles which are apparently sound in themselves and from a design conforming thereto so far as the available data will permit. Much that was hoped for has not come to fruition, and justifiable expectations have been disappointed.

The commercial ports of India are not numerous, in spite of the enormous extent of its seaboard. They can almost be counted upon the fingers of one hand, and most of them are of comparatively recent development. Up to the year 1875, there was nothing at Madras of the nature of a harbour, either natural or artificial; there was simply an open roadstead on an exposed, sandy coast, swept by storms and occasional monsoons of extreme violence, during which vessels were far safer out at sea. In fact, even at the present time, throughout the entire distance of 2300 miles extending round the

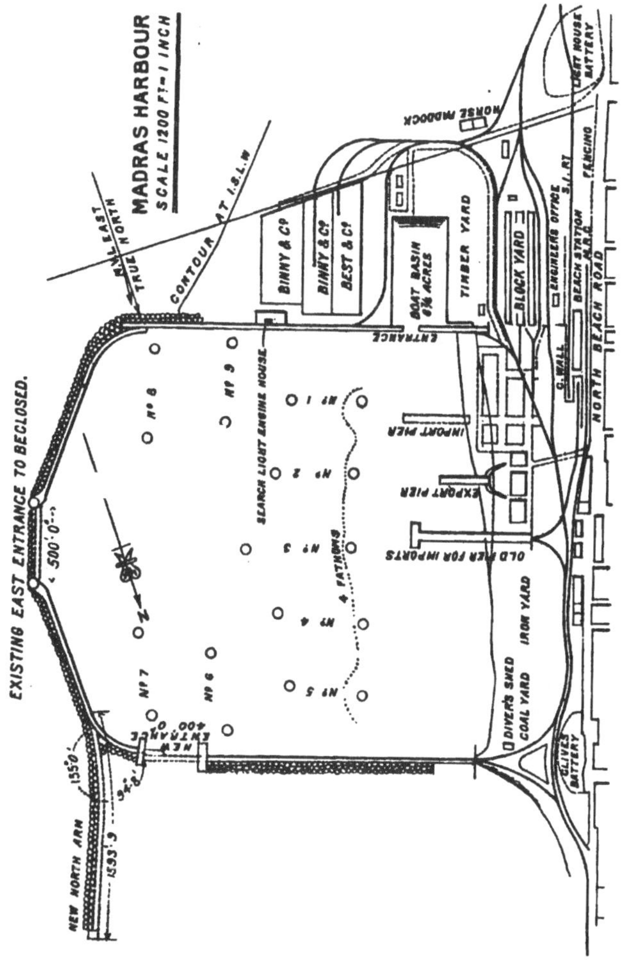


FIG. 30.—Madras Harbour.

peninsula from Calcutta to Bombay, there exists no mainland port capable of affording adequate shelter to shipping in times of cyclonic disturbance; still less of accommodating it, uninterruptedly, at the quayside.

At Madras, the unloading of vessels is carried on by lighters, or barges, of about 10 tons capacity, which traverse the distance between ship and shore. The transfer of cargo in the open was and could only be effected at considerable risk, and the first step to be undertaken in any attempt at an amelioration of the conditions was manifestly the creation of a sheltered area, within which barges could move freely and without danger.

There were no natural features to lend assistance to such an undertaking, and purely artificial dispositions had to be made. The initial design was prepared by the late Mr Wm. Parkes, and consisted of two breakwaters projecting perpendicularly to the coast-line, with rectangular returns terminating centrally in pier heads, 450 feet apart. Subsequently, the plan was modified by the Marine Department, so that the return angles became

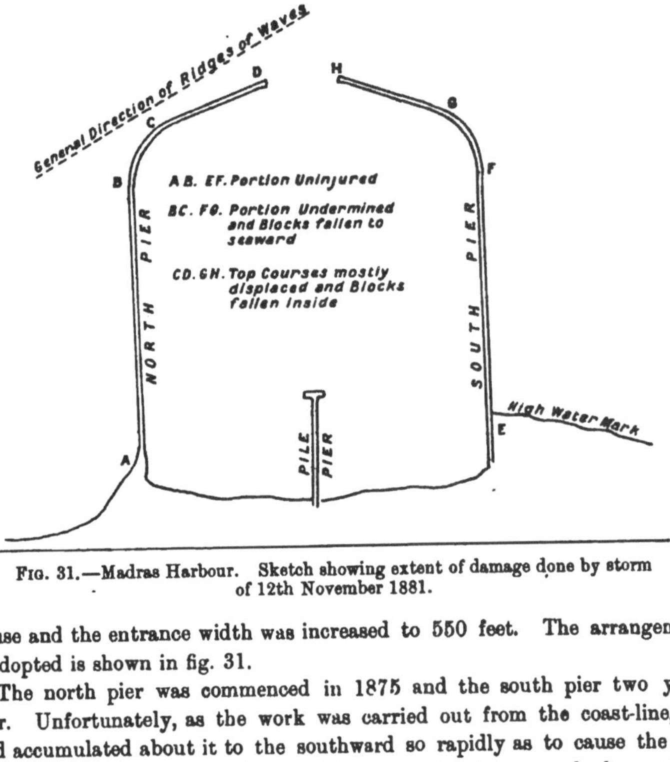


FIG. 31.—Madras Harbour. Sketch showing extent of damage done by storm of 12th November 1881.

obtuse and the entrance width was increased to 550 feet. The arrangement as adopted is shown in fig. 31.

The north pier was commenced in 1875 and the south pier two years later. Unfortunately, as the work was carried out from the coast-line, the sand accumulated about it to the southward so rapidly as to cause the line of foreshore to keep pace almost with the work. It was only by pushing forward with despatch in the intervals between the monsoons that the walls were eventually got ahead of the sand drift. They reached their respective pierheads by the year 1881.

At this epoch a disastrous cyclone occurred. On 12th November 1881, the sea swept over the breakwaters from both sides of the harbour, damaging the work to an enormous extent. Blocks of 27 tons a-piece were dismantled and flung into the inclosure, the walls were undermined by the scour (in places to the unprecedented depth of 22 feet below water) and some of the work even fell outwards under the pressure of the water pent up within the harbour.

In consequence of this disaster, the walls, after repair, were raised an additional 9 or 10 feet in height, and this has so far proved satisfactory, for no case of overflooding has since occurred.

The incident afforded an opportunity for a review of the design with the object of ascertaining whether any modifications were desirable. In 1883, a local committee convened at Madras to consider the official report of Sir John Hawkshaw, Sir John Coode and Professor Stokes, recommending a basis of reconstruction, advocated the adoption of an improvement scheme of their own, with a new entrance facing north-east. In assigning their reasons for wishing to abandon the original entrance, the members stated that:

"No matter what the direction of the wind, the unceasing swell on this portion of the coast rolls in with the crests of the waves parallel, or very nearly so, to the coast-line. In no case is it believed that the angle exceeds 30° to the general line of the coast. The result is that seas enter the present mouth freely, and, owing to the small length of the harbour, are not dispersed before reaching the shore at its base. The action is, of course, greatly intensified during storms, and particularly with the wind from the east. At such times, the sea inside the harbour, though not so high as outside, is certainly of a dangerous character, being exceedingly broken. Taking these and other facts into consideration, the committee have to record their opinion that unless means be found for closing entirely the present entrance, no radical cure will have been applied to the chief defect of the work as at present designed." [1]

In 1887 they issued a further statement.

"It is agreed on all hands that, owing to the frequently disturbed state of the water, the facilities for landing and embarking passengers, cargo, etc., offered by the harbour, are very much restricted, nor would it be feasible, for the same reason, to use, without serious interruption, wharves or jetties along the shore-line, or to keep in safety within it such improved lighters, tugs, and other harbour craft as would greatly increase its value as a trading port. Much cargo is said to be lost overboard in the process of transhipment, and, for want of tugs, no sailing vessels use the harbour at all.[2]

"The present or east entrance we believe to be the easiest and safest for ingress or egress, but not only does it admit the sea in the manner described, but we are of opinion that the time is not very far distant when the depth at the entrance will be so far reduced as to become too shallow for the larger class of vessels frequenting the port.

"The alternative is an opening in the north-east corner with a covering arm. This is the plan favoured by the Madras Board, and to this we have given our most careful consideration.

"The opinions of the captains of steamers frequenting the port differ materially. Some see considerable difficulty and danger in taking an

[1] *Official Papers, Madras Harbour*, 1902, p. 39. [2] *Ibid*, p. 53.

entrance so placed; others see none. We give it as our opinion that, although it may not be so easy of ingress, and ships may be detained outside more frequently than at present, the increased difficulty is not sufficient to condemn it."[1]

The recommendation was, however, set aside by the Secretary of State, and the work proceeded in accordance with the original design.

The allusion to the silting of the eastern entrance indicates another difficulty of the situation. Both prior to and since the completion of the works, the entrance has shown manifest indications of shoaling, at the rate of about one foot per annum. Although disquieting, this is not a cause of immediate anxiety, in that there is still at the present date something like 34 feet depth of water to meet the requirements of vessels which do not reach

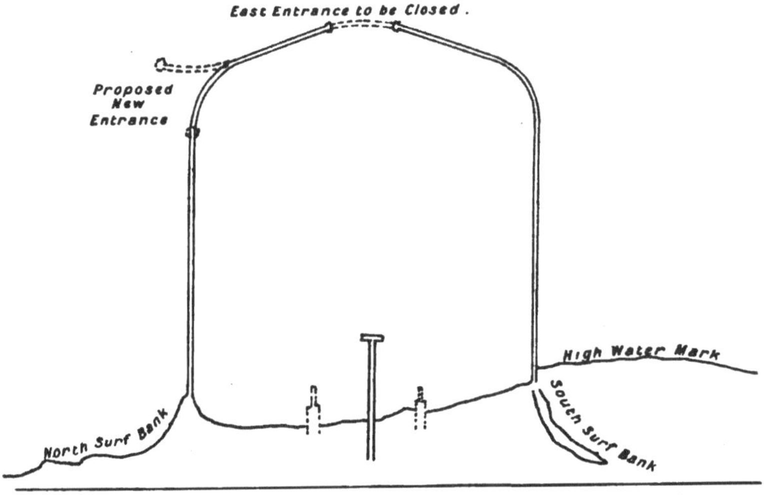


FIG. 32.—Madras Harbour. Proposals by Madras Special Committee, 1887.

that draught by 10 feet or more. Still, in view of future increments, the matter called for attention. Dredging operations have been put in hand and the evil has been checked, one month's work sufficing to remove a year's deposit.

Although shelved for the time, the project of a north-east entrance revived, and in 1902 practically the same recommendation as that put forward in 1883 was endorsed by an advisory committee, consisting of Sir Charles Hartley, Sir George Nares, and Sir William Matthews. An extension to the covering arm was added, making it 1600 feet long. The plan which is shown in fig. 30 has now received official sanction and is in course of execution. It is expected to be completed by the year 1910 or 1911.

The objections which have been urged against the scheme are the following:—(1) That it will entail the removal of part of the existing breakwater—

[1] *Official Papers, Madras Harbour*, 1902, p. 54

a difficult and uncertain operation; (2) that the covering arm will be costly on account of the strength required to resist a broadside attack; (3) that the arm will favour the travel of littoral drift, and lead to shoaling in the neighbourhood of the new entrance; and (4) that the latter will be navigable with difficulty in a heavy sea. None of these objections, however, can be held to possess sufficient force to overrule the conclusion carefully arrived at, that the existing arrangement is unsatisfactory and no longer maintainable. They possess no insuperable drawback in themselves, and the scheme is undoubtedly the best modification which can be devised in the light of local experience.

For much of his information respecting the Port and its history, the author is indebted to the courtesy of Mr F. J. E. Spring, C.I.E., the chairman and engineer of the Harbour Commissioners. The following interesting observations are direct from the pen of that gentleman:—

"Madras Harbour makes no pretence of offering, nor, owing to its situation, can ever pretend to offer, shelter to shipping in very severe weather. For, though the artificial breakwaters are finished at a height of 12 or 14 feet above sea level, they cannot, at that height, shelter the sides of modern steamers from winds, the force of which may avail to break the mooring chains. There are in the harbour nine berths for large steamers, with permanent bow and stern buoys, the bow buoy chain being 3¼ inches diameter and the stern buoy chain 2¾ inches diameter. But in heavy winds the vessel's own stern lines sometimes break; accordingly, the moorings are so arranged that each vessel may swing round clear of every other vessel and of the breakwaters.

"The breakwaters, however, very greatly mitigate the roughness of the water within the harbour, and so enable the barges and lighters locally in use —of from 2 to 10 tons burden only—to carry cargo in all weathers between ship and shore. The water is not, however, always smooth enough—that is, from twenty to thirty days in the year it may not be smooth enough—to allow ships to lie alongside jetties.

"It is in view of obtaining smooth enough water within the harbour for ships to lie alongside permanent jetties that the scheme of harbour improvement, now in hand and to cost some £375,000, has been initiated. The scheme in question is expected to be completed by the year 1911, and it will then be time enough to consider whether the volume of trade is likely to be such as to warrant the expenditure of perhaps £200,000 more involved in the erection of permanent jetties for ships' use. Madras is not, and probably never will be, a terminal port, where ships can discharge and take in full cargoes. By far the greater number of vessels touching there either have only a few hundred tons to put out and take in, or discharge full cargoes and go elsewhere for return cargoes; or, again, come from other ports for whole or part cargoes. Few vessels, therefore, seem likely to need the convenience of quays erected at an expense which must put a permanent burden on the trade of the port and a fairly heavy tax on the tonnage of goods using such quays, if the dues necessitated by the extra expenditure are confined to goods using them.

"But Madras may be made an improved port for shipping if a better class of barge can be used than the 2 to 10 ton lighters now in use. At present no private owners, nor the Port Trust, can be expected to invest money in such plant as tugs and 100-ton barges, which are practically certain to be wrecked when a cyclone occurs, as it must occur once in some eight or ten years' time. The existing small craft come ashore and not much harm is done, but bigger craft would be wrecked to a certainty. Therefore the Port Trust authorities have recently excavated a 6½-acre basin for such vessels, as well as for the Port's own dredging-plant, the floating timber trade, etc., and they have it in contemplation to provide a few self-propelling 100-ton lighters, which, pending their replacement by private enterprise, will be worked between ship and shore by the Port authorities. There are now in use two steel jetties, one 800 feet long, for imports, and the other 350 feet long, for exports. Another export jetty, fully equipped with modern hydraulic cranes, is about to be erected, as well as some wharves at the harbour breakwaters, alongside of which coal-ships can lie and discharge during, probably, 300 days in the year. The new north-eastern entrance, under its sheltering arm, and the closing of the old entrance, will enable work to be done with far greater comfort and convenience and less relative motion between ship and lighter or ship and coal-wharf than is now possible."

Respecting the shoaling difficulty, Mr Spring adds:—

"The sandy shore of the thousand or so miles of the eastern coast of India is continually being acted upon by surface waves dashing upon it, for part of the year with a north-western set, and for part of the year with a south-western set. The resultant of these two sets has a north-westerly trend, and the effect is that wherever an obstruction juts out from the shore, *e.g.* such an obstruction as is offered by a harbour arm, there is necessarily an accretion of sand to the south of it, and the contrary, in the form of erosion, to the north of it. The accretion is greatly assisted in its formation by the effect of the wind upon the sand thrown up on the shore by the surface waves; for it will be understood that in a tropical climate the sand is very quickly dried by the intense solar heat. The prevailing winds, blowing for months together in one direction, bear the dried sand landwards, and pile it up until its level is from 3 to 5 feet above the level of high water. Such an accretion, to the extent of some hundreds of acres, has found itself on the south side of Madras harbour, affording valuable land for various port purposes. It is in this accretion that the 6½-acre boat basin, previously alluded to, has been excavated.

"The sandy accretion has now extended itself nearly out to the extremity of the south harbour arm, and the sea current, running for several months in the year northward past the entrance, can be seen laden heavily with sand in suspension, which, naturally, is being dropped in the entrance. The entrance has thus, in the last ten years, been steadily shallowing at the rate of about one foot annually. But lately it has been found, since the Port authorities have become possessed of a modern 1000-ton suction dredger, that a month's work in the year easily overtakes the annual silting.

"When the old entrance is closed in 1911, the sand will undoubtedly tend to flow in the current, past the closed entrances and northwards to the end of the new sheltering arm, there to be dropped in the still water formed by that arm near the new entrance. But there seems no reason to fear that modern dredging methods will not be equal to coping with the difficulty."

**Whitby Harbour.**—The town of Whitby, lying at the mouth of the river Esk, affords an instance of a fishing harbour maintained chiefly by tidal scour. There are two piers at the entrance, the west pier originally projecting considerably further seaward than the east pier. The meeting of the flood-tide, however, with the river current, produced an eddy just within the west pierhead, leading to slack water and shoaling.

Thus, a vessel desirous of entering the harbour had to go nearer the east pier in order to avoid the bar; in so doing it was in danger of losing steerage-way, owing to the strength of the flood-tide (which flowed eastward at a considerable rate), and tended to drift beyond the east pierhead before making the entrance. In order to minimise the trouble of the bar, some large stones were placed N.N.W. of the west pierhead below the water level, at which ships might safely enter the harbour. These stones acted as a groyne, tending to prevent the formation of the bar by arresting the eastward progress of the sand deposit.[1]

The remedy ultimately adopted consisted in prolonging the east pier until both pierheads were in a line parallel to the set of the tide. By this means the width between the pierheads was diminished by nearly one-half, the bar disappeared for a time, and heavy seas which formerly entered the harbour were perceptibly reduced. The projection of the east pier, however, caught the waves from the north-west, instead of allowing them to pass the entrance as before. These waves, passing into shallow water, stirred up the sandy bottom, becoming heavily charged with material. They swept along until they struck the inner face of the east pier extension, whence they rebounded within the harbour, and, reaching slack water, the sand which they carried was deposited.

Besides silting up the harbour, the decreased width of the mouth made the entrance exceedingly dangerous, as, owing to the rapid cross-flow of the tide, a vessel had great difficulty in shooting in between the pierheads when running before a north-west wind. She might strike the east pier-end or drift on to the rocks beyond; or, if she effected an entrance, she might collide with the inner face of the extension.

The state of affairs was, therefore, far from satisfactory, and a further scheme of improvement has been decided upon and is about to be undertaken. Messrs J. Watt Sandeman and Son, the engineers, have favoured the author with the following observations, both in regard to the defects of the present harbour and the proposed remedial measures.

[1] Vide Austen on Whitby Harbour, *Min. Proc. Inst. C.E.*, vol. clvi. p. 264.

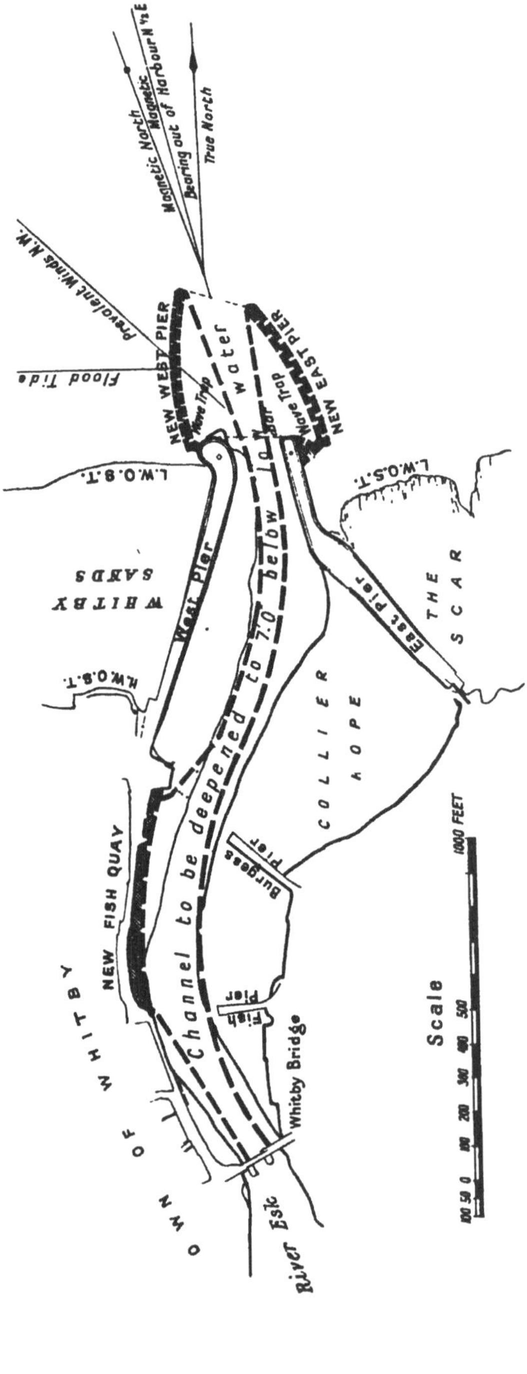


FIG. 33.—Whitby Harbour. Showing proposed improvements.

"*Defects of the present harbour*: (1) The shoal depth, narrow breadth, and irregular course of the entrance channel.

"At low water, the depth of the channel between the pierheads is about 2 feet, and from the pierheads to the harbour quays it averages only about 1 foot. This shallow depth causes detention to the largest class of herring-boats of from 2½ to 5 hours, which greatly diminishes the value of the fish, particularly in hot weather.

"The breadth of the channel varies from 50 to 80 feet. Its course, instead of being midway between the piers, is close to the outer arm of the east pier, and at the inner end of that arm it takes a sharp bend. But the most dangerous feature is the sand-bar just outside of the pierheads, which dries several feet at low water, and which, owing to wave action from pre-

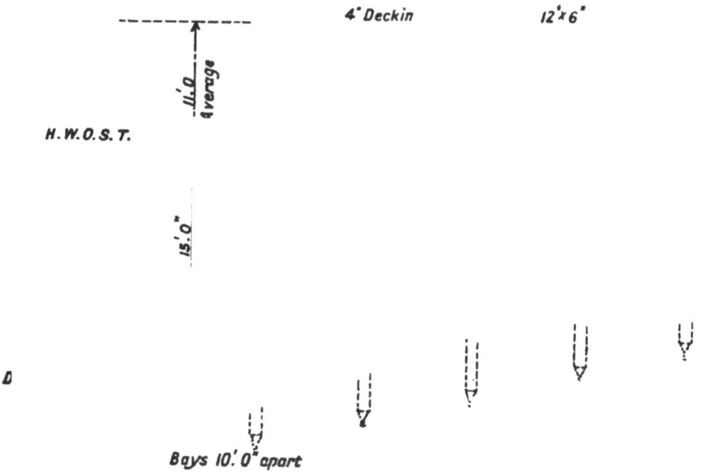


FIG. 34.—Whitby Harbour. Proposed fish quay.

vailing northerly winds combined with the flood-tide, is continually tending to close the entrance channel, and, at times of drought in the river, does so.

"Upon this bar, during storms, seas break so heavily that it is impossible for boats either to enter or leave the harbour, and this is also the case for some time before and after low water, even during moderate gales.

"(2) The range of waves into the harbour. This is so great, that even during slight gales boats cannot lie in the outer harbour, and are obliged to go above the bridge for safety.

"*Proposed Works of Improvement.*—The following works are the least which it would be necessary to undertake in order to remedy the defects enumerated, and to render Whitby an efficient fishing port:

"(1) Two new piers extending from the present pierheads to a permanent sea depth of 7 feet at low water.

"The necessity for such piers is, first, to enable that depth of water to be maintained in the entrance channel, chiefly by reducing the travel of sand from the west foreshore, and, secondly, to reduce the admission of waves and

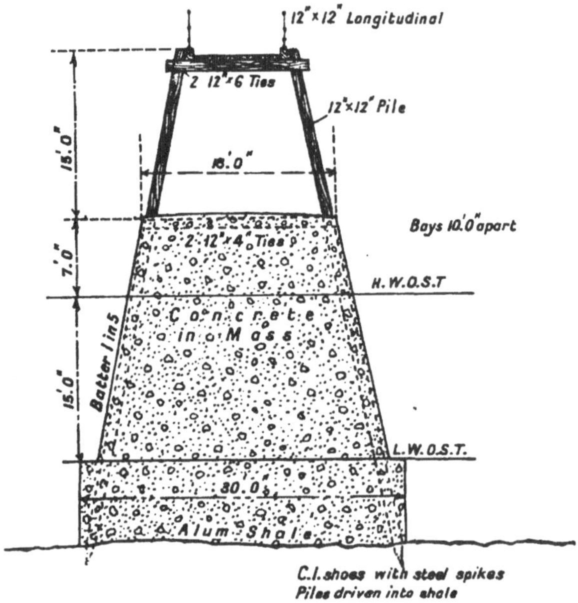


FIG. 35.—Whitby Harbour. Proposed piers.

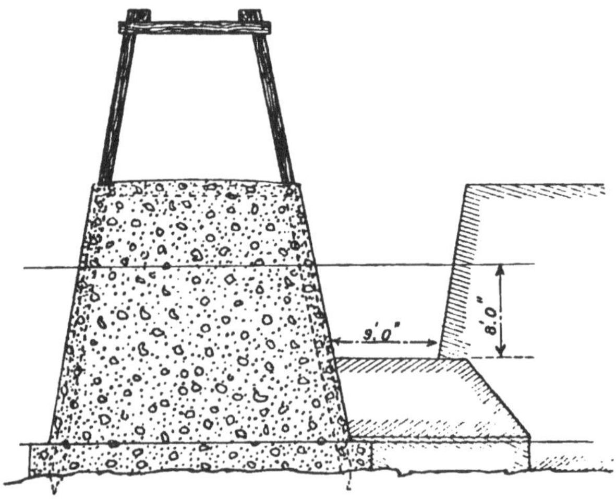


FIG. 36.—Whitby Harbour. Front view of proposed wave trap.

consequently the range of sea into the harbour, and so render the harbour safer to be taken by ships and boats during storms.

"(2) Dredging a channel to a depth of 7 feet at low water from the new pierheads up to the bridge, and widening the same for a length of 700 feet in front of the new fish quay.

"(3) The construction of a new fish quay, 700 feet in length.

"There is not at present sufficient quay space in the harbour for the landing and sale of fish, and the existing quays would not admit of a depth of 7 feet at low water being dredged alongside of them.

"The prominent position of Whitby gives it a great advantage for sailing boats over embayed harbours, such as Scarborough and Hartlepool, where boats often lose much time by being becalmed; while at Whitby, when there is any wind at all, boats get it immediately outside of the pierheads.

"The position of the railway at Whitby, alongside of the harbour and at the level of the quays, is another great advantage as compared with North Shields, Scarborough, and other harbours, where fish have to be carted uphill, at great expense to the railway."

**Danish Island Harbours.**[1]—The Danish Isles and the Peninsula of

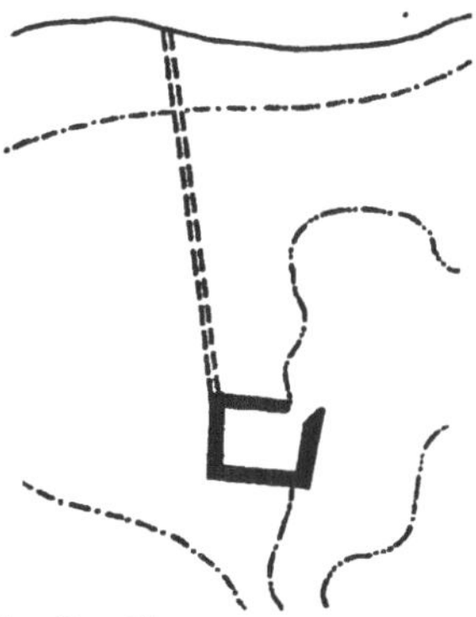

FIG. 37.—Plan of Arnager Island Harbour.

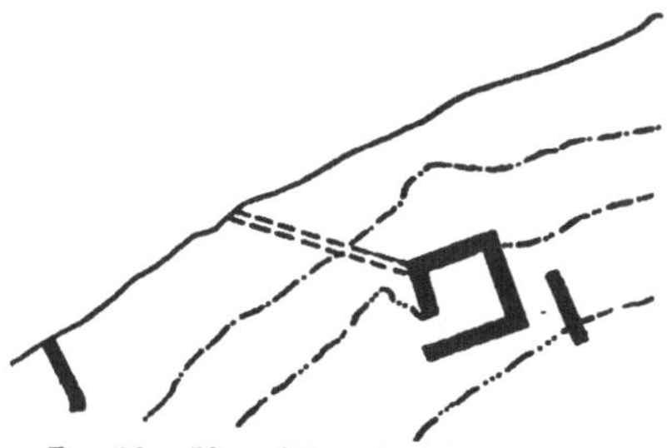

FIG. 38.—Plan of Snogebæk Island Harbour.

Jutland have an area of only 14,850 square miles and a shore-line of about 3274 miles. On a part, perhaps about $\frac{1}{7}$th, of this length, particularly on the south coast of the Island of Bornholm in the Baltic, on the north coast of Zealand facing the Cattegat, and on the north and west coasts of Jutland, facing the Skagerrack and the North Sea, construction of harbours is rendered difficult by the littoral drift. Here the island harbours are situated. Two of these, at Arnager and Snogebæk on Bornholm, were built in a tentative way, in 1883 and 1888, whereas the third one at Hundested on Zealand was formed in 1893 by the transformation of an originally land-connected harbour. All three, shown in figs. 37–39, were built by Mr H. Zahrtmann. The basins inclosed by riprap moles are from 4 feet 6 inches to 8 feet deep, and cover areas of ·27, ·2 and 1·62 acres respectively, or, including the outer basin of the latter, 2·16 acres. From the moles, open

[1] P. Vedel on "Island Harbours," *Trans. Am. Soc. C.E.*, vol. liv. Part A., Proc. Int. Eng. Conf., 1904.

viaducts, wooden or composite, from 330 feet to 660 feet long, lead to the shore, and are divided into from thirteen to twenty bays, which span openings from 20 feet to 30 feet wide.

On the coast of Denmark, tides are insignificant, hardly perceptible in the Baltic, not exceeding 1 foot in the Cattegat, and rising only to 4 feet 6 inches in the North Sea near the Southern Boundary. Hence, any movement of material which takes place is not due to tidal action, but to the action of the waves, combined, perhaps, with that of local currents, and, whichever of these agencies be considered, attributable to the effect of the wind.

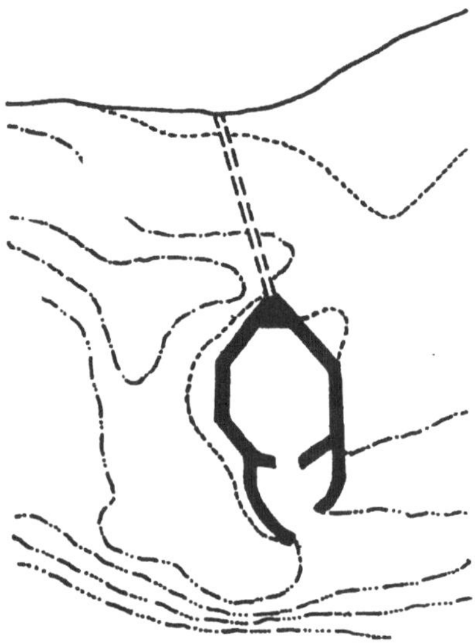

FIG. 39.—Plan of Hundested Island Harbour.

The object aimed at by the three fishing ports seems to have been accomplished fairly well. At neither Arnager nor Snogebæk has material accumulated to an alarming degree; it is pure quartz sand, the size of the grains being ·45 and ·25 mm. respectively. At Hundested the drifting material is more heterogeneous, consisting of a mixture of quartz sand with grains of ·25 mm. in diameter, gravel, shingle, larger pebbles, and good-sized boulders. Some accumulation has taken place inside the 5 m. contour, and banks have formed at the south-east mole and at the shore south-east of the port; but a state of equilibrium seems to have been reached, in which these two banks play an important part.

# CHAPTER III.

## SURVEYING, MARINE AND SUBMARINE.

Soundings—Methods of Procedure—Appliances: Line, Chain, and Pole—Sutcliffe's Apparatus—Maintenance of Alignment—Variations in Water Level—Tide Gauges—Current Observations—Floats—Localisation—Plotting Positions—Diving Operations—Bells—Diving Dress and Equipment.

**Marine Surveying,** while generally and appropriately considered as constituting a special department of maritime work, with a field and purview of its own, has at the same time certain of its operations so closely associated with the ordinary routine of harbour engineering that some reference to them, if not absolutely imperative, becomes, at least, eminently desirable.

It is not contemplated, however, to enter into any detailed explanation of the principles underlying the carrying out of a hydrographical survey, nor even to describe in outline the series of operations involved in the preparation of a chart of any portion of the coast-line for purposes of harbour design. Such matters, affording scope for no little mathematical investigation and requiring much elucidation of particular problems, are to be found treated in geodetic text-books specially devoted to that end; and even supposing that this were considered an appropriate course to pursue, it would hardly be possible to deal with the questions which would inevitably arise, in a manner at once sufficiently comprehensive and succinct for inclusion within the limits imposed by the requirements of this treatise.

The operations more immediately concerning the engineer in the actual constructional and maintenance work of a harbour are, (*a*) the taking of soundings, and (*b*) the determination of the direction and velocity of tidal and fluvial currents. Our observations, therefore, will be confined to these points.

**Soundings.**—The taking of soundings is a very common operation in navigation, but the appliances used in that connection and the methods in vogue are by no means identical with those characteristic of harbour practice. In the latter sphere much greater precision and accuracy are essential than can be afforded by the somewhat rough and ready appliances employed in connection with shipping.

Manifestly the simplest way of taking a sounding, that is, of ascertaining the depth of water at any spot, is to lower a pole or weighted line until the bottom is reached. If the pole or line be graduated to linear measure, the depth can be read directly therefrom. This method is perfectly satisfactory when performed from a stationary base, as, for instance, when the operator is standing on a quay wall or on a boat which is moored, and in this way it is, of course, only applicable to single dips.

When it is desired to take a series of dips along a given line in any direction, the base, if afloat, must obviously be movable; and while a boat, no doubt, may be alternately moved and moored so as to fulfil the condition stated above, yet the process would be slow and tedious. It is evidently preferable to adopt some method of taking soundings in close and uninterrupted sequence while the boat is in continuous motion.

FIG. 40.—Wire Sounding-line.

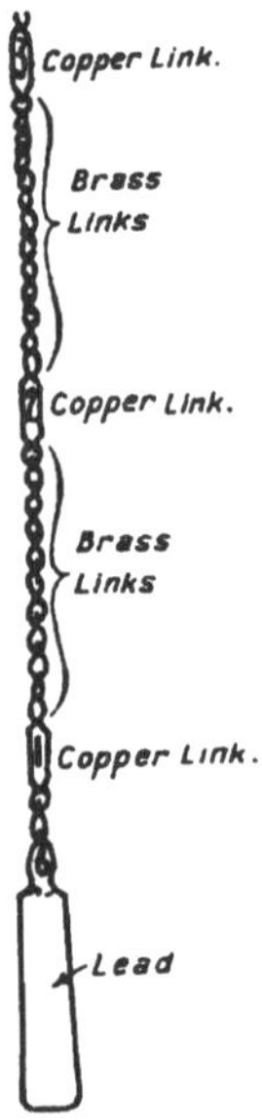

FIG. 41.—Chain Sounding-line.

Lowering a pole or line under these circumstances would lead to inaccurate readings, as, by the time the bottom was reached, the travel of the boat would have produced considerable inclination in the instrument, so that it was no longer vertical. Seamen get over this difficulty by throwing or heaving the lead, which weights the line, some distance ahead of the boat, giving it time to reach the bottom before the boat passes perpendicularly over it. The reading is then taken at the moment of verticality, as near as can be judged.

The method is somewhat rough and crude, but, except in deep water, it gives fairly reliable results. The hand-lead, ordinarily used for the purpose, has a length of line not exceeding 30 fathoms, which is more than ample for any engineering requirements. The length of line for purposes of harbour soundings need hardly exceed 50 or 60 feet, and very often much shorter lengths than this will serve.

The **line**, if of hemp or wire, is graduated by tags of different texture, shape, and colour, so as to be identified by night as well as by day (fig. 40). Strips of woollen material, cotton, leather, serge, and even string, are utilised to give variation. If a chain line be used, copper-stamped links are substituted for the tags (fig. 41). A **chain** is preferable to a hemp line on account of the excessive shrinkage of the latter, amounting, when new, to as much as 5 per cent. during the course of a day's work. New lines should, in fact, be avoided for this reason, and all lines should be well wetted just before use. Finally, they should be tested frequently—before and after each line of soundings, if possible—by some standard length, which, for all practical purposes, may be marked on the boat itself.

A seaman's lead is usually an octagonal bar, weighing from 8 to 10 lbs. For engineering work, a flat disc weighing about 5 lbs. is more suitable, as there is less tendency for a lead of this shape to sink into mud or other soft material. If, however, it be desired to penetrate through the mud to firm ground, a bar or ball must be employed.

For soundings in shallow water, in depths, say, not exceeding 20 or 25 feet, a **pole** is sometimes used. As regards convenience in handling, white pine, which is light, forms the best material from which to make it. The pole is either circular in section, or oblong, about 2 inches by 3 inches, with hollowed faces, painted and graduated in feet and quarter-feet; sometimes in feet and inches. It should be shod with a flat-bottomed shoe for the reason stated above, and the weight of the shoe should be just sufficient to assist in sinking the pole to the bottom, and no more. The manipulation of a long pole, however, is by no means an easy matter, and it is only or mainly used for minor purposes.

The drawback attaching to all these appliances—line, chain, and pole—is their liability to miss some prominent protuberance in the bottom due to the isolated nature of the dips and the distance which lies between them. Except in perfectly still water, it is necessary to keep the boat moving at a certain rate in order to steer it and prevent it from being deflected out of its course by the current. It is difficult, therefore, under ordinary circumstances, to take soundings with the lead line at intervals of less than 10 feet. And a good deal may lie hidden in 10 feet. As a matter of fact, soundings are frequently taken at much greater distances. Furthermore, there is the effect of wave motion, which interferes very materially with the accuracy of the readings.

For these reasons, and for others which it is unnecessary to enumerate, a more complete and reliable system of recording depths in connection with harbour work is highly desirable, and a number of attempts have been made to supply apparatus which will conform with the requirements of the case. Chemical and electrical agencies have been proposed, and tested with varying, but generally unsatisfactory, results. They are too sensitive in action and too delicate in adjustment for use in exposed situations amid unstable surroundings. Whatever possibilities they may contain, at any-rate they have not yet been put into a working form, and mechanical appliances still seem to supply the only practicable means of dealing with the problem.

For sounding work in estuaries, harbours, and coastal inlets generally, the most serviceable and efficient machine with which the author is acquainted is one designed and patented by Mr Fielden Sutcliffe of Liverpool. Having had occasion to use it many times, the author feels in a position to speak authoritatively on its capabilities, and he has no hesitation in testifying to its value. The following is a description of the machine.

**Sutcliffe's Sounding Apparatus.**—The apparatus illustrated in figs. 42, 43, consists of three parts: the Sounding Machine (shown fitted up on a boat), the Horizontal Distance Measurer, and the Section Plotter.

The *Sounding Machine* consists of a wheel with a grooved rim, on which is wound the sounding line, a fine steel wire having the lead attached to its free end. The wheel is mounted in a frame which is arranged for clamping to the gunwale of a boat at its starboard quarter. The sounding-boat is also equipped with a sprit and a leading block fitted over the starboard bow.

On the back of the wheel is a spiral reel, on which a second line, called the "Preventer Line," is wound. The free end of this line, after being passed forward and through the sprit block, is taken back to the lead to which it is attached, the line thus forming a right-angled triangle. The function of the line is to prevent the trailing of the lead astern of the boat when the latter is moving. It not only ensures the verticality of each dip, but it also enables the lead to be maintained within a few feet of the bottom, so that dips may be taken frequently and regularly. Any casting forward of the lead is entirely obviated. The wheel and the reel are so proportioned relatively to each other and to the horizontal line from the wheel to the sprit block, that they each pay out or take in the requisite amount of their respective lengths to maintain the lead-line vertical at all depths of its range.

The wheel measures 10 feet in circumference at the bottom of its rim groove; consequently, the length of sounding line paid out per revolution is 10 feet, with fractions of a revolution in proportion.

At the front of the wheel frame there are two scales, along which a pointer is caused to travel at a rate proportional to the vertical travel of the lead. These scales are adjustable; one may be set so that the pointer indicates upon it the absolute depth below the surface level, while upon the other is indicated the depth relative to any assigned datum line. Indications are also afforded by the rim of the wheel, which is graduated and provided with a fixed pointer.

In taking soundings with this machine, the operator first sets the lead at the surface of the water with zero on the wheel at the fixed pointer, and the scales adjusted to the movable pointer.

He then grasps the rim of the wheel with his left hand, releases a catch with his right hand, and allows the wheel to revolve until the lead strikes the bottom; then, reversing the motion of the wheel, he reads from the pointers the depth indicated at the instant he feels the sounding line become taut. Continuing the movement a little further, he raises the lead clear of the bottom in readiness for the next dip.

The operator is assisted in the picking up of the slack line by the reaction of a coil spring on the wheel axle.

The *Horizontal Distance Measurer* consists of a drum on which is wound a length of fine steel wire. The drum is mounted, with its axis vertical, in a bracket fixed at the stern of the sounding-boat. As the boat moves away from its starting-point, to which one end of the wire is attached, the drum rotates and pays out the wire, the unwinding being regulated by a hand-wheel and screw acting upon a band brake. The revolutions of the drum

Fig. 42.—Sutcliffe's Sounding Apparatus. [*To face p.* 46.

are made to indicate on a counter the length of wire paid out, or to act in conjunction with the Section Plotter.

The *Section Plotter* is a device for recording on paper a line of soundings to scale, without the necessity for any intermediate bookings. By its means a band of paper is made to traverse at right angles the track of the movable pointer of the sounding machine at a rate proportionate to the paying out of the distance line. Thus the pointer, the motion of which across the paper

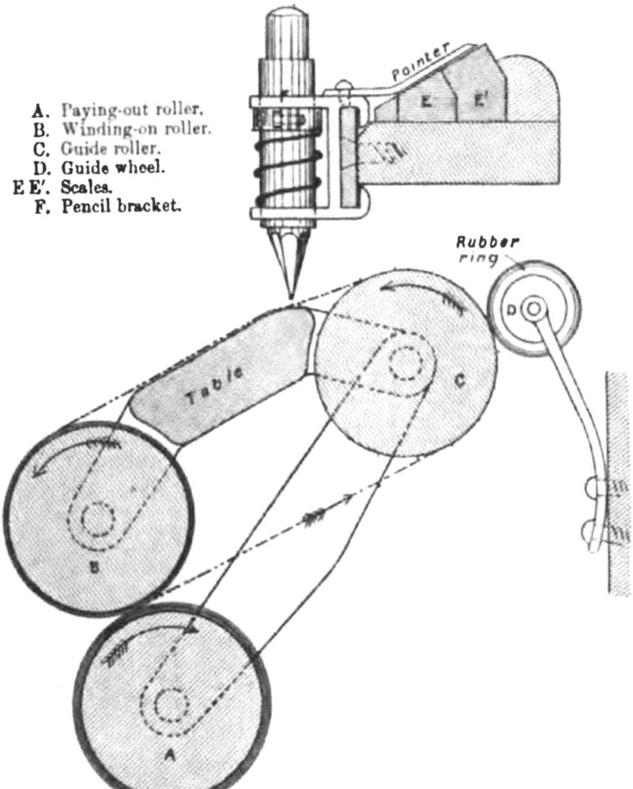


**FIG. 43.—Mechanism of Section Plotter in Sutcliffe's Apparatus.**

represents the amount of rise and fall of the lead, acquires relatively to the paper a second movement corresponding to the horizontal travel of the lead. The pointer carries a marker, sprung just clear of the paper, and the operator, on taking a sounding, has only to tap the marker with a finger of his right hand in order to record the sounding on paper, by means of a dot which denotes the position of the lead on the section plane, both vertically and horizontally.

There are a number of ingenious details by which the apparatus is rendered

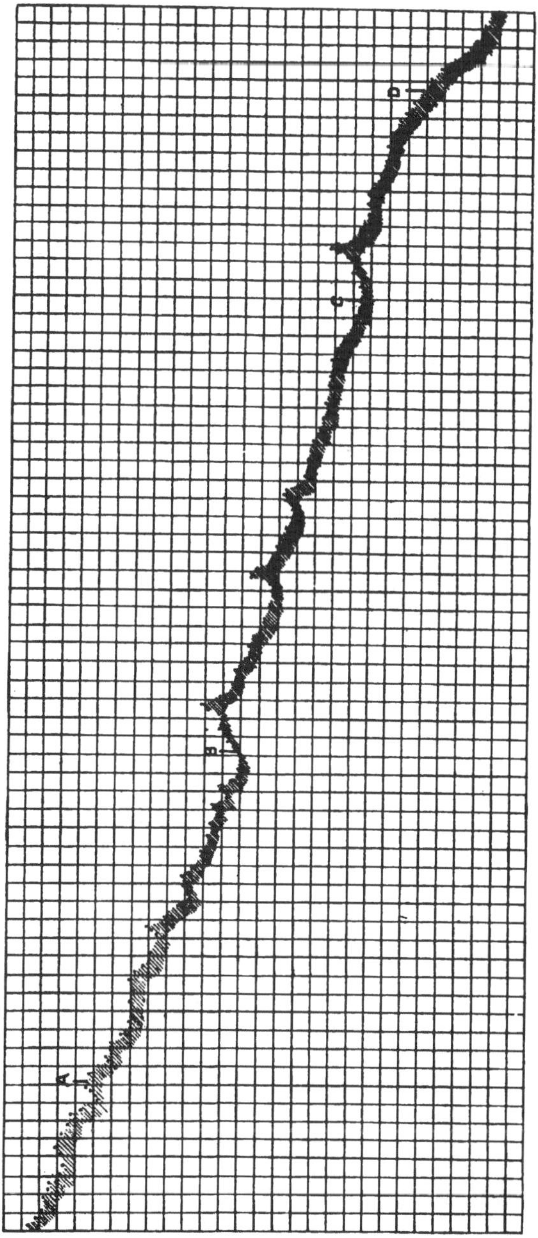


FIG. 44.—Line of Soundings taken by Sutcliffe's Apparatus. The dots only were indicated by the machine; the etching added subsequently

fully efficient for the purposes stated above. It would, however, take too long to enter into these, and we must dismiss the subject at this point in order to return to the main theme under consideration.

In the example (fig. 44) of a line of soundings taken and plotted by Sutcliffe's Apparatus, the letters A, B, C, D represent the position of cross sights, by means of which the operator verifies his distances.

During the process of taking soundings in a tideway, two points demand the constant attention of the operator. One is the *alignment of the dips*, and the other is the *mutation of the water level.*

**Alignment** is difficult to maintain in a cross current, and the boat needs to be carefully watched to see that it does not drift out of its course. If a line of soundings be taken from the shore or quay out towards the open, a couple of poles or other suitable uprights, one at the water's edge and the other some 30 or 35 feet back, should afford adequate guidance. When the bank is steep, it may be necessary to give the rear pole greater elevation than the other in order to be able to range them both with the eye from a lower level. If a line of soundings lie between two fixed points, such as stakes fixed into opposite banks of a river, the distance between them not being very great, a rope may be stretched taut from one to the other. In this case, by providing the rope with tags at regular intervals, the exact distance of each sounding can be recorded. When there is only one fixed point, distances may be read off a cord or rope, similarly tagged, and paid out from the boat as it proceeds. Failing this, the position of the boat at each dip must be fixed by angular measurement from the shore, with the aid of the sextant or theodolite.

**Variations in the Water Level** should be noted at regular and stated intervals (say every ten or fifteen minutes) by an observer stationed at a tide-gauge adjacent to the site of operations. The operator in the boat also notes the dips which correspond to the same intervals of time, and, by subsequent comparison with the tide-gauge readings, the proper correction can be made by which all the soundings are referred to one datum line, either local and temporary, or established and general.

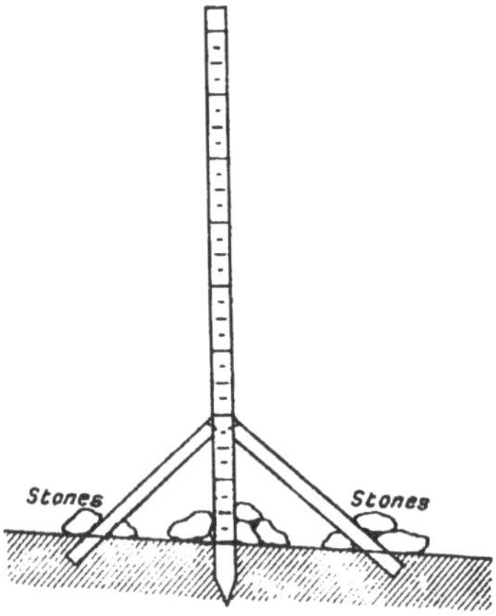

Fig. 45.—Temporary Tide-gauge on Beach.

A **Tide-gauge** is an appliance for the purpose of indicating changes in the sea surface level. In its simplest form it consists of an upright stake or post (fig. 45) driven into the shore or bank, and graduated to linear measure. In some situations a single post may suffice to indicate the whole tidal range; in other cases a number of posts may be necessary, extending across a sloping shore from high water level to low water level and forming a series of steps.

In cases where there is any swell, the gauge may consist of a rod or indicator, with a float at its lower end inclosed in a tube, the bottom of which is perforated as shown in fig. 46. Such an apparatus may be affixed to a quay or other vertical wall.

In important localities it is customary for tide-gauges to be more

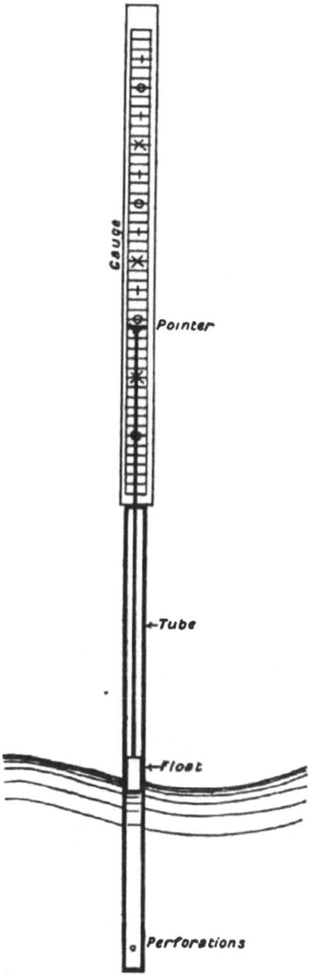


FIG. 46.—Tide-gauge for use in rough or choppy Water.

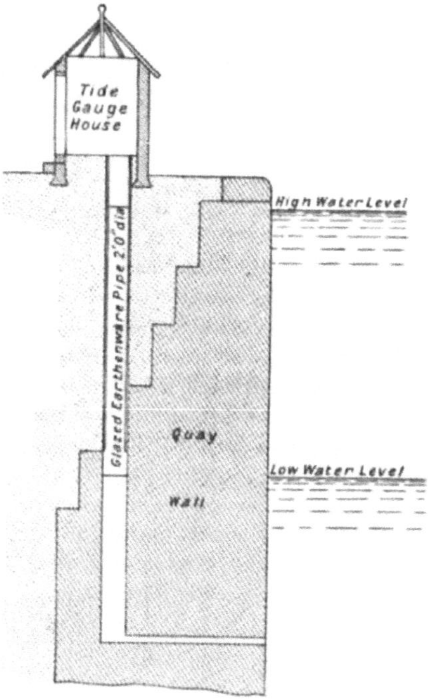


FIG. 47.—Tide-gauge House on Quay Wall.

elaborately constructed and to possess self-registering apparatus. A well or tube in free communication with the sea is fitted with a float supporting a graduated upright stem, which passes upwards to a scale and pointer. For self-registering purposes the float is connected by means of a chain or cord with a movable pencil suitably situated so as to mark the surface of a paper-covered cylinder, which is rotated by clockwork.

In taking soundings, it is necessary to have a tide-gauge immediately adjacent to the scene of operations, and when the scope of these is extensive, several gauges at various points will be required, because fluctuations in the water level are frequently local, and they are by no means uniform.

Examples of permanent tide-gauge stations are shown in figs. 47 and 48: one situated at the edge of a quay wall, and the other on a river bank. In muddy places a flushing bank is provided for cleansing the gauge well and maintaining free access of water to it.

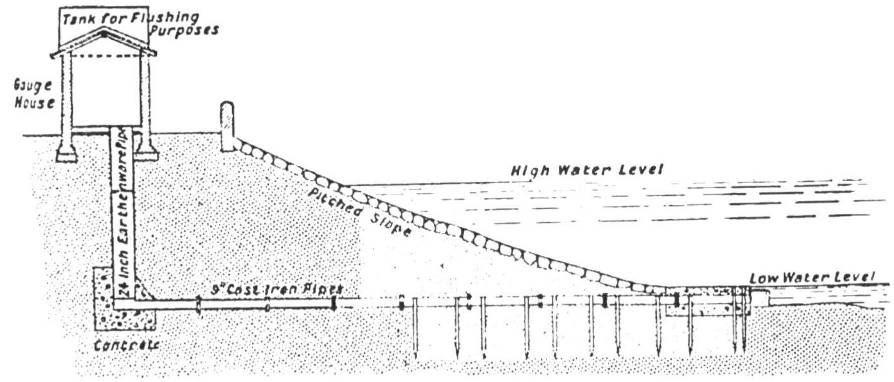


FIG. 48.—Tide-gauge House on River Bank.

**The Determination of Currents.**—It is important to the engineer to know the directions taken by tidal currents at various times during the day, and to observe their relation in regard to the configuration of the coast-line and the maintenance of navigable inlets. To acquire this knowledge, he has in many instances to fall back upon personal observation, and one of the earliest steps in connection with the laying out of harbour work will be to acquire the requisite data in regard to current flow.

**Floats.**—The most obvious method of observing the set or direction of a current is by means of some floating object. Any substance drifting upon the surface of the water affords a means of recognising the trend of tide or stream. Yet it must be pointed out that so simple an expedient—despite its apparent reliability—is not without very serious drawbacks, and that its universal efficacy is by no means to be taken for granted.

In the first place, paradoxical as it may seem, the topmost layer of the water may flow in a different direction to the lower layers or main body. Fresh water has a less specific gravity than salt water, and does not readily mix with it. A fresh-water stream, encountering a tidal inset, will therefore flow over it for some distance before becoming incorporated therein. The wind also is capable of exerting so powerful an influence on the surface of a body of water in motion as to make it move directly counter to the under portion. Then again, there are cross channel currents producing a spiral action, in such a manner that the particles of water move across the bottom in a totally different direction to that which they take along the surface.

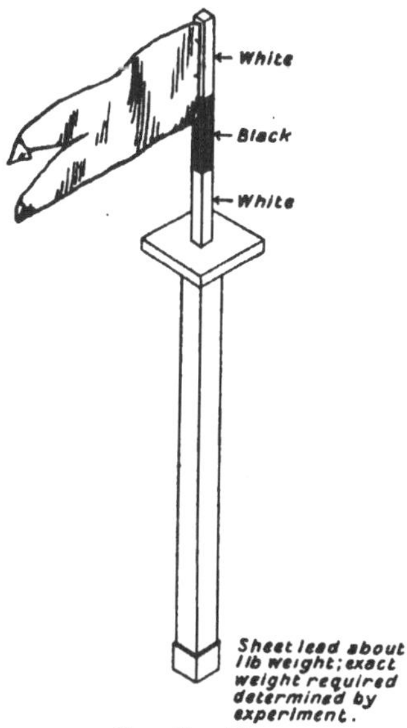

FIG. 49.—Float.

And not only is direction affected, but velocity is very much more involved. The velocity of a moving stream varies very materially throughout its depth. The maximum lies about midway between the surface and the bottom, at both of which points there are sources of retardation, due, in the former case, to aerial movement, and, in the latter, to friction with the ground.

This being so, floats for determining the flow of currents should extend some depth into the water so as to partake of the influence of as many layers as possible. Furthermore, for purposes of observation, they must project sufficiently above the surface level to be accurately observed, without exposing too great a surface to wind pressure.

A circular or square pole with a wooden cylinder or prism at its lower end, weighted so as to float vertically, affords a suitable form of instrument. Such an indicator is shown in fig. 49.

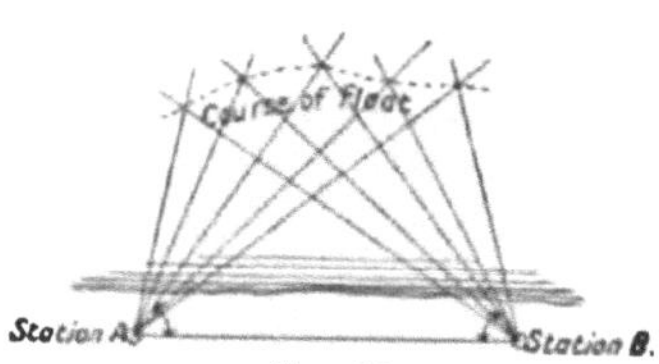

FIG. 50.

√ There are two ways of taking the necessary observations, which fix the location of the pole at any desired point of time.

In the first method, two operators, each with theodolite or sextant, are stationed at a fixed distance apart along a base line on the shore (fig. 50). At concerted signals each operator measures the angle subtended by the line joining the other operator and the float. The intersection of the lines forming these two angles respectively determines the position of the float at the time of observation.

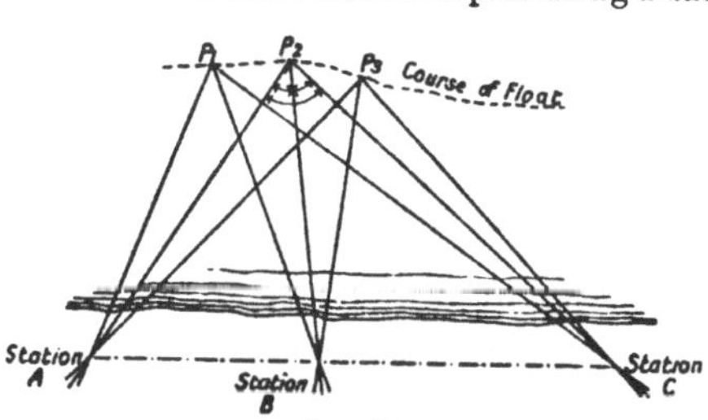

FIG. 51.

For the second method (fig. 51) one operator will suffice, but he must generally be provided with two sextants. Following closely in a boat the course of the float, at any assigned moment and as rapidly as possible, he takes the readings of the two angles which the float makes with three fixed objects ranged along the shore frontage, conveniently situated as near as possible abreast of his position and preferably co-linear or nearly so.[1] These angles having been plotted on a piece of tracing paper, the latter may be adjusted over a plan so that the three lines pass through the fixed points on the shore. When this condition is fulfilled—and there is only one position corresponding to any pair of angles,—the point may be pricked through.

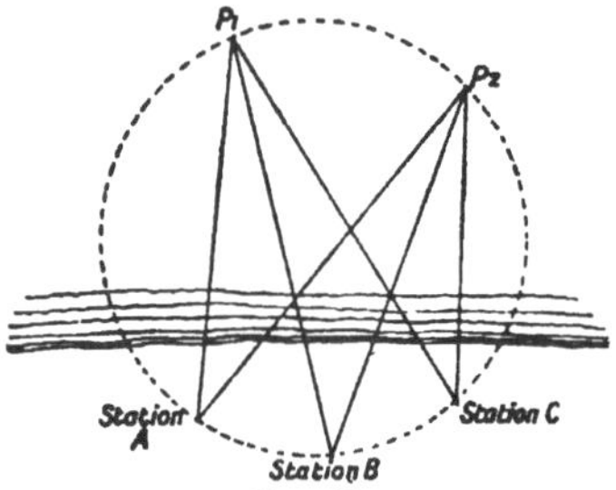

FIG. 52.

A *station-pointer* may be used in place of the tracing paper. This instrument has three long flat arms, or splayed straight-edges, all radiating from a common centre. A graduated circular arc on the middle arm, with vernier indices on the side arms, enables the instrument to be accurately adjusted to any given combination of angles. This performed, the instrument is laid upon the plan so that the straight-edges pass through each of the given stations. The point of intersection of the arms is then pricked through as before.

It has been remarked that for this method of locating the position of the float, two sighting instruments are generally necessary. It is manifest that the observations must be as simultaneous as possible. Any hurry in reading the first angle prior to adjustment for the second would lead to error. It is preferable, therefore, to fix both angles and defer the readings of either until that has been done. As a check, the angle containing the two subsidiary angles may also be read, but this involves the provision of a third sextant, with, probably, the aid of another operator.

Besides the staff or pole, other forms of float are available. On the river Avon an empty 5-gallon oil-drum has been used, sunk to almost complete immersion and ballasted by a basket of stone attached below it in the manner shown in fig. 53. On top of the drum is fixed a sighting mark consisting either of a semaphore, flag, or disc.

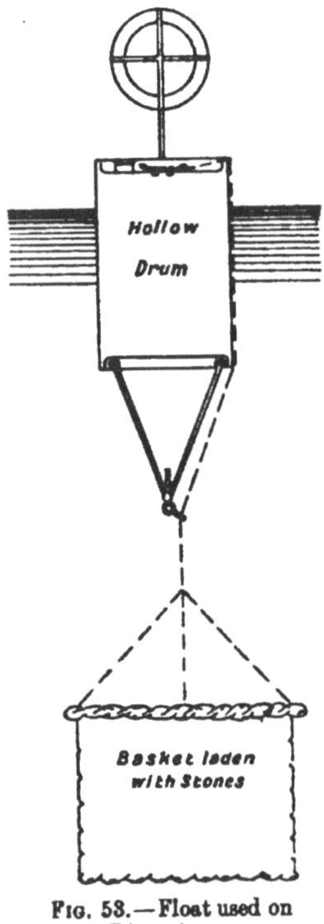

FIG. 53.—Float used on River Avon.

[1] The method fails if all four points lie on the circumference of a circle (fig. 52). All angles in the same or equal segments of a circle are equal. Thus the angles $P_1$ = the angles $P_2$.

For purposes to which no great accuracy is essential, any convenient buoyant object may serve, such as an empty keg or barrel, and in positions where a small object can be easily seen, an angler's float will do admirably. Small pieces of cork or wood suggest themselves as equally utilisable. An orange makes a good float under convenient circumstances of visibility. Its specific gravity is very little less than that of water; hence it floats nearly wholly immersed, exposing little or no surface to atmospheric action.

**Diving.**—Perhaps the most interesting, not to say romantic, feature of harbour engineering work is the use of diving apparatus in connection with the preparation of submarine foundations. A very large proportion of the operations necessary to the satisfactory stability of breakwaters and quays has to be conducted under water, and much of it would be almost impracticable without the aid of diving-bells and diving-dresses. Natives of the East Indian Seas engaged in pearl fisheries do, it is true, remain under water for appreciable periods without any special apparatus for the supply of air, but the strain is very great, causing bleeding at the nose, mouth, and ears, and if unduly prolonged, leading to fatal results. Apart from these physical effects upon the agents engaged, there is much interruption to the operations, which, in the case of structural work, would be inimical to its satisfactory accomplishment, and the haste with which the operations have to be performed would be incompatible with the exercise of care and accuracy. A regular and constant supply of atmospheric air to workers below the surface enables them to remain on duty for some time without any serious discomfort, and for this reason alone the use of air chambers and diving suits has become an integral accompaniment of all maritime operations.

The **Diving-Bell** (fig. 54) is a metallic chamber of sufficient capacity to accommodate any number of workers, from one man up to a dozen or more. The chamber, which is formed of mild steel plates, carefully rivetted together and caulked so as to be absolutely water-tight, is suspended from the jib of a crane or overhead traveller, or from any lifting appliance afloat or ashore, and so raised and lowered as the case may be. In the interior are seats and footrests for the occupants during ascent and descent, and shelves for tools. Signalling gear is provided, and many bells are now fitted with electric light and with telephonic apparatus. In addition to this equipment, there are the necessary air-valves and pipes for maintaining a supply of fresh air at the required pressure. At the top and sides of the bell are observation lenses, affording a view of the environment. The bell is ballasted with cast-iron kentledge, placed in a special chamber, to enable the structure to sink without endangering its equilibrium. The size of bells varies very considerably. Two instances may be quoted as examples. The diving-bells used on the Dover Harbour Works were 17 feet long by 10½ feet wide by 6½ feet high, inside measurement. At Marseilles the dimensions were 66 feet in length, 22 feet in width, and 6½ feet in height.

Diving-bells may be kept in comparatively free communication with the upper air by means of a cylindrical tube, carried up above the surface of the

FIG. 54.—Steel Diving Bell.

[*To face p.* 54.

water and surmounted by a special chamber, known as an air-lock. This chamber has two tightly fitting doors; one giving on to the open, and the other in to the tube. After entrance to the chamber from above or below, the door is closed and the pressure equalised with that of the bell or the atmosphere as the case may be. The air-lock thus forms a convenient means for the transmission of material to and from the base of operations—for the passing out of excavation and the taking in of stone and cement; and it becomes, in fact, essential, if the progress of the work is to be uninterrupted.

In many cases, however, owing to the bulk and weight of the foregoing apparatus, submarine operations have been mainly, and even entirely, carried on with the aid of individual divers, each equipped with a helmet diving-dress and capable of acting in perfect independence of any submerged chamber. Where the locality is free from currents, there can be little doubt that this method of working is preferable in many ways; but a current exceeding 3 miles per hour constitutes not only a trying force for the diver to contend against, but it throws considerable strain on the air-pipe and, to a less degree perhaps, upon the life-line. It must be borne in mind that a diver, when immersed, is a very buoyant object, and that he necessarily finds it difficult to withstand any powerful lateral force. It is comparatively easy for him to be swept off his feet, and even a moderate flow makes his foothold far from secure. The author recollects with painful vividness how one diver, engaged on submarine work at Liverpool, was suddenly sucked through a culvert to his death. There was gross negligence which contributed to the fatal result, but the incident illustrates the uncertainty of a diver's equilibrium and the great risk he oftentimes runs.

Apart from this drawback, it must be admitted that work, as a rule, can be more expeditiously performed by men moving in perfect freedom over a large area than is possible when they are confined within a narrow space, where there is a limit to the number of men employable and the likelihood of their impeding one another's movements.

**Diving-Dresses.**—The diver's outfit comprises the helmet, the dress, the air-pipe and life-line, and the air-pump.

The *helmet*, which is spherical in appearance, is of highly planished tinned copper, as also is the breastplate or corselet to which it is connected, though gun-metal is sometimes employed for the latter. Connection is effected by means of segmental screw neck-rings of gun-metal, the joint being rendered perfectly water-tight by turning the helmet through an angle of 45 degrees. The breastplate is moulded to the shape of the shoulders on which it sits, sometimes with a padded bearing. It receives the collar of the india-rubber dress over a series of brass screws through corresponding eyelet holes. Gun-metal flanges and wing nuts form a secure and impervious connection. The headpiece is fitted with side and front lights in the form of round or oval plate glasses, set in brass frames, with stout wire-guards. The front glass is detachable by unscrewing, or hinged to open. Air should be introduced into the helmet in such a way as to pass closely over the surfaces of these glasses, so as to prevent the condensation of the diver's breath upon them. The other fittings of a helmet are the inlet and outlet valves of the air supply, the latter of which is equipped with a regulator, so that the diver can control his supply of air to a nicety. The inlet valve is so constructed that air is allowed to enter freely, but cannot possibly escape that way, and, in the event of damage occurring to the supply pipe, by closing the outlet valve, the apparatus would retain sufficient air to enable the diver to return to the surface.

The *dress* is in one complete piece, made of solid sheet india-rubber between double-tanned twill. It is fitted with vulcanised india-rubber cuffs and collar, the former being sufficiently close fitting to the wrists to prevent the entrance of water, and the latter pierced with holes to correspond with the clamping screws of the breastplate. In English practice the number of these holes is about a dozen; in French practice, three. The cuffs have generally to be expanded with metal expanders, shaped like shoe-horns, to admit of the passage of the hands, but, in some cases, a bead is moulded on the edge of the cuff, which enables it to be rolled back over the hands. Should the cuffs not prove sufficiently water-tight, the writer has found it a good plan to bind the wrists with a band of moistened chamois leather before the cuffs are put in place.

As it is no uncommon occurrence for a little water to enter the dress through leakage, or occasionally through allowing the outlet valve to be open rather too widely, the diver, before putting on the dress, removes his outer garments and dons a guernsey, drawers, and stockings, as protection from wet and also as padding to his body. For deep or cold water these habiliments may be doubled or trebled. He wears a pair of canvas socks over the feet of the dress to protect it when walking about without shoes, and, if his work is likely to lead him into rough and rocky places, an outer suit of canvas overalls is desirable.

The *boots* are strapped on at the last moment before descending. They are either of specially stout leather, heavily shod with lead, or cast in brass with leather uppers. Additional weight is generally provided for the body of the dress by loading the breast and back with lead pads slung across the shoulders.

The diver's personal equipment is completed by a leather waist-belt containing a knife in a sheath. India-rubber gauntlets may be added, but in this country most divers work without them.

The *air-pipe* is made of the best india-rubber hose with a core of either hardened steel wire, tinned to prevent rusting, or of brass or copper wire. The pipe may be made to float or sink by adjusting the weight of metal. It should be tested to a pressure of 200 or 300 lbs. per square inch. After being screwed up to the helmet, the pipe is led and secured under the diver's left arm, so as to be conveniently at his command, and thence it passes upward to the pump. A life-line of stout cord is fastened round the diver's body. Both life-line and air-pipe are paid out together through the hands

FIG. 56.—Submersion of Diving Bells at Folkestone Harbour Extension Works.

[*To face p.* 56.

generally of a single attendant, though they are sometimes in charge of two men. The life-line also acts as a communication cord, according to a code of preconcerted signals, but the most modern outfits are furnished with special speaking-tubes, or with telephonic apparatus, as also with an electric glow-light.

The *pump* is usually double acting, worked by a couple of men, with either single, double, or triple cylinders, according to the depth of water and the pressure required. It is furnished with a gauge indicating both these data.

The qualities needed in a professional diver are not exceptional, but preference will naturally be given to men of nerve and intelligence. The first descent, no doubt, is always more or less a trying experience from its very novelty. The sense of helpless confinement in the midst of a strange and artificial environment, a feeling of oppression, and the increased pulsation all tend to render the initial trip below water (as the writer's experience went) somewhat uncomfortable, if not a source of trepidation. The disagreeable sensations, however, pass away with acclimatisation and practice. Almost anybody in health may make a descent in perfect safety; but for regular and continuous work under water, full-blooded men with short necks are not desirable subjects; neither are those suffering from palpitation or from poor and languid circulation; nor intemperate and generally unhealthy men. Diving is said to be good for the lungs owing to the compressed air affording an increased supply of oxygen and deepening the respiration.

A diver of ordinary powers may descend to a depth of 100 feet with impunity, and may even reach 150 feet without ill effects; but deeper descents are not easily made, and are rarely recorded. The greatest depth to which any diver has descended by authentic testimony is 210 feet, at which point the pressure on his body was 90 lbs. per square inch in excess of atmospheric pressure. Harbour work very seldom entails diving in water exceeding 10 fathoms in depth.

The following table shows the pressures sustained over and above the ordinary atmospheric pressure at varying depths:—

| Depth. Feet. | Pressure. Lbs. per sq. in. | Depth. Feet. | Pressure. Lbs. per sq. in. |
|---|---|---|---|
| 10 | 4¼ | 80 | 34¾ |
| 20 | 8½ | 100 | 43½ |
| 30 | 12¾ | 120 | 52¼ |
| 40 | 17¼ | 140 | 60¾ |
| 50 | 21¾ | 160 | 69¾ |
| 60 | 26¼ | 180 | 78 |

Care should be taken in descending and ascending not to move too rapidly—more particularly in ascending. The rate of movement should not be greater than 2 feet per second for depths less than 80 feet.

The following regulations usually observed in regard to divers' working hours—except, of course, in cases of a temporary nature or of special urgency—afford an idea of their capabilities. A shift consists of four hours nett, not counting the time taken by the diver to dress, which he does in his own leisure. He is allowed a period of fifteen minutes during each shift for rest, and another fifteen minutes at the end for undressing. One or more shifts per day may be worked according to the needs of the case.

The minimum number of attendants required for a single diver is three—one for the signal line and air-pipe, and two to work the pump. For two divers an additional man is required to look after the second signal line and pipe. Pumpers and signalmen may relieve one another at their respective duties.

The following notes on the care of diving apparatus are extracted from instructions issued by Messrs Siebe, Gorman & Co., in connection with their goods.

"After the day's work is over the air-pipes should be thoroughly dried and the gun-metal joints carefully cleaned before being packed away. The diving-dress should be cleaned, and, if wet inside, turned inside out and hung up in the shade to dry; the dresses, if used in salt water, should be washed at least once a week in clean fresh water. The underclothing should also be kept dry and well aired.

"When in store, the pump and its fittings must be kept clean and free from verdigris, and, if likely to be out of use for some time, it should be occasionally oiled and the handles turned two or three times, in order to prevent the piston leathers getting hard. If the pump has been lying by for a considerable time, then it would be well to have it taken to pieces by a good fitter and examined to see that it is in proper working order. When a piston rod works loose, the screws at the top of the stuffing-box, in the case of the double-acting pumps, should be turned a little with a spanner. Only good olive and neat's-foot oil mixed should be used for lubricating.

"Should the diving-dress, from constant use or accident, get leaky, it is easily repaired by laying two or three coats of india-rubber solution on each side of the seam, rubbing it with the finger as much as possible and allowing each coat to dry before the next is applied; the sides of the seam may then be laid down, and two or three coats applied in the same manner to the channel of the seam, when the prepared twill (which should have an extra coat laid on and dried) may be immediately applied and well pressed down by the hand. Superfluous solution may be removed with a piece of india-rubber, but it is better to lay it on the proper width so as not to require cleaning off. Diving-dresses should never be packed away in a wet or damp state; they must be thoroughly dried, both inside and out, before so doing, otherwise they will mildew and become so rotten as to be of very little service afterwards. The following represents an easy and efficient mode of drying the diving-dress:—Take two pieces of wood each about 8 feet long, nail or screw them together in the form of a St Andrew's cross, place them inside the dress, and pass another piece through the arms to keep them distended; the dress

FIG. 57.—Divers at Work. [*To face p.* 58.

can then be set in an upright position until it is dry. In case the diver urinates in the dress, it should be turned inside out, washed with clean water, and then allowed to dry.

"Should the helmet have been lying by for some time, the valves must be unscrewed and examined to see that they are free from verdigris; they may be slightly greased with tallow, and new springs should be fitted if necessary. All the screws of the helmet breastplate should be kept clean, and occasionally wiped with an oily rag."

# CHAPTER IV.

## PILING.

Use of Framework in Maritime Structures—Association of Piling therewith—Varieties—Bearing Piles—Sheeting Piles—Materials for Piles—Timber—Varieties—Destructibility and Preservation—Metal—Concrete—Reinforced Concrete—Typical Systems—Pile-driving—Sustaining Power—Various Data.

**Structural Principles.**—Maritime structures are, generally speaking, based on one or other of two distinct systems of construction. First, there is the compact, solid mass, capable of withstanding the attacks of the elements by means of sheer intrinsic inertia, and, secondly, there is the framework structure, composed of an association of members or parts, all slender in themselves, but so contrived and connected as to afford one another mutual support, and at the same time able to discharge special individual functions.

Typical of the former system is the breakwater, mole, or quay, built as a solid mound or mass of rubble, masonry, or concrete, or a combination of these, the ideal being a homogeneous monolith, without break or joint. This type is really an adaptation of nature's own system exemplified in rugged cliff and massive headland.

The framework structure, on the other hand, is a strictly scientific design, utilising the minimum of material to the maximum advantage. It is based on the same theoretical considerations as those which govern the synthesis of all trusses, whether in the form of bridges, roofs, or other openwork. Its principal source of weakness lies in the jointing together of the various parts, for under the violent alternations of impact and recoil, which are characteristic of marine forces, there is every disposition for the joints to become loosened through excessive vibration. There is, moreover, another disadvantage attending those structures which are composed of unprotected metal and timber, viz., their liability to corrosion and decay. Both these considerations militate greatly against the realisation of any great degree of durability and permanence, and render structures of the second class inferior in certain respects to those founded on the former system, while, at the same time, they obviously involve much greater expenditure in the way of maintenance and repair.

There are, however, circumstances under which framework structures become inevitable, and many others where they are undoubtedly desirable.

Thus a solid pier inevitably deflects the course of a littoral current, thereby diverting navigable channels into unknown directions, and bringing about physical results which it is not possible to forecast with any certainty. A columnar pier, on the other hand, offers very trifling obstruction to current flow, and practically leaves the coastal régime unaltered. On these grounds, it has been deemed politic, at Zeebrugge for instance, to construct in open-work the portion of a projecting mole which immediately adjoins the shore, while that portion which lies beyond the range of the littoral current, or which is not likely to offer any injurious opposition to the motion of the sea, is built in the solid.

Framework, as adapted to maritime situations, consists of two distinct parts: the supporting columns or piles and the superstructural trussing. With the first of these we propose to deal in this chapter. On the second it will be necessary to touch but lightly, as the principles upon which it is based are common to all branches of engineering work, and in no sense can it be considered as a special feature of harbour engineering operations.

**Piling** is the term applied to all columnar members driven vertically, or nearly so, into the ground to form a foundation for constructional purposes. It includes two varieties: first, **sheeting piles**, which are employed to inclose or confine an area, and secondly, **bearing piles**, which act as isolated supports.

*Sheeting piles* are often much wider than they are thick, and are set with their edges in close contact, so as to form a continuous wall or partition. In order to achieve this result, they are driven in bays of moderate length, between leading or guide piles, to which horizontal walings are affixed. *Bearing piles* are more equilateral in cross section, and are driven quite separately, or in clusters. Sheeting piles are provided with a knife edge at their lower extremities; bearing piles have either pointed or butt ends.

The materials from which piles are made are extremely varied, and include timber, iron and steel, concrete and ferro-concrete.

*Timber piles* are, perhaps, those which have been most extensively used up to the present time. They have been adapted to purposes both of a temporary and of a permanent nature. For the former class of work, they are still in universal demand, but for the latter class they are now only utilised when considerations of economy outweigh all others. For jetties and piers destined to wear and rough usage, the durability of material, composed of metal and mineral which is practically indestructible, gives it an enormous advantage over perishable fibre; but for temporary work, such as gantry staging, cofferdams and the like, the cheapness and adaptability of timber confer upon it qualities relatively superior.

The character of the timber employed in harbour work depends upon the probable or estimated duration of its services. When utilised for permanent structures, only the best, hardest, and soundest timbers are admissible. In other cases, softer and less durable wood will suffice, provided it be kept under constant supervision and renewed whenever necessary.

So far as soundness and strength are concerned, there are few trees which are incapable of supplying logs and balks of a thoroughly satisfactory character. In harbour work, however, durability is the crucial consideration, and the conditions attaching to that qualification are much more exacting than those which govern the choice of suitable timbers for constructive purposes elsewhere. The alternations of exposure to the atmosphere and submersion in the sea, due to tidal fluctuations, constitute in themselves a most fertile source of decomposition, such as is not experienced in any other environment nor associated with any other branch of engineering. And, as if this were not sufficient, there is allied therewith a most pernicious and deadly subjection to the mechanical attacks of insectile[1] borers, which infest the waters of most ports.

In addition to the question of durability, however, there are the subsidiary, but no less essential, considerations of available scantling, cost, and facility of supply, each of which demands the careful attention of the engineer.

It is not proposed to enter into any lengthy dissertation of a botanical nature on the very great variety of trees which are available for engineering purposes; it will be sufficient to confine our attention to details of a practical kind in connection with those comparatively few species which have obtained wide and general recognition in connection with maritime work. These may be enumerated briefly in three groups:—

I. Greenheart, Mora, and the Eucalypti. These woods are extremely durable and highly repellent of insects.

II. Teak and Oak. These are also very durable, but subject to insect attack.

III. Beech, Elm, and Pine. These are moderately durable, and they succumb easily to insects.

By far the most important group to the harbour engineer is the first. To this we must pay greatest attention, leaving the other groups, though they comprise timbers of more extensive use, to be but briefly noticed.

**Greenheart** (*Nectandra Rodiæi*) is an American product, the tree being a native of Guiana and the adjacent states of the South American continent, where it grows very abundantly in tracts lying within a hundred miles of the coast-line. It is a wood of extreme hardness and durability, with a very fine and compact, though uneven, grain. Its resistance to crushing is enormous, but it is very brittle and it splits under the least provocation. Before sawing, logs have to be bound very tightly with chains and wedges on each side of the projected cut; otherwise there is great danger of splitting, and a crack once started is prevented from extending with difficulty greater than that which characterises avoidance of the danger in the first instance.

Greenheart contains a poisonous oil, which renders necessary considerable circumspection on the part of carpenters and others engaged in dressing it.

[1] Objection may be taken to the use of this word in this connection. It is difficult, however, to find an accurate generic name for these pests. The term insect is applied under license which is justifiable, since no confusion is likely to arise from its use.

A splinter in the flesh almost invariably produces blood-poisoning, and the merest scratch should be promptly sucked and washed in clean water.

The weight of greenheart ranges from 60 to 75 lbs. per cubic foot, so that it has practically no flotation. This characteristic facilitates its manipulation for piling purposes, as it sinks readily into position. It can be obtained in balks from 12 to 24 inches square and up to 70 feet in length. It has an ultimate compressive strength, in short prisms, of 8 to 8½ tons per square inch, and a beam of unit dimensions, *i.e.* 1 inch square in section and 1 foot between supports, will fail at loads ranging from 950 to 1500 lbs., centrally and concentratedly applied.

The colour of greenheart ranges from green to almost black.

**Purpleheart** is a wood of the same kind, from the same locality, with a difference only in colour, as indicated by the name. It is perhaps a little tougher and slightly more durable, but, on the other hand, it is not so readily procurable. Balks can be obtained up to 30 inches square.

**Mora** (*Mora excelsa*) is also a native of Guiana, but is a light-red wood, with several distinguishing characteristics. It shares the strength and durability of greenheart, while it differs from it in possessing great toughness and in lacking any disposition to split or splinter. It is rather lighter in weight, too, than greenheart, weighing from 57 to 68 lbs. per cubic foot.

The Eucalyptus family is a numerous one, and indigenous to the Australian continent.

**Jarrah** (*Eucalyptus marginata*) is a timber found in abundance in Western Australia, and, from its resemblance to mahogany, it is sometimes called Australian mahogany. It is hard, heavy, and close-grained; very liable to warp and split. It is also beset with clefts filled with resinous matter, which is sometimes found to be in a state of decay. The fibres also contain an acid having a pungent odour. The tree grows to a height of 200 feet and more, but sound logs are limited to 40 or 45 feet in length and 12 to 24 inches square.

The weight of jarrah is just about equal to that of an equal volume of water. It has little more than half the crushing strength of greenheart, and the ultimate transverse strength of a unit beam (1 inch square and 1 foot clear span) is between 500 and 650 lbs., concentrated at the centre.

**Karri** (*Eucalyptus diversicolor*) is a hard, heavy, straight-grained wood, with some claims to toughness. It is somewhat stronger than jarrah, but less durable in damp situations; though when totally and continuously immersed, it is said to last well.

The **Blue Gum** (*Eucalyptus globulus*) and the **Stringy Bark** (*Eucalyptus obliqua*) are two varieties of the same species, which have latterly come into use and have demonstrated considerable merit for staging purposes in connection with the improvement works at Dover Harbour.

The former is so named from the characteristic glaucous blue tint of the young plant, though the colour of the mature wood is a golden yellow or brown. Both trees grow to an enormous height and girth, and furnish tough, strong wood, extremely durable under favourable circumstances, and more particularly in dry and open situations. Piles, 100 feet to 120 feet long and 20 inches square, have been obtained in Tasmania. Stringy bark, according to some authorities, weighs about 70 lbs. per cubic foot, and blue gum about 77 lbs.; others place the figures at 60 and 65 lbs. respectively. Some variation of weight in different specimens is, of course, inevitable. The transverse strength of unit beams (see p. 63) may be taken at anything from 450 to 850 lbs.

It will be noted that all the timbers in this group have a very high specific gravity, and this property is found to be very useful in connection with driving piles in water of any depth. The lighter kinds of wood have necessarily to be weighted at the lower ends, in order to cause them to assume an upright position suitable for driving.

As regards durability in marine situations, it cannot be claimed that any of the foregoing timbers are absolutely immune from the attacks of insects. On the contrary, there is distinct evidence that boring has occurred in each kind of wood, though it is apparent that there is no great attraction in these timbers when others are present in the neighbourhood. Greenheart appears to be least susceptible, possibly on account of the poisonous oil which it contains. At certain ports it exhibits no sign of any depredation whatever, but this may be due to the absence of the inimical agencies. Altogether as a class, the timbers are the least vulnerable of any which can be applied to marine work, and in many instances they have demonstrated extremely high resisting powers.

The second group includes timbers which, though durable enough in themselves, are much more subject to insect attack.

**Teak** (*Tectona grandis*) is a native of India, Burmah, Siam, and Java. It is a firm, durable wood, fine and straight in grain, and easily worked, though possessing a tendency to splinter. It contains an aromatic oil of a resinous nature, which, on exposure, coagulates to such a degree of hardness as to spoil the cutting edges of tools. The tree often attains a height of over 100 feet and sometimes a girth of 10 feet. It is usually imported in logs from 25 to 40 feet long and from 10 to 20 inches square. The weight of teak varies from 41 to 52 lbs. per cubic foot, and the transverse strength of a unit beam lies between 600 and 700 lbs.

**Oak** (*Quercus*) is found on both the European and American continents, as also—less commonly—elsewhere. The best is grown in Great Britain. The wood is firm, with a fine, straight grain, comparatively free from knots, and it is readily cleavable. Logs vary from 10 to 40 feet long and from 10 to 24 inches square. The longer logs come from America. Oak is heavier than teak, weighing from 49 to 61 lbs. per cubic foot; but it is not quite so strong—about 50 to 100 lbs. less in ultimate transverse strength. Oak contains an acid which corrodes iron, and is therefore destructive of bolts and other fastenings.

Both the above timbers are admittedly assailable by insects, but they offer greater resistance and attain a higher degree of exemption than do members of the third and last group.

**Elm** (*Ulmus*) and **Beech** (*Fagus sylvatica*) are two well-known timbers, to which the term durable is only applicable provided the conditions be those of total immersion or continuous dryness. The weight of elm is about 35 lbs., and the weight of beech about 48 lbs. per cubic foot. As regards strength, beech has the superiority, being half as strong again as elm. The mean ultimate transverse load on a unit beam of elm is 400 lbs.; that of beech, 600 lbs.

**Pine** and **Fir** include a number of varieties of timber, some of which, such as pitchpine and Oregon pine, are highly serviceable to the harbour engineer for temporary staging and dams. Their durability under exposure to water is not very great, unless it be assisted by some treatment, as creosoting, which also affords protection to a certain extent against insects. These timbers must needs, however, be under constant supervision and inspection, and it is certainly not desirable to set them in positions difficult of access nor to place too great confidence in their capabilities of resistance.

**Pitchpine** (*Pinus rigida*) is obtained from the southern states of North America. It is a highly resinous wood, reddish or reddish brown in colour. The resin in its pores renders it hard and difficult to work, but also increases its durability. The strength of pitchpine is often reduced by the practice of "bleeding" the growing tree, that is, tapping it for the turpentine which it contains. Logs are obtainable from 10 to 18 inches square and up to 60 or 70 feet long. The commonest sizes for piling purposes are from 12 to 15 inches square and from 40 to 50 feet long.

**Oregon Pine** (*Abies Douglasii*) comes from the north-west of North America. It has a light reddish colour. It is obtainable in logs up to 20 and 24 inches square and up to 100 feet in length. It is not so strong as pitchpine, but, affording larger sizes, is useful in certain situations.

**Destruction of Timber.**—The utility and value of timber being so greatly affected by its liability to destruction and decay, it is necessary to consider the sources of deterioration and the possibilities of their avoidance or cure.

Insectile ravages claim first attention, as they constitute the most serious and pressing danger to which timber piles are exposed. Woods of the utmost durability in regard to chemical changes succumb only too rapidly from purely mechanical causes.

*Teredo Navalis.*—This animal, one of the most pertinacious assailants of marine timber structures, is a member of the family *Pholadidæ*. It is found in all British seas and, indeed, frequents the majority of the seaports of the world. It has a decided preference, however, for clear salt water, and deliberately avoids water which is muddy or sewage-polluted, or even fresh. The process of its depredations appears to be as follows. Its eggs, drifting in the water, adhere to any exposed woodwork against which they happen to be washed by the sea, and there remain till ripe for hatching. On leaving its egg, the young teredo attacks the wood in its immediate vicinity by boring or tunnelling into it, principally in the direction of the grain. The boring implements are two strong, sharp black teeth, which can be disclosed for inspection by applying pressure at the back of the creature's head. The holes, or galleries, increase in size with the growth of the animal, and they are lined throughout with a chalky secretion forming a thin, hard, smooth shell. It is no uncommon experience to find holes ½ inch or ¾ inch in diameter, and the teredo has been known to attain a length of as much as 2 feet, though the average length is not more than 7 or 8 inches. Its operations seem to be chiefly confined within the tidal range: that is, between highest high water mark and lowest low water mark; but it also attacks timber at any moderate depth. At times it works with extreme rapidity. Some of the Memel fir piles of the old pierhead at Southend showed signs of the teredo within six months after completion, and in twelve months' time they were reported to be seriously injured. Fir and alder appear to furnish the most favoured fields for operation; oak and teak are less susceptible; greenheart and jarrah have a general reputation (not strictly maintainable) of being free from attack. Greenheart has been used for piles at the mouth of the Mersey without the slightest sign of deterioration of any part, even after the lapse of many years; but at Bombay the same wood has been freely ravaged.

Fig. 58.—*Teredo Navalis.*

The *Pholas dactylus* is another member of the *Pholadidæ* family. On the whole, it evinces a more pronounced taste for mineral substances, such as limestone and sandstone; but it also turns its attention to woodwork, which it honeycombs by boring a number of holes very closely together. Compared with the teredo, however, it is of small importance, so far as timber, at any-rate, is concerned. The animal attains a length of 4 or 5 inches.

Another boring tribe of similar habits and tendencies is the *Xylophaga.* Its members are small in size, and they do not line their excavations.

The *Chelura terebrans* (*Amphipoda*) is a small crustacean resembling a minute shrimp, both in shape and colour. It is very small (fig. 59), not more than ¼ inch in length. In addition to feet, it is provided with a pair of limbs, near the tail, which it employs in leaping. The *chelura* destroys wood by cutting or tearing it away in thin flakes, working inwards from the exposed surface. It manifests a decided partiality for pure, clear sea-water, and is consequently more often found along the open coast than in inclosed basins and harbours.

Fig. 59.—*Chelura Terebrans.*

Fig. 60.—*Limnoria Terebrans*

The *Limnoria terebrans* (*Asellidæ*) is another lilliputian, whose length seldom reaches to more than $\frac{1}{8}$ inch. In appearance it is not unlike a grain of rice. It is mainly troublesome on account of the vast numbers in which it infests certain localities, and, as it is indifferent to the foulness or otherwise of the water, no harbour precincts can be considered free from its presence. The *limnoriæ* are active mainly about and below high water of neap tides, depredations proceeding rapidly until the whole of the timber-work is eaten away. Large balks of unprotected fir have been completely destroyed in three years, and even creosoted timber has perished within a decade.

The *Tanai vitalis* belongs also to the *Asellidæ* family. It preys upon vegetable fibre with powerful claws, rending it to pieces. The length is about $\frac{1}{8}$ inch.

The attacks of the white ant (*Termes*) in tropical countries do not call for detailed mention. They are not particularly associated with maritime situations, and submerged timbers are, of course, not affected in any way.

**Decay of Timber.**—Apart from the mechanical destruction of timber, there is the question of natural decay, which is due to one or other of two distinct forms of decomposition, known respectively as *dry* and *wet rot*. The former, which is a process of fibrous disintegration, accompanied by the growth of a parasitic fungus, is attributable to, and certainly accelerated by, the absence of adequate ventilation. The woodwork attacked is mostly that which is situated in confined and stuffy places, to which air has insufficient access—conditions not generally allied with harbour work.

Wet rot, on the other hand, has a much more general and appropriate connection. It is the most characteristic disease, in fact, to which marine timbers are liable. It arises from, and is promoted by, frequent alternations of dryness and moisture, and these conditions are obviously prevalent along the water's edge. Every time a log becomes immersed and dries again, a fresh portion of the fibre is converted into soluble matter, which, in due course, is abstracted and lost. Furthermore, the continual evaporation of moisture from the pores of the wood results in putrefaction, the progress of which, once commenced, is often rapid.

Wet rot will attack indifferently any part or substance of a log, whether it be heartwood or sapwood; whereas dry rot is generally to be found in the latter only. The disease, moreover, is contagious, and affects adjacent timbers which may not in themselves be exposed to the same predisposing causes.

**Preservation of Timber.**—Having enumerated and described the inimical agencies, we come now to the means used to combat them. Expedients, as diverse as they are multitudinous, have been tried from time to time with a view to increasing the durability of timber, both as regards preserving it from internal decay and protecting it from external attack.

The commonest preservative for structural work is **paint**. Applied to seasoned timber completely deprived of free sap, the method is one of the most efficacious which can be devised; but it calls for frequent and regular renewal, and this, in the case of submerged work, is an insurmountable obstacle to its adoption. Moreover, sea-water tends to soften paint, and the chafing of floating objects against the surface of the wood soon wears away its protective coating. The same objections apply to other substances, such as *tar*, *verdigris*, and *paraffin*, which have either been used or proposed as a substitute for paint.

The best and only really effective agency for increasing the longevity of timber-work in contact with moisture is the process known as **creosoting**. It also acts as a deterrent to sea-worms, though not to the extent of rendering the wood invulnerable. The process consists in coagulating the alburnum in the pores of the log, so that the latter become filled with an antiseptic, bituminous substance, which excludes air and moisture, repels the lower forms of vegetable and animal life, and prevents putrefaction and rot.

**Creosote** is an oily liquid contained in the second distillation of tar, from which the ammonia has been expelled. Its composition is somewhat variable; but in order to be effective, it should contain over 40 per cent. of naphthalene, about 4 or 5 per cent. of carbolic acid, and as little pitch as possible. It is essential to the efficacy of the treatment that, as a preliminary, all moisture be abstracted from the interior of the timber.

Soft woods, such as fir and pine, may be simply immersed, direct from the drying-house and while still warm, in an open tank of hot creosote. Logs treated in this way will absorb from 8 to 9 lbs. of creosote per cubic foot, and this quantity is generally sufficient for inland purposes.

For marine work, however, and especially for piles and the timber-work of jetties, impregnation to the extent of 10 or 12 lbs. per cubic foot is requisite. In order to achieve this result, the timber, after being dried, is placed in a vacuum. Creosote, at a temperature of about 120° Fahrenheit, is then introduced into the containing cylinder, under a pressure of about 175 lbs. per square inch. By this means, with suitable woods, the amount of creosote absorbed may reach a maximum of 16 lbs. per cubic foot.

Hard, compact woods, such as oak, do not, under any degree of pressure, absorb more than 3 lbs. of creosote per cubic foot; but in their case this is found to be sufficient.

*Boucherie's process* consists in impregnating timber with a solution of sulphate of copper (1 per cent. by weight) in water. The usual course of procedure is to cap one end of a log in a water-tight manner and then to allow the liquid to penetrate the pores from the other end, so displacing the sap, under a head of 30 or 40 feet, which produces a pressure of 15 to 20 lbs. per square inch. The extent to which penetration takes place can be tested by means of prussiate of potash: whenever this substance comes in contact with sulphate of copper, a brown stain is left.

Timber is **kyanised** by immersing it in a saturated solution of corrosive sublimate (perchloride of mercury) contained in a wooden tank. The strength of the solution varies from ⅔ to 1 per cent. by weight, according to the porosity of the timber.

**Burnettising** is the term applied to a process of treatment with a solution of chloride of zinc, containing 2½ per cent. by weight of the chloride.

Ordinary immersion will suffice, but the impregnation is expedited by using pressure.

None of the last three methods has proved so effective, or come into such general use, as creosoting. In fact, it is doubtful whether any of them is of the least benefit in warding off insectile attack, and this, in maritime situations, is an object no less important than the preservation of timber from decay.

The only apparently completely successful way in which timber may be guarded in this respect is by means of some external covering excluding the wood from actual exposure.

**Sheathing** is a protective device which consists in enveloping a pile in a covering of metal, earthenware, or other material impenetrable by insects. A thin covering of copper plates has proved satisfactory in repelling worms from piles, when the covering has extended from below the mud level to above high water-mark; otherwise the insects intrude themselves between the metal and the wood. The method is obviously an expensive one, and therefore not likely to commend itself for general adoption. Zinc has been tried as a substitute for copper, but it is soon corroded by sea water. Muntz metal is another substitute, but its application has been too limited for definite judgment of its powers. Studding with broad-headed scupper nails is an old expedient, the principal drawback of which is its troublesomeness, and, of course, its expense.

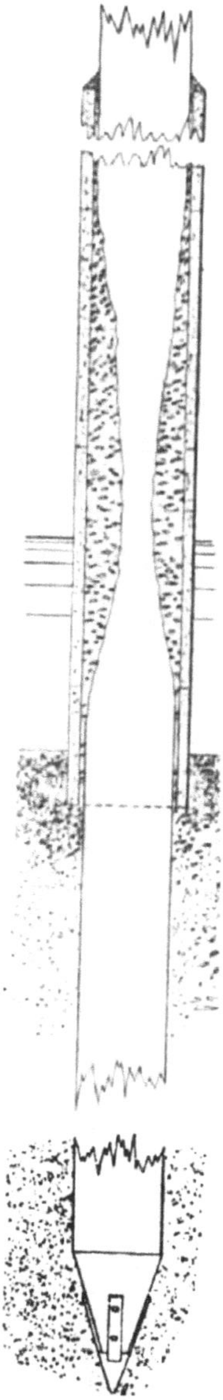

FIG. 61.—Application of Reinforced Concrete Slabs to Decayed Timber Pile.

Earthenware pipes, such as ordinary drain-pipes, and cylindrical casings of wire netting bedded in concrete, are efficient preservatives of piles in situations free from shocks, collisions, and erosion. The space between the pipe and the pile must be filled in with sand or cement grout. A simple coating of Portland cement has been tried, but the film is too thin and easily cracked. Lately, a system of facing wooden piles with reinforced concrete slabs has been promoted by Mr Cooper Poole, the harbour engineer of Southampton. The slabs are primarily intended for application to piles which are in such a state of dilapidation as to call for renewal or repair. At each corner of the pile a small angle iron is spiked on to the timber so as to form a guide for the slabs. These last are connected up and allowed to sink into the mud until they take a bearing, when the inclosed space is filled with concrete. The system, however, is applicable also to piles which are whole and perfect, as a preservative. On the Pacific coast a wrapping of jute burlap, in combination with a preparation of paraffin, powdered limestone, and kaolin, is reported to have achieved successful results.

From the foregoing details, it is obvious that the use of timber piles, though convenient, is attended by a number of serious disadvantages. There can never be any complete sense of security in reference to the part they play in permanent structures, and the increasing scarcity of logs of a suitable size, together with the difficulty of obtaining them at a moderate cost, has led to the introduction of piles composed of metal entirely or of metal and concrete combined.

**Metal piles.**—Metal piles are ordinarily either of wrought iron or steel. The pointed or driving end is frequently cast, but, generally speaking, cast iron is of too brittle a nature for use in the shank of a pile, unless special precautions be taken in driving, or the ordinary method of impulsion by a falling weight be replaced by some other system. Thus, with screw ends, cast iron tubes or pipes are often used instead of timber logs (which are equally available), the means of forcing into the ground being rotation round the vertical axis. This constitutes, however, a method of treatment so distinct and exceptional that it may be regarded as not affecting the general question.

For the sake of dismissing it from further consideration, it is convenient to introduce here a few explanatory words concerning the system of screw piles. The screw end consists of a broad blade, forming, in most cases, little more than a single turn or a turn and a quarter. It has the property, therefore, of furnishing a base of much greater area than that afforded by the ordinary pile, and on this account is useful for foundation work in compressible strata, where it is desirable to spread the load over as large an area as possible. Moreover, there is an absence of vibration in the process of driving, which is a distinct advantage. The piles are driven by means of a capstan head or a drum of large diameter temporarily bolted on to the shank, and raised from time to time as the rate of driving requires. In the former case, capstan or lever bars are used; in the latter, a winch, to which is led a wire rope wound round the drum, supplies the motive power. In primitive and isolated cases, animal labour has been utilised.

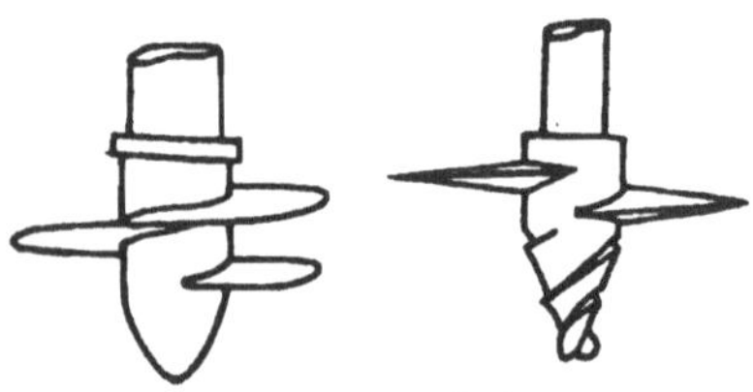

FIG. 62.—Screw Pile Bases.

**Steel or wrought iron piles** partake of all the recognised forms emanating from manufacturers' rolling-mills. Channel and joist sections are most common. Such piles, though available for solitary positions, are more generally found in close association, as sheet piling. When this is the case, a certain, and by no means negligible, amount of mutual interdependence and support is afforded by binding intimately together the adjacent edges of the piles. This can be done by forming a series of grooves with the aid of rivetted connections, as exemplified in the figures shown, which represent typical sections patented by the Friestedt Co. of Chicago, U.S.A.

The interlocking arrangement is extremely useful in forming a water-tight inclosure for dams. Hydraulic pressure against the outer face will generally prevent the passage of water, but where any leakage manifests itself, it can easily be checked by sprinkling ashes, sawdust, or any light material of a similar kind, upon the outer surface, whence it will be sucked into the defective joint.

The driving of these piles is effected in the ordinary way by means of a falling ram; only, it is necessary to interpose a wooden "dolly"—a 6- or 8-

FIG. 63.

FIG. 64.

FIG. 65.—Interlocking Steel Sheeting Piles.

foot length of greenheart timber—between the cap of the pile and the underside of the ram. The cap of the pile is a removable block or plate of cast steel, several inches thick, temporarily secured in position with the aid of bolts and removed after the operation of driving is finished.

Metal piles, though indestructible by insects, are subject to corrosion, with results equally disastrous in the long-run. The effects of oxidisation are most to be dreaded in the case of the outstanding piles of piers and jetties. All ironwork immersed in salt water, and especially when alternately wet and dry, undergoes chemical changes subversive of its strength and durability. Hence the manifest necessity of providing it with some protection akin to that which is accorded to timber piling.

Of the methods in vogue for the prevention of corrosion in iron or steel, two stand out in greatest prominence—painting and galvanising. The former of these is only of the nature of a temporary perservative, and has to be re-applied at regular and frequent intervals; the latter cannot be renewed in the case of *in situ* structures, and, though the initial treatment is understood to be more effective than painting, yet the environment of the seacoast is extremely detrimental to its efficacy.

The coatings applied to ironwork under the head of paint comprise those which are composed of red lead and those which have oxide of iron for their base. The latter of these has been advocated on the ground that it removes the tendency to galvanic action produced by two diverse metal substances in contact with one another in the presence of moisture. Other coatings are mineral or vegetable tar, black varnish, siderosthen, and various bituminous solutions. It is obvious that only the surfaces of piles which lie above the water level can be treated with these applications after erection.

For cast ironwork, and especially for cast iron pipes, no better preservative could be devised than the *Angus-Smith treatment*, which consists in dipping the pipes while hot into a liquid mixture of coal-tar, pitch, linseed oil, and resin.

Iron and steel are *galvanised* by dipping them into a bath of molten zinc so that a veneer of the latter metal covers them completely. To effect this treatment properly, the surface of the metal treated must be absolutely clean and free from scale and grease. The process is effective against ordinary atmospheric influences, provided the zinc covering be maintained intact. If a crack or perforation occurs, corrosion sets in and proceeds rapidly. Against sea air and water, galvanising does not afford much protection.

It is obviously no simple matter, therefore, to find a satisfactory and reliable method for insuring the permanence of iron and steelwork in maritime situations, and particularly in the case of piling, where the work is so difficult of access. The desired result, however, has been achieved by the ingenious expedient of enveloping the metal in concrete, and this brings us to the system of combined steel and concrete which now generally goes by the name of reinforced concrete.

**Reinforced concrete** consists essentially of a core or internal network of metal, completely embedded in concrete, so that no part of the metal is exposed to, or in contact with, any external atmospheric or aqueous influences. As applied to piling, the system has many and important advantages. Reinforced concrete piles are not subject to oxidisation, decomposition, or decay. Experience has demonstrated that steel bedded in Portland cement concrete does not rust even when immersed in water, and that a rusty bar so treated manifests no increase in corrosion. Moreover, reinforced concrete piles do not offer the least incentive or attraction to sea-worms or insects; they are fireproof as well as waterproof; their durability is beyond question; they cost less than long greenheart piles, and little, if anything, more than creosoted pitchpine; they can be jointed, and lengthened or shortened at will; and, finally, their compressive strength and supporting power is very great.

Reinforced concrete piles vary considerably in design, according to the individual ideas of numerous inventors. It will only be necessary, however, to refer to a few of the better known examples, which are distinctly applicable to harbour work. The circumstances of foundation piles for inland structures and for piers and jetties are by no means identical. It cannot fail to be evident that a pile driven wholly into the ground, as in the former case, does not need to possess the same lateral stiffness which must essentially appertain to a pile only partially buried, and subject, moreover, to the incidence of forcible impact throughout a very considerable part of its length.

Thus, for landwork, concrete piles may be formed by simply drilling or boring a hole within an iron shell or tube, and filling the latter with concrete, the shell in many cases being withdrawn as the work proceeds. This method, of course, is quite inapplicable to piling in water.

The **Hennebique bearing pile** (fig. 66) contains a series of long, round bars, generally from four to eight in number, set parallel to, and arranged symmetrically around, the longitudinal axis of the pile. These bars are connected together and maintained in position by bonds, or ties, of iron wire and distance pieces. The bars vary from 1 inch to $1\frac{5}{8}$ inches in diameter, and the wire is usually $\frac{3}{16}$ inch thick. The distance pieces, which are about $\frac{1}{2}$ inch in diameter, with forked ends, are set at a normal distance of 10 inches apart, but at and near the top of the pile the distance is reduced to 2 inches. The toe of the pile is a pyramidal block of cast iron, into which wrought iron straps have been inserted. The upper ends of the straps are bent inwards towards the centre of the pile. The longitudinal rods of the pile are continued as far as the casting, being deflected to the required splay.

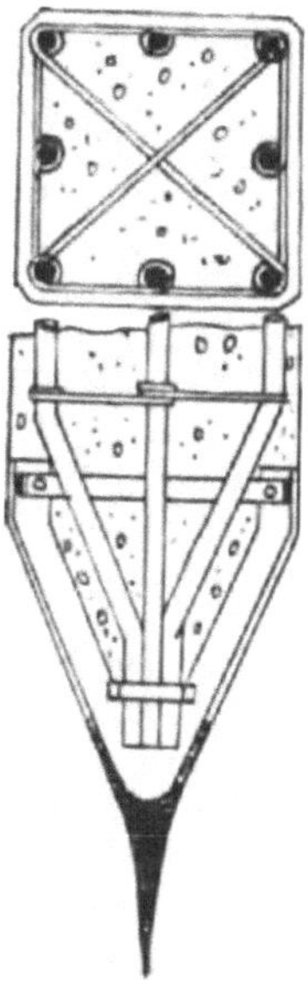

FIG. 66.—Hennebique Pile.

The **Hennebique sheeting pile** is made on the same lines as the bearing pile. There are three rows of longitudinal rods, arranged in pairs, and connected, as before, at 10-inch intervals with iron bands or clips. The ends of sheet piles are wedge-shaped, with a downward splay towards one side. This is extremely useful, during the driving process, in keeping a pile in continuous contact with its neighbour, towards which the resultant pressure on the splayed edge causes it to be urged. To complete the connection, piles are moulded with cylindrical grooves in each of their sides, one of which possesses a short spur or projection, capable of engaging in the groove of an adjoining pile. When two consecutive piles have been driven, their combined grooves form a cylinder which, after being cleansed by forcing water through it under pressure, is grouted with cement.

A slight variant on the Hennebique pile is the **Mouchel hollow pile** (fig. 67). It has the longitudinal rods, wire ties, and distance pieces of the former, but, with the object of saving material and reducing weight, it is concreted with a core, which, being withdrawn, leaves the pile hollow. Diaphragms at intervals strengthen the concrete work. The Mouchel pile is light and easy to handle. The reduction in strength is such as to be practically inappreciable, and does not affect the utility of the pile.

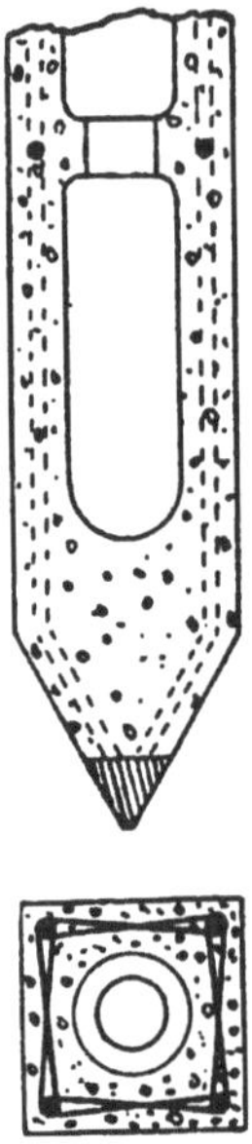

FIGS. 67.—Mouchel Hollow Pile.

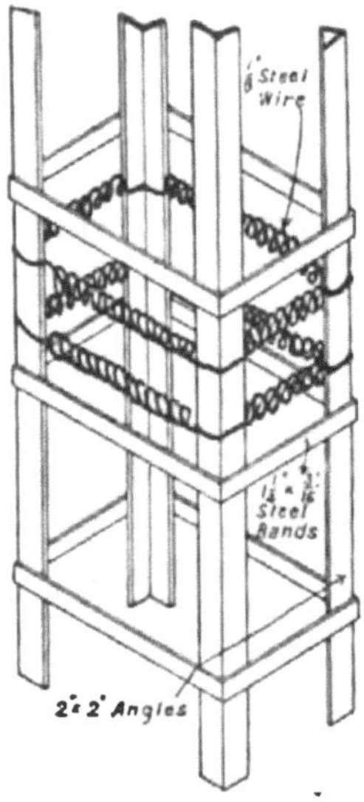


FIGS. 68.—Johnston Pile.

The **Johnston pile** (fig. 68) differs from the preceding in that the longitudinal rods are replaced by angle bars at the corners of the pile. These are bound together by flat bands and coiled steel wire.

The **Chenoweth pile** (fig. 69) is constructed on totally different lines. A sheet of iron mesh is bent round a longitudinal axis in the form of a continuous spiral, forming a cylinder which is surrounded and filled with concrete.

The **Williams pile** (fig. 70) consists of a central rolled steel joist surrounded at intervals by steel wire hoops and having cambered longitudinal stiffeners.

**Moulding.**—Reinforced concrete piles may be moulded either vertically or horizontally. For the former method, it is claimed that it results in greater uniformity in density throughout any horizontal layer, while the latter method is characterised by greater convenience. The advantage gained by vertical moulding is of questionable validity—there is no reason why horizontally moulded piles should not be absolutely homogeneous—and, in any case, it cannot be said to compensate for the greater trouble of moulding in that way and the higher cost involved.

In *horizontal moulding* a box is formed of the dimensions of the pile, but without a top—that side being reckoned the top which comes uppermost when the box is laid flat on the ground. The sides of the mould are well soaped or oiled to prevent adhesion; then a layer of concrete is deposited in the bottom to the extent of the outermost covering of the metal: that is to

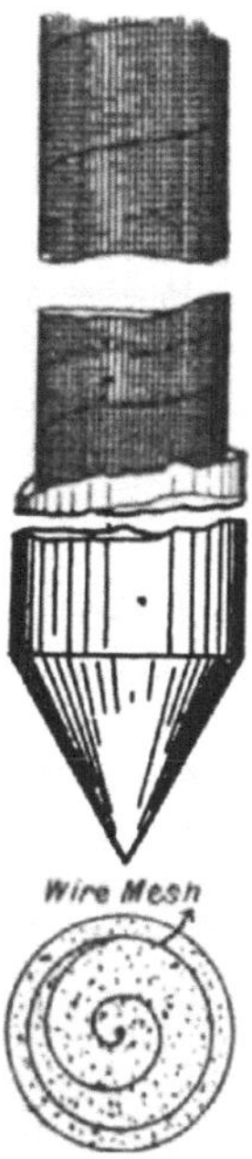


FIGS. 69.—Chenoweth Pile.

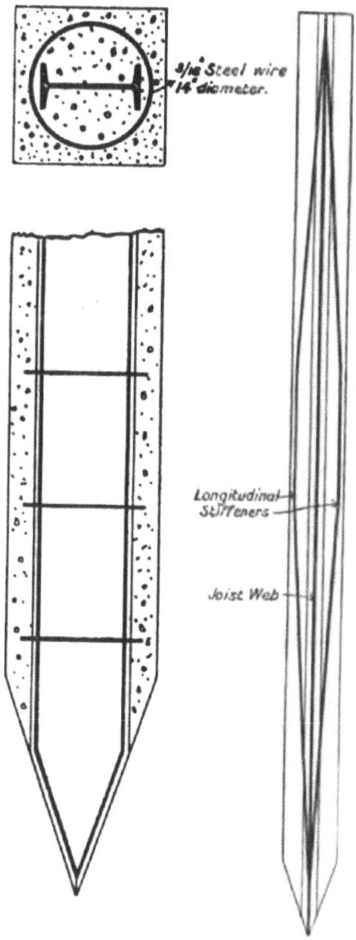


FIGS. 70.—Williams Pile.

say, 1 or 2 inches, as the case may be. This is very carefully rammed and consolidated before the metal framework is laid upon it. The latter operation requires great care. The framework must be set perfectly true to the axis of the pile, and the shoe, with its bevelled sides, must be accurately adjusted and brought into close contact with the ends of the frame. The box is then carefully filled with concrete in a series of thin layers, deposited without a break, each layer being well punned and the concrete pressed into all corners, angles, and recesses. The top, or fourth side of the pile, is formed by striking the edges of the box with a straight-edge, so that the concrete just comes flush with them. The pile is left for a week in the mould, then the mould is removed and the pile allowed to harden, either in water or while constantly wetted. A month or six weeks elapses before the pile is ready for driving.

To facilitate lifting, a bolt hole is cast near the top of the pile. The bolt and a shackle enable the pile to be swung easily into position. The green pile, however, is not handled in this way, but by means of chain slings passing round the pile, the sides of which are protected by deals at the points of contact.

For *vertical moulding*, the box is set upright and the metal framework first placed in position with the shoe downwards. Concrete is then filled in to the mould and around the metal, as carefully as in the previous case. The pile is built up in a series of layers from 4 to 6 inches in depth, the fourth side of each layer being formed by a batten fixed across the open face by fitting into grooves or being otherwise secured to the box, the whole height being treated in this way. The remaining operations are as before.

The materials used for reinforced concrete piles must be the best of their respective kinds. The concrete particularly calls for special attention. The proportions used lie between one part of Portland cement to four or five parts of aggregate, the latter compounded of gravel and sand in the ratio of 2 : 1. In one system (the Williams') the aggregate consists of clean shingle, which will pass through a ¾-inch gauge but not through a ¼-inch gauge, mixed with half its volume of sand. In Hennebique work the gravel is also sifted through two sieves. The first has apertures 1 inch square; the other has four uncrossed meshes per linear inch. The residue from the first sieve is thrown against the second, and equal parts taken of that which passes through the second sieve and that which fails to do so. After the pile has been removed from the mould, it is well to give it a coat of pure cement wash. This closes the outermost pores and renders the pile more highly impervious. The non-porosity of a reinforced concrete pile is obviously essential to its durability. It is only by the complete exclusion of moisture from the embedded steelwork that the latter can be maintained in a serviceable condition.

**Pile-driving.**—Piles are forced into the ground, or driven, by means of piling machines, which are actuated by hand or steam power. The exceptional use of the screw pile has already been noticed (p. 70). The impelling force is commonly a heavy weight or ram, which is allowed to fall within vertical guides from any desirable height. In the hand or ringing machine, the weight rarely exceeds one-third of a ton, and the fall, 4 feet. In other appliances the weight and fall range from 15 cwts. and 10 feet to 3 tons and 4 feet respectively. A heavy weight and a low fall are preferable to a light weight and a considerable fall, owing to the greater oscillation resulting from the latter arrangement and the consequent jar in the delivery of the blow, which thus tends to injure and split the pile. In concrete piles the absence of vibration is of primary importance. Indeed, such is the care which has to be exercised to prevent rupture, that the pile head is capped in a very elaborate manner. A cast steel helmet completely envelops the head, its interior being filled with sawdust and sacking. Between the helmet and the ram of the pile-driver is also interposed a wooden dolly, so that a very considerable proportion of the momentum of the blow is absorbed before it reaches the pile.

A much more efficient implement, where conditions admit of its employment, is the steam hammer. Blows can be delivered with greater rapidity and effect. Timber piles driven by an ordinary weight machine to the utmost capability of the ram have responded readily to the steam hammer and have been forced to a considerably increased depth. Steam hammers are of two types. In the first, the piston is maintained in constant contact with the pile head, while the blow is administered by means of a heavy cast-iron cylinder, moving up and down under steam pressure. An average cylinder will weigh a ton and its stroke will be 3 feet. In the second type the cylinder is affixed to the head of the pile and the hammer is attached to the piston. The disadvantage attaching to machines of the steam hammer type is the leakage of moisture from the cylinder, which softens the head of the pile under impact, and reduces it to a pulpy state. This necessitates cutting and dressing a fresh head, otherwise the power of producing penetration is much impaired.

In driving through sand and sandy gravel, very excellent assistance has been derived from the use of the water-jet. A pipe led down the side of the pile to be driven, transmits water under pressure to the ground in advance of the pile, and maintains the former in a state of fluidity until the required depth has been obtained. Immediately after the withdrawal of the pipe, the sand consolidates firmly round the pile and there is no further tendency to sinkage even under load. Piles treated in this manner rarely have pointed ends, as a butt end affords greater bearing area without appreciably increasing the difficulty of driving. Indeed, the perpendicularity of a butt-ended pile is more easily maintained.

The limit of driving varies so strikingly according to local requirements that no precise figure can be assigned to it. Obviously, a pile may support a light load with ease where a heavier one would cause sinkage. With a ram of one ton weight falling through 10 feet, the pile may justifiably be considered adequately driven when eight or ten blows fail to produce a depression of more than $\frac{1}{4}$ of an inch. This will indicate the attainment of thoroughly firm ground, and any further attempts at driving will only tend to shatter the pile. Pile-ends become "broomed" or splintered under an excessive amount of impact. Apparently easy driving, after a check, may be due to this cause, and there is no means of ascertaining the fact except by withdrawing the pile.

**Sustaining Power.**—Piles, if completely embedded and driven to the limits stated above, may be loaded safely to the extent of half a ton per square inch of the area of the pile section. Those in soft, muddy ground, sinking freely, say 1 inch per blow, at the end of driving operations should not be loaded with more than a tenth of a ton per square inch, and even then some slight settlement may be anticipated.[1]

Illustrative of the first class, the following instances may be cited of piles driven to refusal in various situations at the port of Liverpool:—

| Nature of Superstructure. | Sectional Dimensions of Pile. | Sectional Area of Pile. | Length of Pile. | Total Load on Pile. | Load per sq. inch of Pile Area. |
|---|---|---|---|---|---|
| | | sq. ins. | | cwts. | |
| Riverside Warehouse . | 12" dia. | 113 | 40' to 50' | 1568 | 13·88 |
| Quay Cargo Shed . | 15" × 15" | 225 | 40' to 45' | 2296 | 10·20 |
| ,, . | 12" × 12" | 144 | 40' to 45' | 1170 | 8·12 |
| Riverside Warehouse . | 14" × 14" | 196 | 40' to 45' | 2177 | 11 11 |

At the port of New York, the conditions in many instances are such as to be typical of the second class. Along the North River, for example, where

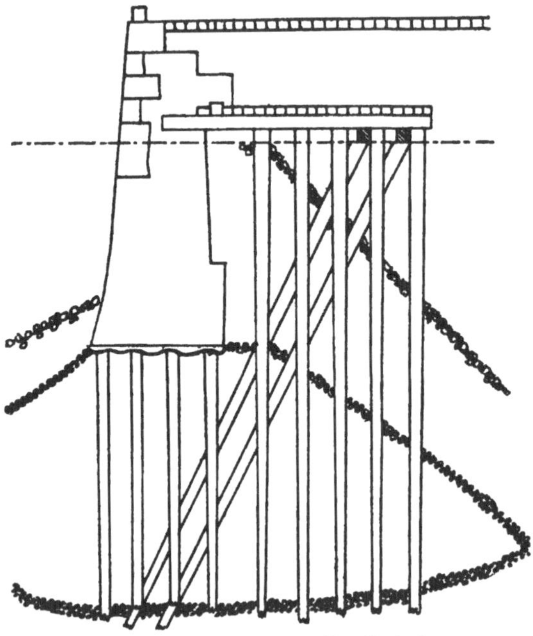

FIG. 71.—Piled Foundation to New York Quay.

most of the transatlantic liners are berthed, a firm stratum cannot be reached by piles 80 feet long. Such piles, therefore, are dependent on the friction of

[1] This figure relates to piles of ordinary size, say 12 inches square in cross section. As a matter of fact, the supporting power depends upon the surface exposed to friction, and therefore is governed by the sectional perimeter of the pile.

mud against their sides to support both themselves and the load they carry. And although, under circumstances of this kind, great sustaining power could hardly be expected, it is recorded that loads equivalent to 40 tons per pile have been safely carried. This figure, indeed, is very much in excess of the limit previously specified. At one-tenth of a ton per square inch a circular pile, 18 inches diameter in the butt (such as is commonly used at New York), would only support 24 tons and a 22-inch pile not more than 36 tons.

Mr J. A. Bensel, engineer-in-chief of the Department of Docks and Ferries at New York, has carried out some experiments on the limiting capabilities of the piles employed there.[1] A platform was built near the foot of Seventeenth Street, North River, and the piles to be tested, all about 80 feet long, were driven in four groups within the area of the platform, and arranged as

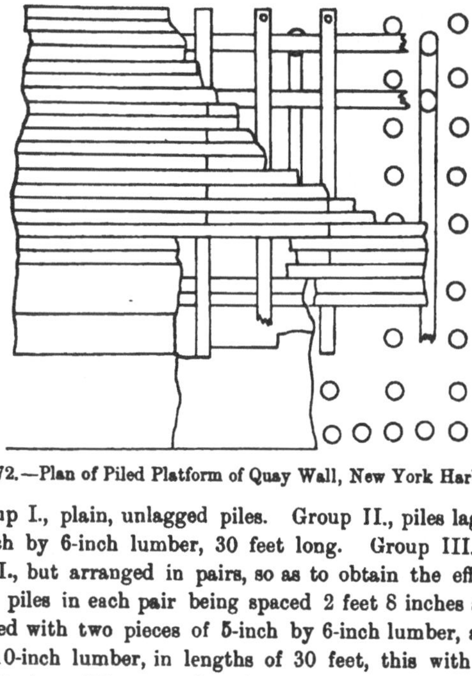

FIG. 72.—Plan of Piled Platform of Quay Wall, New York Harbour.

follows:—Group I., plain, unlagged piles. Group II., piles lagged with four pieces of 5-inch by 6-inch lumber, 30 feet long. Group III., piles lagged as in Group II., but arranged in pairs, so as to obtain the effect of greater proximity, the piles in each pair being spaced 2 feet 8 inches apart. Group IV., piles lagged with two pieces of 5-inch by 6-inch lumber, and two pieces of 4-inch by 10-inch lumber, in lengths of 30 feet, this with the object of obtaining results for a different style of lagging.

The testing platforms were loaded with granite and concrete blocks eighteen days after the last pile had been driven, thus affording the mud an opportunity of consolidating round the piles. A boring, taken at the site of the platform, indicated mud of uniform character to a depth of 100 feet below mean low water-line. The consistency of the mud at the top was such as to admit of the piles sinking by their own weight through 10 or 15 feet when lowered gradually; a little further down the mud attained the consistency

[1] Bensel on Dock Work in New York Harbour, *Proc. Int. Eng. Cong. St Louis*, 1904.

of wet modelling clay. The depth of water was 22 feet below mean low water level, and the piles were driven by a hammer weighing about 3000 lbs., having a uniform effective fall of about 8 feet.

The experiments were spread over a period of fifty-four days, when they came to an abrupt conclusion owing to the failure of the platform under the wash occasioned by the passage of a steamship. The last observations taken showed the maximum settlement of any test pile to be about 3½ inches, and that the settlement of the working platform in its vicinity was 1¾ inches. It is to be noted that this latter settlement took place under no load beyond the weight of the piles and the timber upon them, and that, therefore, the maximum settlement under load of any test pile was practically only 1¾ inches. A settlement of this amount appears to be not uncommon in local piers formed of similar piles, even before the structure is finished, or has received any other load than its own weight.

Mr Bensel, from various considerations indicated in his report, concludes that the ultimate bearing power in the unlagged piles of Group I. might be taken at 20 tons per pile, and that in the remaining groups of lagged piles the ultimate bearing powers would be 30, 20 and 30 tons respectively, per pile.

The following table shows details of the observations made during the experiments. The tons are given in American units of 2000 lbs. Roughly, their equivalent value in English units of 2240 lbs. may be arrived at by deducting one-tenth. This modification applies also to the figures quoted in the preceding paragraph.

PILE TESTS AT NEW YORK HARBOUR.

| Group. | Pile. | Diameter of point. | Diameter of butt. | Approximate weight of pile. | Total number of blows to drive. | Total penetration into mud. | Surface contact of pile with mud. | Average penetration for each of last five blows. | Load supported during latter twenty-seven days of test. | Settlement under load as in preceding column. | Settlement in working platform during same period. | Remarks. |
|---|---|---|---|---|---|---|---|---|---|---|---|---|
| | | ins. | ins. | lbs. | | ft. | sq. ft. | ins. | tons. | ins. | ins. | |
| I. | 1 | 8 | 18 | 4200 | 26 | 48 | 139 | 8 | 18·3 | ⅛ | ... | Unlagged piles 5 ft. 6 in. apart; in pairs, 11 ft. 6 in. apart. |
| | 2 | 8 | 17 | 3900 | 19 | 49·6 | 134 | 9¼ | 18·7 | ... | ... | |
| | 3 | 7 | 14 | 2700 | 13 | 50·9 | 127 | 13¼ | 18·7 | ⅛ | ... | |
| | 4 | 6½ | 17 | .3640 | 11 | 45·7 | 123 | 14¼ | 18·7 | ... | ... | |
| II. | 1 | 6 | 17 | 3600 | 46 | 49·6 | 220 | 4 | 31·8 | 13/16 | 5/16 | Piles spaced as above and lagged with four 5-in. × 6-in. pieces 30 ft. long. |
| | 2 | 6½ | 15 | 2940 | 58 | 50·1 | 219 | 3⅝ | 31·8 | 1 9/16 | 5/16 | |
| | 3 | 7½ | 16 | 3450 | 46 | 51·0 | 233 | 4 | 31·8 | 11/16 | 5/16 | |
| | 4 | 8 | 16½ | 3700 | 69 | 49·3 | 238 | 3 | 31·8 | 13/16 | 5/16 | |
| III. | 1 | 7 | 17 | 3700 | 70 | 50·7 | 247 | 3⅛ | 28 0 | 1 1/16 | ⅞ | Piles lagged as in Group II. and arranged in two lines of two pairs each, the space between each line being 11 ft. 6 in.; between the pairs in each line, 5 ft. 6 in.; and between the piles in each pair, 2 ft. 8 in. |
| | 2 | 8 | 17½ | 4060 | 70 | 49·8 | 243 | 3⅛ | 28·0 | 1⅛ | ⅞ | |
| | 3 | 7½ | 18½ | 4370 | 64 | 49·0 | 241 | 3 | 28·0 | 1⅛ | ⅞ | |
| | 4 | 9 | 18 | 4450 | 80 | 49·9 | 250 | 2⅝ | 28·0 | 1 1/16 | ⅞ | |
| | 5 | 6 | 22 | 5700 | 90 | 52·6 | 257 | 2 13/16 | 28·0 | 1 9/16 | ⅞ | |
| | 6 | 7½ | 21 | 5500 | 67 | 49·4 | 247 | 2¾ | 28·0 | 1 11/16 | ⅞ | |
| | 7 | 5½ | 22 | 5660 | 72 | 49·7 | 243 | 2¾ | 28·0 | 1⅛ | ⅞ | |
| | 8 | 9 | 19 | 4870 | 83 | 48·8 | 245 | 2½ | 28·0 | 1½ | ⅞ | |
| IV. | 1 | 5 | 16½ | 3270 | 59 | 50·7 | 241 | 3⅝ | 34·6 | ⅝ | ¾ | Piles spaced as in Group II. and lagged with two pieces 5 in. × 6 in. and two pieces 4 in. × 10 in. all 30 ft. long. |
| | 2 | 10 | 18½ | 4870 | 77 | 49·4 | 247 | 2½ | 34·6 | ⅝ | ¾ | |
| | 3 | 7½ | 19 | 4600 | 44 | 47·1 | 281 | 4⅜ | 34·6 | 1 11/16 | ¾ | |
| | 4 | 7 | 17½ | 3890 | 65 | 47·8 | 233 | 2⅞ | 34·6 | 1 9/16 | ¾ | |

The following data[1] relating to piles driven at Portsmouth Dock Yard extension constitute a typical record of ordinary experience in driving piles into very firm ground. The ground consisted of a regularly stratified, argillaceous sand, containing perhaps half its bulk of pure clay. The beds were fine, some of them not being half an inch thick, wholly impervious to water across the stratification, and very slightly, if at all, pervious in the direction of stratification.

All the piles were of fir 17 feet 4 inches long and 13⅜ inches square. Six of them were driven by hand by five men with a monkey weighing 15 cwts., and six by steam with a monkey weighing 22 cwts. The maximum fall in the first case was 21¾ feet, and in the second case 14¼ feet.

| Pile. | Time in Driving. | No. of Blows. | Max. Fall. | Final Depression. | Total Penetration. | | Remarks. |
|---|---|---|---|---|---|---|---|
| | hrs. | | ft. ins. | ins. | ft. ins. | | |
| 1 | 14 | 228 | 19 0 | ¼ | 12 9 | Hand machine. | Piles in all cases except two driven to the stage of pronounced or incipient splitting. In cases 9 and 12, as much as 3 feet and 4 feet 6 inches respectively had to be cut off the top of the piles from this cause. |
| 2 | 12 | 187 | 20 1 | ½ | 11 8 | | |
| 3 | 15¼ | 179 | 21 6 | $\frac{5}{16}$ | 12 3 | | |
| 4 | 15 | 169 | 21 10 | ⅜ | 12 4 | | |
| 5 | 18¾ | 270 | 21 9 | ¼ | 13 11 | | |
| 6 | 14 | 181 | 21 7 | ½ | 13 2 | | |
| 7 | 3¼ | 265 | 7 8 | ¼ | 10 6 | Steam machine. | |
| 8 | 2 | 255 | 14 3 | $\frac{3}{16}$ | 14 3 | | |
| 9 | 1½ | 280 | 9 5 | ¼ | 14 8 | | |
| 10 | 2½ | 180 | 10 9 | ½ | 14 0 | | |
| 11 | 2 | 228 | 10 8 | ⅛ | 13 9 | | |
| 12 | 3¼ | 130 | 9 10 | ⅜ | 11 0 | | |

Formulæ professing to give the exact sustaining power of piles are numerous, radically different in form, and conflicting in results. They are to be found in all engineering pocket-books, and little advantage would be derived from quoting them here. Some are extremely complex, embodying elements which have little or nothing to do with the capabilities of a pile to sustain an imposed load. When all has been said, it must be evident that the true test of sustaining power is the resistance offered to the final blow. The length, weight, and modulus of elasticity of the pile, are factors possessing no practical value, and a simple formula, linking up the weight of the ram and its fall with the resulting depression, should give all that is required. Major Saunders's formula is certainly based on these lines, but, unfortunately, it does not adapt itself to all cases. Thus, with a depression of $\frac{1}{10}$ inch under the last blow of a 2000-lbs. ram falling 9 feet, the safe load becomes 270,000 lbs. To sustain such a load, a pile, in the author's estimation, should be not less than 15½ inches square, and, by Rankine's rule, 16½ inches. Any pile, therefore, of less dimensions would be incapable of supporting so heavy a load with reasonable regard to safety, whereas, as a matter of fact, many 11- and 12-inch piles have been driven to comply with the standard stated in the formula. The values given by the formulæ of other authorities for similar conditions are as follows:—Haswell, 30,000 to 60,000 lbs.; Weisbach, 26,800 to 28,000 lbs.; Wellington, 32,727 lbs.; Trautwine, 32,460 to 97,380 lbs.,

[1] *Min. Proc. Inst., C.E.*, vol. xliv. p. 204.

the range in each case being due to the limits in the coefficient or factor of safety, which must always remain a matter of conjecture and arbitrary selection. The multiples, in fact, lie anywhere between $\frac{1}{2}$ and $\frac{1}{12}$.

Another and no less essential point to be noted is that the values obtained by these formulæ relate solely to loads imposed upon piles which are completely embedded in the ground. In so far, therefore, as a pile acts merely as a foundation for a pier column, the foregoing estimates of its resistance to pressure are strictly and legitimately applicable. But when a pile is only partially embedded in the ground, the calculations for its stability are of a dual nature: first, as a pile pure and simple up to the surface of the ground, and secondly, above the ground level, as a column or strut.

This aspect of the case calls for careful consideration, because a framework wharf, or pier, may fail through the flexure of its vertical members as much as through the subsidence of their bases. The longer the unsupported length, the less becomes the permissible load. And it follows, as an obvious corollary, that cross and diagonal bracing should be introduced from the lowest level at which it becomes practicable.

Failure by flexure involves an investigation of the relative values of the resistance of a material to tension and compression. Within the limits of the present treatise, it is not feasible to enter into all the details of so complex a problem. Neither in the present connection is it even desirable. Gordon's well-known formula furnishes all the information necessary for empirically determining the limiting load on columns, whether in the sea or ashore, and any further information on the subject should be sought in works dealing specially with columns and structural work generally.

**Gordon's formula** for the determination of the limiting loads on long columns or struts may be expressed as follows:—

$$p = \frac{f}{1 + a\frac{l^2}{d^2}}$$

where $p$ = ultimate load per square inch of sectional area;

$f$ = compressive stress per square inch of the material;

$\frac{l}{d}$ = ratio of length of column to its diameter, or to its least dimension in cross-section;

$a$ = coefficient given in table below.

The ultimate compressive stress ($f$) may be taken as follows, according to the material of which the strut is composed:—

| | |
|---|---|
| Timber | 2 to 4 tons per square inch. |
| Wrought iron | 16 " " |
| Mild steel | 30 " |
| Cast iron | 40 " |
| Concrete (4 to 1) | 2 " " |
| do (8 to 1) | 1 " " |

The values of the coefficient $a$ are given in the subjoined table.

COEFFICIENTS IN GORDON'S FORMULA.

| Material. | Cross Section. | Values of $a$. | | |
|---|---|---|---|---|
| | | Both ends rounded. | Both ends fixed. | One end rounded, one fixed. |
| Timber. | Rectangular or circular | $\frac{4}{250}$ | $\frac{1}{250}$ | $\frac{1}{100}$ |
| Wrought iron | Rectangular / Circular (solid or hollow) | $\frac{4}{3000}$ | $\frac{1}{3000}$ | $\frac{1}{1000}$ |
| ,, | L T + ⊏ ⊜ I ⊔ | $\frac{4}{900}$ | $\frac{1}{900}$ | $\frac{1}{360}$ |
| Cast iron | Circular (solid) | $\frac{1}{100}$ | $\frac{1}{400}$ | $\frac{1}{160}$ |
| ,, | ,, (hollow) | $\frac{1}{200}$ | $\frac{1}{800}$ | $\frac{1}{320}$ |
| ,, | Rectangular | $\frac{3}{400}$ | $\frac{3}{1600}$ | $\frac{3}{640}$ |
| ,, | Cross-shaped | $\frac{3}{200}$ | $\frac{3}{800}$ | $\frac{3}{320}$ |
| Reinforced concrete | Circular (solid) | $\frac{1}{600}$ | $\frac{1}{2400}$ | $\frac{1}{960}$ |
| Reinforced concrete | Rectangular (solid) | $\frac{1}{784}$ | $\frac{1}{3136}$ | $\frac{1}{1254}$ |

In the case of compound columns of concrete and steel (reinforced concrete), it is necessary to find the equivalent sectional area in terms of one material and make the ensuing calculations on that basis. Thus, if $\rho$ be the ratio of the coefficient of elasticity of steel to that of concrete, that is, if $\rho = \frac{E_s}{E_c}$, then an area A of steel is equivalent in resistance to an area $\rho$A of concrete. If A is the area of a reinforced concrete section (including the area of the steel reinforcement), and $A_s$ is the area of the steel, then the equivalent section $A_c$ in simple concrete will be

$$A_c = A + (\rho - 1)A_s.$$

The strength of a reinforced concrete pile can therefore be determined by treating it as a simple concrete pile of augmented area, the equivalent area being determined as above. The value of $a$ in the table corresponds to this method of treatment. The value of $\rho$ may be taken as 15″.

# CHAPTER V.

## STONE: NATURAL AND ARTIFICIAL.

Stone Supplies—Qualities desirable—Density and Hardness—Weight of Stone—Obtainment—Mine Firing—Drilling Operations—Implements—Charging—Tamping—Firing—Fuses and Detonators—Seam Firing—Wedging—Blasting Agents—Description of Quarrying Operations for Breakwaters at Goodwick, Alderney, and Holyhead—Concrete—Its Ingredients—Their Qualities and Proportions—Sea-water in its relationship to Concrete—Model Specification for Concrete in Maritime Works—Japanese Standards.

### Stone.

**Natural Stone.**—One of the most important considerations in connection with the construction of a breakwater is the supply of stone. Even in the case of those breakwaters which consist mainly of concrete blocks, it is eminently desirable, from an economical point of view, to pack the concrete with as many stone burrs, plums, or displacers, as possible. And in mound breakwaters a plentiful supply of rubble is obviously a paramount requirement.

The matter opens out into two branches. First, there is the quality of the stone, and secondly, the cost of obtaining it. The former question involves a consideration of physical characteristics and chemical qualities; the second, the proximity of a suitable quarry and the means of transport.

**Quality.**—In regard to physical characteristics, there are two features of pre-eminent importance—density and hardness. *Density*, or high specific gravity, is essential, because, when immersed in water, a stone loses a very considerable part of its effective weight; and when the sea is in motion, its stability as an inert mass is thereby reduced to a very great extent. Furthermore, if, compared with its weight, the stone possess a very large bulk, it presents a correspondingly large surface to wave action, thus increasing the scope or field of the disturbing force. These two factors of volume and weight must therefore be taken into joint consideration; they show that the smaller the surface area of a stone and the greater its unit weight, the less likelihood there is of disturbance. In other words, the higher the specific gravity, the greater the stability.

A concrete example will perhaps render this fact clearer. Take two blocks of stone of the same size—say exact cubes, each containing one cubic yard—but with specific gravities, represented in one case by 3 and in the other by 2. In air, the weights are 5184 lbs. and 3456 lbs. respectively.

In sea-water, the weights (after deducting the weight of the volume of water, which is the same for both) are 3456 lbs. and 1728 lbs.—a ratio of 2 to 1, representing an increase of 33 per cent. As the exposure to wave-stroke is the same in both cases, it is obvious that, when immersed, the stability of one block has relatively increased from half as much again to twice that of the other.

The second point is *hardness*, or durability. A good stone in this respect is one which is dense, compact, impervious, and free from all susceptibility to disintegration. In maritime situations, stones are subjected to much wear and friction—certainly more so than on land. The swell of the sea keeps those of small size in a state of continual agitation, rolling them over one another and chafing them until they assume that smooth spherical or ellipsoidal form which is so characteristic of pebbles along the beach. Moreover, in stormy weather, shingle, shells, and gravel are taken up by the waves and dashed with tremendous force against any surface upon which the waves happen to break. The effect of continual impact of this kind is to wear away even the hardest masonry. Wave action is supplemented by that of the wind, which blows sand in great volumes with the severity of a sand-blast. The cumulative results of abrasion and attrition are to be observed on any rocky coast, where towering cliffs stand honeycombed and fretted into fantastic shapes, while the strand is strewn with the comminuted fragments of quondam boulders.

The chemical qualities of a stone are not perhaps of such striking importance as its physical characteristics, but they are nevertheless deserving of consideration. The acidity and salinity of sea-water may, and often does, bring about molecular changes in minerals containing soluble salts. Certain compounds of lime are decomposed and softened by sea-water, and they also give rise to the formation of other compounds which tend to destroy the cohesion of the material of which they are ingredients, by producing cracks and fissures. Caustic lime and caustic magnesia, which are to be found in inferior and imperfectly made artificial stone or concrete—more rarely in natural stone,—are causes of disintegration by reason of their expansion under hydration, and also on account of their solubility. Still, on the whole, the chemical aspect of the question assumes a secondary importance, because those rocks which come under the category of minerals available for marine purposes, on account of their physical properties, are mostly, if not altogether, free from unstable constituents. The only exception, perhaps, is granite, which is a composition of three minerals—quartz, felspar, and mica, in a state of physical, not chemical, incorporation. Of these, the quartz is durable beyond cavil—it is practically indestructible; but certain varieties of felspar are liable to decomposition, and the mica is always more or less easily disintegrated. Nevertheless, granites, as a class, have gained a high reputation for strength and permanence, and it is only in very inferior qualities that the imperfections just mentioned manifest themselves, or where any appreciable deterioration is produced by natural agencies.

The heaviest and most durable varieties of stone are, generally speaking, those of igneous origin, such as basalts, granites, and traps, and metamorphic rocks such as quartzite. Many of the harder sedimentary rocks, though suitable in other respects, are unfortunately subject to the depredations of two troublesome molluscs, the *Pholas dactylus* and the *Saxicava*, both of which attack limestone and sandstone. Limestone blocks at Plymouth breakwater have had to be replaced by granite blocks on account of the ravages of the *Pholas*, which has already been mentioned in connection with its attacks on timber structures. Boring its holes in close proximity to one another, it honeycombs masonry work until it brings about its destruction.

The weights and specific gravities of stone suitable for maritime purposes are somewhat as follows. It will be understood, of course, that there is often a considerable range of weight in material of the same class, according to locality, owing to variations in composition and texture.

WEIGHT AND STRENGTH OF STONE.

| | Weight in lbs. per cub. ft. | Crushing Load in lbs. per sq. in. | Specific Gravity. |
|---|---|---|---|
| Granites . . . . | 160 – 190 | 8,000 – 14,000 | 2·5 to 2·97 |
| Basalts and Traps . . | 170 – 190 | 8,000 – 16,000 | 2·65 to 2·97 |
| Limestones . . . | 130 – 170 | 3,000 – 9,000 | 2·03 to 2·65 |
| Sandstones . . . | 150 – 170 | 2,000 – 8,000 | 2·34 to 2·65 |

Granite has been used in the construction of two notable breakwaters in this country—those of Plymouth and Portland. The stone used at Plymouth came from the quarries of Colcerrow and Roughtor in Cornwall and Pewtor in Devonshire. Penryn in Leicestershire, in addition to Cornish quarries, supplied stone to Portland, where a large quantity of the local limestone was also used. Holyhead breakwater was built of Anglesea stone, which, nominally a granite, is really a quartzite. Alderney breakwater consists mainly of the native Mannez stone, a sandstone grit of such extraordinary hardness as to exceed that of the neighbouring Guernsey granite.

**Obtainment.**—Next to the selection of a stone comes the question of the facilities for its obtainment and the cost of conveyance. Certain breakwaters have been so fortunate as to be located in the immediate neighbourhood of a suitable quarry. In other cases stone has had to be transported from some distance. Generally speaking, upon a rocky coast stone is likely to be fairly plentiful and cheaply procurable; on a sandy shore, where stone is not so accessible, other forms of construction, such as fascine work, may commend themselves to preference on economical grounds.

**Quarrying.**—The art of quarrying is one which is often applied to special purposes: some quarries being mainly worked for building blocks, and others almost entirely for setts and road metalling. Obviously, neither of these departments claim any attention here. Stone which is required for breakwater purposes is of an intermediate character—not so small as for macadam, nor so regular as for architectural work. The rubble which is desirable for maritime undertakings is of varying size, and, in fact, is such as results more or less naturally from the simple blasting of rock. Except for copings and string courses, no dressing is required, and the main bulk of the work is executed in blocks of irregular size and shape. In order to obtain these blocks to fairly large dimensions, some discrimination has to be exercised both in regard to the manner of boring the holes for blasting purposes and the nature and amount of the charges employed.

Much, of course, depends upon the disposition of the working face of the quarry and its relationship to the strike and dip of the strata. Natural joints and beds should obviously be taken advantage of to the fullest extent. These features are most irregular and uncertain in the igneous rocks, and therefore call for the aid of some skill and experience in their utilisation.

When blasting operations are projected on a large scale, the system of **mine firing** is adopted, and headings are driven in from a vertical face, or shafts are sunk from the top—the relative economy of these methods being dependent on the height of the quarry escarpment. Drainage and ventilation are more readily assured by the use of headings. In this case galleries are formed of the smallest possible sectional area consistent with the working space required for a man in each; they are arranged zigzag in direction or with one or more abrupt turns, and they terminate in chambers which are filled with explosives. Shafts, on the other hand, are straight and vertical.

Mine firing, which produces huge downfalls of stone—ranging, in many instances, from 100,000 tons to 500,000 tons—results in the dislodgment of so many and such enormous masses of rock that these last have to be again broken up into serviceable sizes by means of smaller charges. The method, therefore, does not altogether obviate the alternative system of small-charge firing, which, in less extensive operations, is more generally adopted.

**Drilling Operations.**—For the purpose of boring the necessary holes, either to receive the blasting charge proper or as a preliminary in the formation of a shaft or heading, various kinds of drilling instruments are employed, including the jumper, the hand-drill, and the machine drill. Of these, the two former involve manual labour; the last is mechanical and automatic. Where the work is sufficiently extensive to justify the initial cost of installation, there can be little doubt as to the superior economy and efficiency of machine drills. They can be worked much more accurately and with greater ease and convenience, there being situations where the application of hand-drilling would prove awkward, tedious, and slow.

The *Jumper* is an implement worked by one or several men. It consists of a long heavy bar of steel, sometimes circular or cruciform in section, but generally octagonal. The length varies from 6 to 8 feet, and, though not commonly the case, the bar is sometimes thickened in the middle in order to give increased momentum to the blow. In drilling a vertical hole, the jumper is lifted and allowed to fall, its uprightness being maintained throughout. It is caught at each rebound and raised again, being given, at the same time, a slight turn. For horizontal work, the drill is swayed backwards and forwards and slowly rotated as before. One drawback of the jumper is its liability to deflection from its assigned direction if it happens to come across a vein of harder material. Guidance is practically absent at the moment of impact.

The *Hand-drill* is a short steel bar of octagonal section, manipulated either by one man, who holds the bar with one hand while he strikes it on the head with a hammer held in the other, or by two, or even three men, one of whom acts as holder and the others as strikers. The command over a hand-drill is more effective in maintaining its alignment than it is in the case of a jumper.

The limiting amount of useful penetration by the hand-drill is about 2 feet, and it is chiefly used for making the short plug-holes, some few inches in depth, which enable large blocks to be split up into smaller pieces. The jumper may be effectively used for holes of from 3 to 4 feet in depth. The rate of progress in either case depends, of course, mainly on the hardness of the rock, and, in the second place, on the diameter of the hole, but it may be taken, on an average, at from 1 foot to 5 feet per hour; the former rate for holes of 2 inches diameter in granite, and the latter for $1\frac{1}{2}$-inch holes in limestone.

At the Kirkmabreck granite quarry, the following rates of work obtain: viz., three men will bore about 7 feet per day of $2\frac{1}{2}$-inch hole, and $8\frac{1}{2}$ feet per day of 2-inch hole. Of plug-holes 9 inches deep, three men will drive 24 feet per day to $1\frac{1}{2}$ inches diameter, and 32 feet per day of holes $\frac{7}{8}$ inch diameter and 4 inches deep. One man alone can do 14 feet per day of $\frac{5}{8}$-inch plug-holes 3 inches deep.

*Machine drills* are either rotary or percussive in action, and are actuated variously by steam, compressed air, water under pressure, and electricity.

Purely *rotary drills* generally take the form of a tube with an annular cutting edge, formed either with hardened steel teeth or with a row of

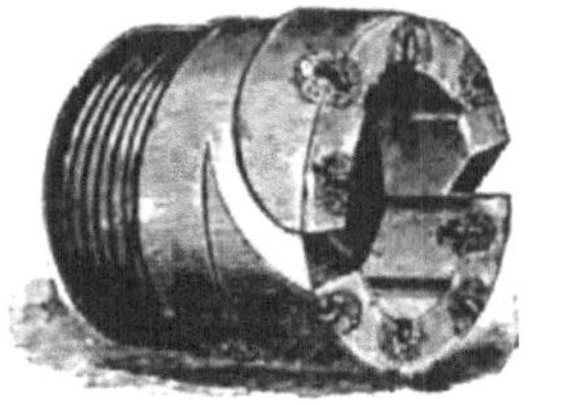

FIG. 74.—Core-bit or Cutting Edge of Rotary Drill, set with Diamonds.

FIG. 75.—Core Lifter.

diamonds. In the Brandt drill, steel teeth are forced against the surface of the rock under enormous hydraulic pressure, while the tube makes from five to eight revolutions per minute. In the ordinary diamond drill, the periphery of the "core-bit," as it is termed, has a number of diamonds embedded in it, and rotation is much more rapidly performed—from 200 to 400 revolutions per minute. The core, which results from the action of the tube, is subsequently broken off and withdrawn by a "core-lifter," which forms part of the internal mechanism of the drill.

The annular form of such drills lends itself to the supply of water to the point of incision, an adjunct which is of decided advantage in all forms of drilling, whether by hand or mechanism. In this respect hydraulic motive power may serve a double purpose, the waste water from the pressure cylinder acting also as a lubricant and dust preventer.

*Percussion drills*, which also have a subsidiary rotary movement, conform to the principle of the manual drill in that they are driven forcibly against the rock by steam or other pressure. The essential parts are a cylinder and piston, the latter of which receives the pressure alternately on each of its faces and acts as a combined hammer and drill, or perhaps more closely resembles the jumper. The drill rod proper is solid throughout and attached to the end of the piston. It is provided with a cutting edge or bit, of I, X,[1] or Z shape. The bit requires sharpening every 2 to 4 feet of penetration. The pressure employed is about 60 to 70 lbs. per square inch. About 300 blows are delivered per minute, and the rate of progress ranges from 3 to 10 feet per hour when the diameter of the hole lies between 1 and 2 inches. One man suffices to operate a machine, which may comprise several drills, but two or three men are required to transport it, and two are generally in attendance. In granite, two men working a steam drill can do about three times the amount of work which would be done by hand in the same time.

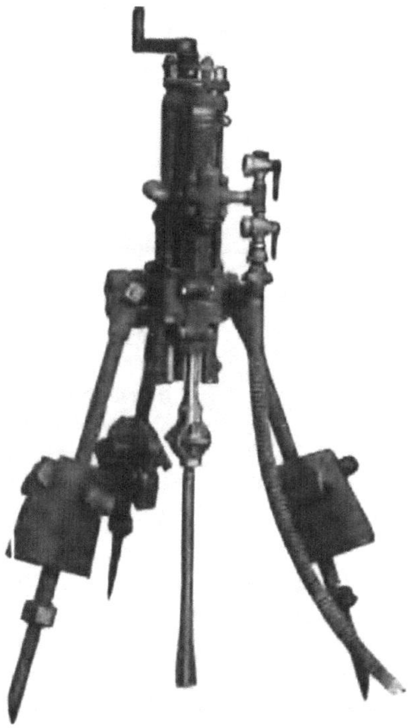

Fig. 76.—Ingersoll Percussive Drill.

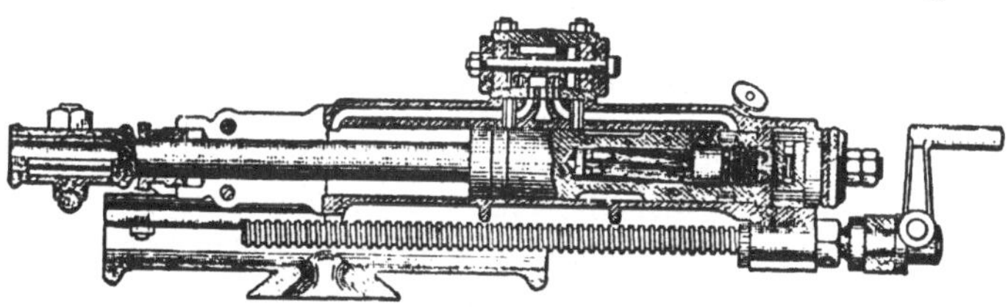

Fig. 77.—Section of Ingersoll Drill.

[1] An X shape is preferable to an exact cross, as it affords less likelihood of the same grooves being struck repeatedly.

The cutting edges, or "bits," of percussive drills, the widths of which range downwards from 4, or even 5 inches in mechanical drills and from 2 inches in manual drills, are slightly wider than the shanks of the bars on which they are worked, in order to ensure the necessary clearance in driving. For the same reason, the diameter of the drill is diminished as the depth attained is increased, at the rate of about $\frac{1}{16}$ inch every 18 inches drilled by hand, and about $\frac{1}{8}$ inch every 2 feet drilled by machine. Thus a 20-foot hole commencing with a diameter of $3\frac{1}{4}$ inches at the top would become reduced to 2 inches by the time the bottom was reached.

FIG. 78.—Drill Steels.

**Charging.**—After driving has been completed to the required extent, the hole is cleared of débris and moisture prior to the insertion of the charge. The amount of the charge is calculated on the same principle as that which underlies the preparation of the borehole, viz., that the explosive acts in the direction of the line of least resistance, or along the shortest route from the charge to the nearest open face; the hole should be considerably longer than this distance, at least twice as long. On the basis stated, we have the following formula:—

$$\text{Charge in lbs.} = (\text{line of least resistance in feet})^3 \times \text{coefficient},$$

in which the coefficient depends upon the nature of the rock and of the charge, being only definitely determinable by actual experience. For ordinary blasting powder in granite it is approximately ·04, and in softer stone ·03· For higher explosives it will be less, in proportion to their specific power. Thus, for dynamite the coefficient becomes ·005 or ·004· Gunpowder exerts, according to its precise composition, an explosive force of from 18 to 40 tons per square inch. For blasting purposes only the lower power is used, and a cubic yard of quarry rock ordinarily requires a charge of from $\frac{1}{4}$ lb. to 2 lbs. according to its nature and position; in tunnels and shafts as much as 6 lbs. per cubic yard has been used.

Blasting powder may be deposited in the borehole in bulk, but high explosives are usually made up in the form of cartridges, and the diameter of these is arranged so as to fit the hole exactly. The charge will consist of as many cartridges as may be considered necessary, carefully pressed into contact with one another by means of a blunt-ended wooden tamping-rod. The topmost cartridge is fitted with the cap, or detonator, required to produce explosion.

**Tamping.**—The next step after charging, prior to explosion, is tamping. This consists in packing the hole with granular or plastic material, so as to completely confine the charge and its gaseous products. For powder, dry, tough clay and powdered brick make good tamping material. By reason of its rapid action, dynamite does not require tamping to such an extent as is necessary for slower explosives. Even water will serve the purpose in deep vertical holes. Mud, sand, brickdust, and clay, are all used in connection with high explosives. Care must be taken in pressing home the first portion of a tamp, so as to avoid prematurely exploding the cap. Only a blunt wooden tamping-rod should be used.

**Firing.**—There is a difference of procedure in regard to the circumstances attending detonation. Low explosives, such as gunpowder, expand progressively by combustion, the gases accumulating until the resistance to expansion gives way. High explosives, on the other hand, act instantaneously, but they require a sharp initial explosion to develop their action fully, and this is provided by means of the detonator.

The *detonator* in general use is a small, solid-drawn copper tube, closed at one end and partly filled with an explosive compound (fulminate of mercury and chlorate of potash in varying proportions) which is capable of producing

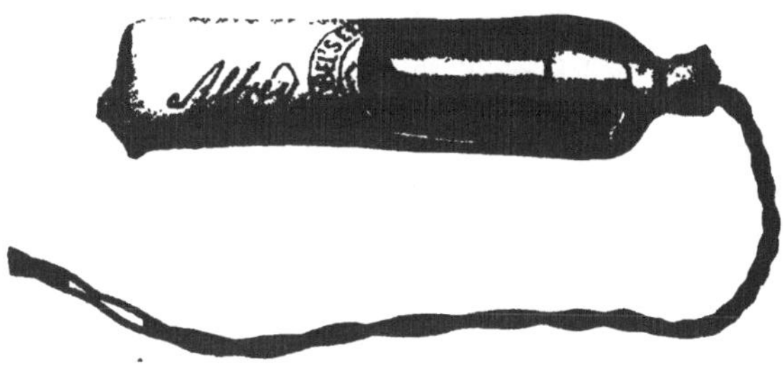

FIG. 79 —Electric Detonator Fuse inserted in Cartridge.

intense local force and heat. The detonator itself may be exploded either by means of a combustible fuse or by the electric current.

The ordinary (safety) combustible fuse burns at the rate of 2 to 3 feet per minute. There is also a "lightning" fuse for simultaneous firing, burning at the rate of 150 feet per second.

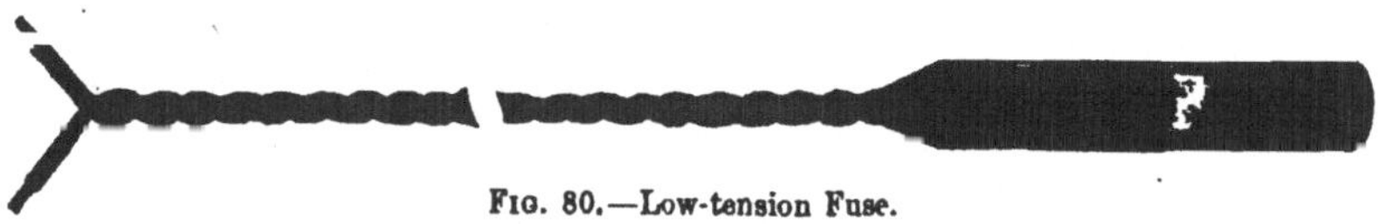

FIG. 80.—Low-tension Fuse.

Electrical discharge, of course, takes place instantaneously. There are two methods of firing. One, called the *low tension* (fig. 80), consists in sending an electric current through a thin platinum wire contained in the detonator, whereby it is made red-hot and so ignites the inflammable material in which it is incased. In *high tension* fuses (fig. 81), the wire terminals are not connected, but a spark is passed between them, effecting the same result.

FIG. 81.—High-tension Fuse.

**Seam Firing.**—It has already been remarked that the efficacy of a blast depends to a very large extent upon the judicious selection of the site and direction of the bore-holes. The natural faults to be found in rock manifestly lend themselves to an economical disposition of the disruptive forces, and it is often a wise plan to try the effect of a small preliminary charge by way of ascertaining to what extent any latent lines of weakness are developed. Thus, in a quarry at Penmaenmawr, in North Wales, where the rock is a hard, compact trap, it is the practice to bore a hole, say 20 to 23 feet deep, and charge it with only 5 lbs. or so of powder. On explosion, certain cracks are produced, the traces of which are followed and enlarged by additional holes until a final charge can be suitably placed for bringing down the whole mass. Similarly, at Goodwick, Pembrokeshire, a hole 20 feet deep and 2¾ inches in diameter at its extremity would be sprung several times by means of single gelignite cartridges of 1¼ inches diameter, producing a pear-shaped cavity capable of receiving the larger quantity—50 or 60 lbs.—required for complete dislocation.

Another method adopted in a granite quarry at Kirkmabreck, Kirkcudbrightshire, where the seams are fairly regular, extending along the face-line for some distance, is to open out a seam by means of plugs and wedges until it is wide enough to admit of the lodgment of a charge of powder. A chamber for the charge is prepared by shaping a couple of boards to fit the seam and setting them temporarily a couple of feet apart. Tamping is tightly rammed against the outsides of the boards, which are then withdrawn. The method of charging is to insert the fuse after the deposition of a third of the powder, the remainder being added on top and covered with a layer of lightly pressed hay followed by tamping above. The first foot of tamping is lightly rammed, the rest compactly.

**Wedging.**—Where the rock is of good quality and is required in the form of large, sound blocks for ashlar work in copings, facings, and string courses, a system of obtainment by wedging is adopted in preference to that of blasting, which may produce unsuspected planes of weakness as well as undesirable cracking and cleavage. In this case, a series of plug-holes, about 1½ inches in diameter and 9 inches deep, are driven along the line of some natural joint. Plugs and feathers are inserted into these holes, and driven by a succession of blows from a 26-lb. hammer until the seam has been sprung to

Fig. 82.—Quarrying for Fishguard (Goodwick) Breakwater. The cliffs before blasting.

Fig. 83.—A Mine Explosion at Fishguard.

[To face p. 94.

an extent admitting of the use of plates and wedges. These last are driven simultaneously, and as the fissure produced widens out, it is kept well packed with hand rubble. When the block is detached and ready to come away, an iron dog is attached to the stone, from which a chain passes to a crane, the tension of which, aided by men with crowbars and levers acting directly upon the block, causes it to become completely dislodged. It can then be converted into convenient sizes by plugging as before, and be dressed to requirements.

When the height of the quarry face is considerable, it is desirable to place planks and pieces of old timber upon the ledges of rock and upon the floor of the quarry, to avoid undue breakage of blocks falling from the upper layers.

**Blasting Agents.**—The number of explosives available for working purposes in a quarry is legion. From a practical point of view, despite the dissimilarity of their ingredients and methods of production, they fall into two classes, viz., (1) those in which a high local intensity is produced, causing much shattering and splintering into small fragments, and (2) those in which the expansive power is more widely and less violently exerted, resulting in disruption and dislocation rather than shattering. The first class is represented by dynamite, the second by gunpowder.

The basis of dynamite is *nitro-glycerine*, which forms a number of compounds possessed of similar attributes, but varying in power. Nitro-glycerine is a fluid combination of glycerine and of nitric and sulphuric acids. The majority of its various combinations, therefore, have a plastic, gelatinous nature, but the first to be noticed below has not this characteristic.

*Dynamite* consists of nitro-glycerine with the addition of a granular absorbent, which may either be an inert substance or, in itself, an explosive. The material more specially employed is a silicious infusorial earth occurring in Hanover, and called "kieselguhr." Commonly, the proportions are 75 parts, by weight, of nitro-glycerine to 25 parts of earth. If cartridges remain immersed in water for any length of time, the nitro-glycerine exudes and the charge deteriorates. Moreover, the substance is affected by changes in temperature and freezes at a higher temperature than the freezing point of water, so that some trouble is incurred in winter-time in thawing cartridges, the operation requiring much care and circumspection. The effects of firing are, as stated above, an intense rapidity of explosion producing extreme local shattering.

Other combinations of the same character are:—

*Blasting Gelatine*, containing 93 per cent. of nitro-glycerine and 7 per cent. of nitro-cotton. This is probably the most powerful blasting agent known at the present day. It is also very little, if at all, affected by immersion in water.

*Gelatine Dynamite*, somewhat inferior in strength to the foregoing, is a compound of nitro-glycerine, nitro-cellulose, and nitrate of potash.

*Gelignite* contains nitro-glycerine, nitro-cotton, nitrate of potash, and wood meal. It is rather more powerful than ordinary dynamite.

*Forcite* is a mixture of nitro-glycerine with cellulose, the latter being gelatinised by heating in water under considerable pressure. Nitrated cellulose is also used in admixture with oxidising salts.

*Gun cotton*, which is cotton dipped into a mixture of nitric and sulphuric acids and itself an explosive, gives rise to the following, amongst other, compounds:—

*Tonite* is finely divided, or macerated, gun cotton, combined with an equal weight of nitrate of baryta. There are two varieties—the white and the black. The former is very shattering in its action, and is therefore chiefly applicable to the breaking up of extremely hard stone, such as quartz. *Black tonite*, containing a larger proportion of baryta and some charcoal, is more disruptive.

Chlorate of potash forms the basis of two well-known explosives, viz., *Rack-a-rock* and *Cheddite*. The former consists of compressed cartridges of chlorate of potash, impregnated with dead oil, either alone, or in conjunction with bisulphide of carbon, or mixed with nitro-benzol. Cheddite, an admirable product of more recent date, contains chlorate of potash, naphthaline, and castor oil.

It is needless to extend the list further. There are many other excellent explosives on the market, and fresh compositions are continually being evolved, each with its own special advantages. But while many of them are characterised by extremely high power, resulting in the production of almost incredible downfalls of rock, yet in ordinary quarrying operations where, as has been pointed out, intense local effect is by no means sought after, it is probable that in the majority of cases blasting powder is every whit as serviceable, and certainly more economical.

*Gunpowder*, the earliest of explosives, is a mixture of saltpetre, sulphur, and charcoal, in proportions ranging between 6 : 1 : 1 and 15 : 3 : 2. These are the proportions used for service powder for military purposes Blasting powder is distinguished from gunpowder, properly so called, in that it contains rather less saltpetre and that it is not manufactured with the same particular selection of material and delicacy of treatment. The effective power is therefore lower.

**Quarrying for Goodwick Breakwater.**—The following particulars relating to the quarrying of stone for the breakwater in Pembrokeshire, forming a protection to the Fishguard terminus of the new Fishguard, Rosslare (Great Western) route to Ireland, have been compiled from information kindly supplied by Mr G. Lambert Gibson, the engineer in charge.

When the works were begun in the year 1896, they were carried out tentatively with a small outfit of plant, but with a considerable body of men. The start was a difficult one, the men having to attack the face of precipitous cliffs of an intensely hard and vitreous texture, rising from the sea to heights of one and two hundred feet. The boring of the rock to receive explosives was done entirely by hand; and, owing to the want of foothold, the men had often to be slung by ropes from the top of the cliff. After six years of somewhat slow progress in this manner, more vigorous measures were decided

FIG. 84.—Quarrying for Fishguard Harbour. The cliffs after blasting.

[To face p. 94.

upon. A complete installation of compressed air drilling-plant was put down, and the work of boring was let to a firm of contractors.

Under the new system, both single firing and mine blasting operations were carried on. In the first case, holes 20 feet deep and 2½ inches in diameter were charged with 20 to 50 lbs. of gelignite. In the second case, the method adopted where the cliff was lofty and the rock exceptionally hard, a tunnel some 40 feet long was driven square into the face of the cliff, with two branches, each also about 40 feet in length, right and left of it, the combined galleries taking the shape of the letter T. At the ends of the cross tunnels small chambers were formed, within which were placed a charge of, usually, 7 tons of gunpowder in two boxes; the tunnels were then built up and the charges fired by electricity. Very little noise or shock to the neighbourhood is said to have been caused by the firing, although as much as 113,000 tons of rock have been dislocated by a 7-ton charge. The yield, however, varied considerably, according to the nature of the rock at various places along the half mile of quarry face, and in some cases 9 tons of gunpowder were required to produce a fall of 70,000 tons of rock.

Generally speaking, it was found that where a drill could penetrate 15 feet per day—the average depth of the holes,—the cost of single hole firing was equal to the cost of mine-firing with a working face of 120 feet. In other words, when the height of quarry face exceeded 120 feet, or when the drills failed to accomplish 15 feet per drill per day, mine-firing proved the more economical method.

The rock, having been blasted, was loaded into waggons. Stones from 3 to 15 tons weight were tipped on the sea side of the breakwater; those from 1 cwt. to 3 tons on the harbour side. Stones of 1 cwt. and less were sent to a ballast crusher for use in the concrete blockwork.

The rock-getting and depositing plant consisted of one 120 H.-P. air-compressor engine, nine 8 H.-P. Ingersoll rock drills, five locomotives, fifteen steam cranes of powers ranging from 1½ to 15 tons, and 175 waggons.

The quantity of rock dealt with amounted in all to about two million tons.

**Quarrying for Alderney Breakwater.**[1]—The stone of which the greater part of the breakwater is built is a local stone obtained from the Mannez quarries, and, although a sandstone grit, considerably harder than granite. The quarries were situated a couple of miles away from the site of the breakwater, and had a working face 75 feet high. The hardness of the stone may be gauged from the fact that where one jumper sufficed to bore a hole in granite, two were required for the Mannez stone.

The mode of quarrying was as follows:—Shot holes from 6 to 8 feet deep were drilled at the toe of the quarry face, charged and exploded. When the rock was sufficiently undermined in this manner, a deep hole was drilled down from the top of the quarry to a shelf or bed, of which there were several, inclined at an angle of from 25° to 45° to the face. This hole was charged

[1] Vide *Min. Proc. Inst. C.E.*, vol. xxxvii. p. 86.

with powder, lightly tamped and exploded several times, till a crack was made longitudinally; then, into the crack, large grained powder was poured and exploded, bringing down a considerable mass of rock. By this means the stone was not too violently shaken, and good face stones were obtained. Many of these were 9 feet long by 3 feet 3 inches thick and 15 feet on the bed. The cost of dressing them into shape came to 6d. per cubic foot. The stones selected for rubble hearting weighed from 3 to 15 tons. The greatest quantity of material conveyed to the bank in one day was 3000 tons. Seventy-four tons of powder were consumed per annum.

**Quarrying for Holyhead Breakwater.**[1]—The stone—a quartz rock —was obtained from an adjoining hill known as Holyhead Mountain, and the quarries were distant rather less than a mile from the commencement of the work.

At the outset, quarrying operations were carried on by a system of single-hole firing, but, although many hands were employed, the output proved insufficient for requirements. Blasting on a much larger scale was then resorted to, by sinking shafts and driving headings or driftways to receive large quantities of powder.

The first large mines were in shafts about 6 feet by 4 feet, sunk from the top and of varying depths, according to the height of the face; but when the quarries had been more opened and the face got very high, sometimes the top only was prepared for blasting by shafts, and the bottom by headings of the same size, or somewhat less. Ultimately, headings were preferred to shafts and adopted whenever practicable. They proved more convenient, as the men could work in front instead of at their feet; the men did not get wet from rain, and the ventilation was better. Headings were also less dangerous, as the tamping was less liable to be blown out. On the other hand, shafts were more easily tamped and required smaller charges of powder, the rock being already weakened by the excavation.

On an average, 4 tons of rock were blasted per lb. of powder, the extremes ranging from 5 to 2 tons. Generally the charges varied from 600 lbs. to 21,000 lbs. Experience determined the following coefficients for the formula given on page 90:—

For ordinary shafts, coefficient $= \frac{1}{15}$ to $\frac{1}{20} = \cdot066$ to $\cdot05$;
For ordinary headings, coefficient $= \frac{1}{12} = \cdot083$.

In the exceptionally difficult case of a mine called a "rooter-out," the coefficient became ·1· This was a mine in which there was a natural joint on one side only, so that the rock had principally to be torn away from the solid mass. In such cases the lowest results were achieved, and, further, the stone displaced was usually in large masses requiring further breaking up, while in more favourable cases the stone resulting from a blast was suitable for immediate use.

Figs. 86 and 87 are types of the best and worst kinds of mine respectively.

[1] Hayter on Holyhead New Harbour, *Min. Proc. Inst., C.E.*, vol. xliv.

Fig. 85.—Fishguard Breakwater in course of construction.

[*To face p* 96.

In tamping the headings, the powder was often built in for a few feet with a dry rubble wall, and the remainder of the galleries rammed with clay, obtained from a decomposed porphyry dyke in the quarry. The shafts were more readily tamped with quarry débris, stones and clay thrown in and rammed. Occasionally, especially at the outset, a space was left round the charge, but it is believed that this was of little use.

The cost of quarrying the rock, including driving, powder and sundries, was 4½d. per ton, and the cost of filling into waggons, including blasting the large stones, was about the same. The prices, however, refer to the period of

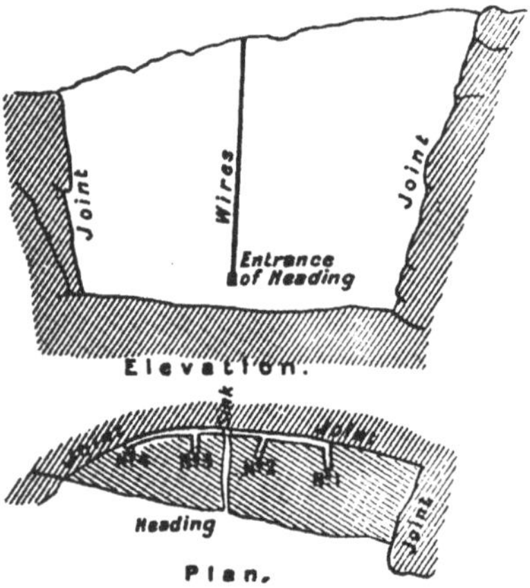


FIG. 86.—Mine at Holyhead.

Height of Face . . 110 feet.
Length of Face . . 140 „
Length of Heading . 89 „
Grip of Heading . . 35 „
Depth of Sink . . 13½ „

| | Chambers. | | | |
|---|---|---|---|---|
| | No. 1. | No. 2. | No. 3. | No. 4. |
| Length . . . | 12½′ | 12′ | 9′ | 2½′ |
| Line of least resistance | 23½′ | 24′ | 25′ | 18′ |
| Charge in lbs. . . | 4,500 | 4,000 | 3,000 | 1,500 |

Prodnce, upwards of 60,000 tons.

Depth of shaft, 44 feet.
Line of least resistance, 21 feet.
Charge, 900 lbs.
Produce, 2000 tons.

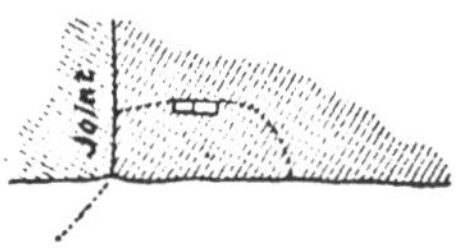


FIG. 87.—"Rooter-out" Mine at Holyhead.

maximum output. The cost of driving the headings ranged from 10s. to 25s. per lineal foot, out of which the miners had to pay, on an average, about 2s. for powder, fuses, etc. The average length of heading driven was 5 feet per week with four men employed.

## Concrete.

The subject of natural stone leads on almost insensibly to the kindred theme of artificial stone, for which an equally valuable, and a practically unlimited, field of usefulness exists. Those parts of maritime structures which are by far the most important and most prominent, are now constructed in concrete, in place of the elaborate masonry which characterised and distinguished the operations of past generations of engineers.

The cause and reason for this is not far to seek. Blocks of stone of large size are difficult to procure, expensive to dress, and equally expensive to convey and set in position. Smaller stones involve a multiplicity of joints. These,

7

in themselves, are a source of weakness, but when their use is inseparably combined with an intricate system of keyage and bonding, they prove doubly unsatisfactory and afford but an indifferent sense of security. What masonry, with its vertical and horizontal breaks and intersections, and its costly chisel-work, fails to ensure, is readily achieved by concrete, easily moulded while in a plastic condition to any required shape or outline, and deposited in position by the simplest means, with a minimum expenditure of time, money, and labour.

Concrete, as the term is generally understood amongst engineers, is an admixture of various mineral substances which, under chemical action, become incorporated into a solid body. Of its ingredients one group is inert, the other is active.

The inert group is called the *aggregate*, and it comprises any number of the following substances: slag, shingle, burnt clay or earthenware, broken stone, broken brick, gravel, and sand.

The active elements are hydraulic lime or cement and water. These constitute what is called the *matrix*.

The uses and applications of concrete are manifold. We have, however, to confine our attention to those points alone which are of pre-eminent importance and value in regard to maritime work, leaving other features and adaptations for treatment elsewhere.

The first thing to be noticed—which has already been animadverted upon in the earlier part of this chapter—is the necessity for an aggregate of high specific gravity. The reasons need not be repeated. From this point of view, heavy materials, such as slag and broken granite, are preferable to broken brick and sandstone.

Secondly, in order to ensure close adhesion, the aggregate should be rough and angular. Porous surfaces are admirable in this respect, but they do not generally appertain to heavy substances. However, the coarse crystalline texture of granite offers a sufficiently marked advantage over the smooth polished surfaces of flints and pebbles to constitute an excellent instance of what is meant by compliance with this requirement.

Thirdly, in order to reduce the number and volume of interstices, fragments of different sizes should be employed so that the smaller material may fill up the voids in the larger. At the same time, it is not desirable to use fragments of greater linear dimension than 4 inches, nor sand so fine as to be dust-like. It is usual to specify that the stone shall pass through a $1\frac{1}{2}$- or 2-inch ring, and that the sand shall be coarse and sharp. If fine sand be used, the grains cohere when watered and impede the introduction of the cement, besides requiring a greater quantity to effect the same complete envelopment.

Lastly, the aggregate should be perfectly clean and free from grease, clay, mud, and any other impurity whatever. Such substances have no adhesive value; they intervene between the parts which should come in contact, and are themselves readily soluble and removable by water, leaving the mass in which they happen to be incorporated in a porous and laminated condition.

The matrix is almost universally **Portland cement**, though hydraulic lime has been, and is still used, and also Roman cement. Hydraulic lime of a special character—the Teil lime—is a favourite with French engineers for sea work. It has been very largely employed in their works on the Mediterranean coast and along the English Channel, and, with one or two exceptions, seems to have answered satisfactorily. The use of Roman cement is limited to situations where rapidity of execution is essential, and where the hardening of the mortar is required to take place within a very short period. Neither hydraulic lime nor Roman cement has anything like the strength and durability of Portland cement.

Portland cement is an artificial product obtained by calcining, at a high temperature, an intimate compound of clay or shale with chalk or other limestone. In this condition, it contains a number of ingredients, of which the principal are lime, silica, alumina, and oxide of iron. These form about nineteen-twentieths of the whole, within the following limits, viz.:—

| | |
|---|---|
| Lime, | 60 to 64 per cent. |
| Silica, | 20 to 24 ,, |
| Alumina, | 6 to 10 ,, |
| Oxide of Iron, | 3 to 5 ,, |

The remaining ingredients are magnesia, sulphuric acid, certain alkalies, and moisture. Of these, the magnesia should not be permitted to exceed 5 per cent., nor the sulphuric acid 1 per cent.

So great a variation has been manifested in the character of the different brands of cements emanating from the numerous manufacturers both in this country and abroad, and so much divergence of opinion has been exhibited in regard to standards and tests to be adopted for reference and comparison, that it was recently felt desirable, and even necessary, to draw up a specification for general use among engineers. This has been done by the Engineering Standards Committee, and the result of their deliberations is embodied in the specification at the end of this chapter. It is not necessary, therefore, at this stage, to enter into the more general requirements of the model specification.

**Effect of Sea-water on Concrete.**—The most vital consideration in regard to the matrix is the effect of sea-water upon concrete. On this point there is scope for much discussion and some ground for difference of view. On the one hand, there is abundant practical exemplification to demonstrate that Portland cement concrete is, in general, a thoroughly sound and durable material, in every way adaptable to maritime situations as elsewhere; on the other hand, there are indubitable instances of deterioration and failure. These instances obviously demand a searching inquiry, for in the absence of definite and authoritative refutation they must inevitably produce a feeling of doubt and uncertainty as to the propriety of using Portland cement in situations where its possible failure would entail consequences of the most serious nature.

The author has already, in another work, devoted some considerable space to a discussion of the subject,[1] and he does not feel that it will be considered incumbent upon him to restate the particulars and details of the investigation which led him to the conclusion that any defects which have exhibited themselves in the behaviour of Portland cement concrete in sea-water are due solely and entirely to inherent deficiencies in the quality of the cement used, and in the materials with which it was mixed, as also, in part, to indifferent and imperfect manipulation.

Briefly stated, the facts in cases of known failure show that disintegration is (1) inaugurated by symptoms of expansion, and (2) subsequently accomplished by the solvent action of the sea.

Now, the constituent elements of sea-water which have any effect upon cement are the sulphates and chlorides, especially the sulphate and chloride of magnesia. On coming in contact with lime, the magnesia in these salts is precipitated as a hydrate, with the simultaneous formation of chloride of calcium and sulphate of lime. Both these latter products are more or less soluble, especially the former, and the washing action of the sea soon occasions their removal.

But, in order that this interchange of constituents may take place, it is essential that there should be present a certain amount of lime, either actually in a free, uncombined condition, or, at least, in the form of a very unstable compound. The silicates and aluminates of lime produced during calcination, at the proper temperature for the manufacture of Portland cement, are all more or less fixed; at any rate there is abundant evidence to show that neither of these combinations necessarily breaks down in a marine environment. It is usual, therefore, to attribute the change to some caustic lime which has failed to combine with the silica and alumina. The supposition is not out of consonance with observed phenomena. The slaking of lime causes expansion, with the formation of a hydrate which is readily soluble in water. One authority,[2] however, suggests that a truer explanation of the expansion and disintegration is to be found in the formation of a "vitreous high-lime compound which slakes or hydrates so extremely slowly that it may be months before visible hydration even commences; but in such cases hydration is accompanied by enormous expansion, the increase in bulk amounting to many times the original size of the mass, sufficient to cause the disruption and total disintegration of the previously set particles in the cement."

The precise nature of this "vitreous high-lime compound" is difficult to identify, but from whatever source the lime be forthcoming, there seems no doubt that it exercises a deleterious effect, and that measures should be taken to ensure its absence wherever possible, and, in the second place, to limit the action of sea-water to the outer skin or surface. This can only be achieved

[1] Dock Engineering, p. 123 *et seq.*
[2] Butler, *Portland Cement*, London, Spon., 1905.

by making the concrete as impervious as possible, so that the bulk of it may be inaccessible to external influences.

Concrete made from sound Portland cement, mixed in proper proportions and thoroughly incorporated, is sufficiently impermeable for all practical purposes. It may not be absolutely water-tight—this is by no means essential[1] and can only be attained by the exercise of considerable trouble,—but it will display amply serviceable resistance to infiltration, which will prove little more than superficial. Even in those cases where chemical changes have taken place, the evidence simply points to the deposition of magnesian salts in the outer pores of concrete from which the calcic hydrate has been removed. The magnesia is an inert substance, and while, in itself, an evidence of decomposition, its presence is attended by no additional ill effects; in fact, it may even be claimed that it exerts a beneficial action in closing up pores which would otherwise remain open for the further penetration of sea-water into the interior of the mass.

In order to secure the highest degree of impermeability, a sufficiency of water must be used for mixing the concrete. An excess, of course, is objectionable, chiefly on the ground that it forms an incompressible volume in the fluid concrete, which passes away in evaporation, tending to leave the concrete porous. But, on the other hand, an insufficiency is attended by the evil that particles of cement may escape hydration, and this is a more vital consideration, in that there is a consequent lack of present cohesion and a source of future disturbance. It is better, on the whole, to water the concrete well rather than sparsely; some proportion of moisture will be absorbed by the environment, the foundation, and adjacent work, and unless the mass be allowed to harden without undue abstraction of moisture, its strength will become impaired.

Speaking from long experience of a wide range of concrete work deposited in a tidal estuary, where the fluctuations of level are very great and where the circumstances are most propitious to the exercise of decomposing influences, the writer is convinced that the dangers attending the use of concrete work in maritime situations are often greatly and needlessly exaggerated. Ordinary care and discretion in the processes of mixing and deposition will prevent any evil consequences, provided, of course, the cement be of unassailable character, conforming in all respects to the requirements of the standard specification.

Another point affecting the use of Portland cement concrete in maritime work is the influence exerted by sea temperature upon its setting properties. The crystallisation or setting of cement is favoured by warmth and retarded by cold. The presence, therefore, of cold or warm currents in the sea exercises a corresponding effect upon the setting time, so that it is not a matter for surprise to find considerable variation at different places, and even at the same place at different seasons, in the period during which concrete work

[1] The remark, needless to say, applies to block work and not to reinforced concrete, the special treatment of which is described elsewhere (p. 76).

hardens. Of course, the more the time is prolonged the greater difficulty will be experienced in preserving the soft concrete from the chafing action of waves; but, on the other hand, it seems to be pretty clearly established that the slower the setting action the greater the ultimate strength attained.

By way of completing this brief review of the subject, a model specification is appended, drawn up from a harbour engineer's point of view, and therefore containing several stipulations of a special character inapplicable to conditions elsewhere. The quality of the Portland cement, however, is strictly in accordance with the terms of the Engineering Standards Committee's specification, the clauses extracted from which are indicated by quotation marks.

## SPECIFICATION.

The *aggregate* shall consist of gravel and broken stone of varying size mixed with sand, the quantity of sand being sufficient to fill completely the interstices in the larger material. The precise proportion of sand is to be ascertained by gauging the volume of water contained in a vessel which has been packed with the maximum amount of gravel and stone it can contain up to, and flush with, the level of the brim. No fragment shall measure more than 4 linear inches in any direction, and every piece must be capable of passing through a ring 2½ inches in internal diameter. The length, breadth, and depth of the larger pieces must not be greatly unequal, *i.e.* there must be an absence of long, flat, slaty slips, as also of smooth, water-worn pebbles. The stone must be heavy, weighing in the solid mass not less than 150 lbs. per cubic foot. Slag from ironworks may be used in place of, or in conjunction with, stone, provided it conform to the same conditions of weight and size and is not brittle or friable in any part. Both gravel and stone or slag must be perfectly clean and free from admixture with any foreign substance, whether mineral or vegetable, and no gravel which has come as ships' ballast will be accepted.[1] The sand must also be clean; sharp, and not too fine, *i.e.* it should all be retainable on a sieve of 32 S.W.G., having 900 meshes to the square inch. Dust and powder, as well as earthy and greasy matter generally, must be rigidly excluded.

Concrete described as $x$ to 1 shall be understood to mean $x$ parts by measure of gross aggregate as detailed above, combined with 1 part of Portland cement.

[Assuming that there is on an average some 35 per cent. of interstitial space[2] in the mixed stone and gravel, and allowing 5 per cent. margin to cover extreme cases, the quantity of sand required will be 40 per cent. The proportion of sand to cement should not exceed 3 to 1. Therefore the minimum amount of cement will be 13 parts in 40.

[1] On account of the liability of a ship's hold to greasiness when used for mixed cargoes, especially in the case of oil in barrels, etc.

[2] Substantiated by experiment.

Totalling the aggregates, we have:—

| | |
|---|---|
| Gravel and stone, | 60 parts |
| Sand, . . | 40 ,, |
| | 100 , |
| Cement, . . | 13 ,, |
| | 113 ,. |

or a limiting ratio for the concrete of $8\frac{3}{4}$ to 1. For general work in bulk,[1] 8 to 1 is a serviceable proportion; for vertical facings, 6 to 1; for quay floors, 4 to 1; for quay steps and landings, 2 to 1.]

*Displacers* or *plums*—large stones and boulders of quality at least equal to that specified for the aggregate—may be inserted in the body of the concrete-forming mass or bulk work, provided that no two stones come within 6 inches of each other and that no part of any stone come within 6 inches of a moulded face. The rock or stone used for the purpose must either be brought fresh from the quarry, or, if old material from paving or building works, it must be thoroughly cleaned by picking and washing so as to free from all mortar, earth, and other accretions. The plums must be sound, hard, compact, and shapely, with no excessive elongation or attenuation and no cracks or flaws, and they must possess rough, preferably rugged, surfaces.[2]

The *matrix* shall consist of Portland cement manufactured by a firm of good standing, and conforming in all respects to the tests and conditions stated below.

"The *cement* shall be prepared by intimately mixing together calcareous and argillaceous materials, burning them at a clinkering temperature, and grinding the resultant clinker. No addition of any material is to be made after burning." [3]

(The standard specification permits the addition of water or calcium sulphate, in neither case exceeding 2 per cent., but only by the express permission of the purchaser.)

As soon as possible after the delivery of the whole of any consignment on the works, *samples for testing* will be taken from ten separate bags or parcels, in different positions. Equal portions of the several samples will be mixed together, and the cement so obtained will be considered as representative of the whole consignment and tested accordingly.

"Before gauging the tests the resultant sample shall be spread out for a depth of 3 inches for twenty-four hours, in a temperature of 58 to 64 degrees Fahrenheit." [4]

*Grinding.*—"The cement shall be ground to comply with the following

[1] 10 to 1 concrete is sometimes used for hearting purposes, but the proportion is somewhat extreme.

[2] No limits of size need be imposed.

[3] This paragraph not transcribed in full.

[4] The limits of temperature throughout are applicable to the climate of the British Isles.

conditions of fineness. One hundred grammes (4 ozs. approx.) shall be continuously sifted for a period of fifteen minutes with the following results:—

"The residue on a sieve 76 by 76 = 5776 meshes per square inch is not to exceed 3 per cent.

"The residue on a sieve 180 by 180 = 32,400 meshes per square inch is not to exceed 18 per cent.

"The sieves are to be prepared from standard wire, the size of the wire for the 5776 mesh being ·0044 inch, and for the 32,400 mesh, ·002 inch. The wire cloth shall be woven (not twilled), the cloth being carefully mounted on the frames without distortion.

"The specific gravity of the cement, when fresh burnt and ground, shall not be less than 3·15 or 3·10 when it can be proved to the satisfaction of the engineer (or of the purchaser) that the cement has been ground for four weeks."

The cement shall be delivered in packages marked with the manufacturer's name.[1]

*Chemical Composition.*—"The cement shall comply with the following conditions as to its chemical composition. There shall be no excess of lime; that is to say, the proportion of lime shall not be greater than is necessary to saturate the silica and alumina present.[2] The percentage of insoluble residue shall not exceed 1·5 per cent.; that of magnesia shall not exceed 3 per cent.; and that of sulphuric anhydride shall not exceed 2·75 per cent.

"The *quantity of water* used in gauging shall be appropriate to the quality of the cement, and shall be so proportioned that when the cement is gauged it shall form a smooth, easily worked paste, that will leave the trowel cleanly in a compact mass. Fresh water shall be used for gauging, and the temperature thereof and that of the test-room, at the time the said operations are performed, shall be from 58 to 64 degrees Fahrenheit. The cement gauged as above shall be filled, without mechanical ramming, into moulds of the form shown in fig. 88, each mould resting upon an iron plate until the cement has set. When the cement has set sufficiently to enable the mould to be removed without injury to the briquette, such removal is to be effected. The briquette shall be kept in a damp atmosphere for twenty-four hours after gauging, when it shall be placed in fresh water and allowed to remain there until required for breaking, the water in

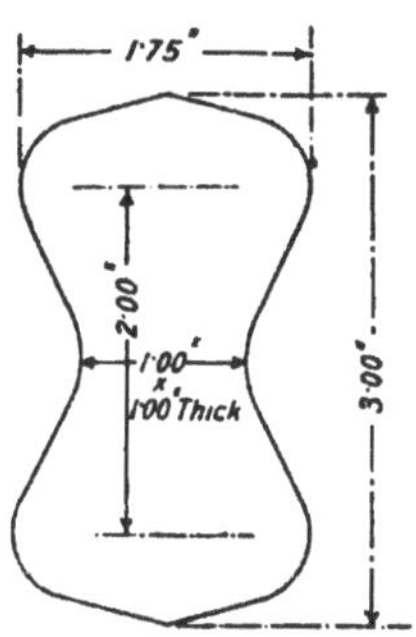

FIG. 88.—Standard Briquette.

[1] Any purchaser wishing to have the cement delivered in sealed bags, or in bags of any certain size, should so specify at the time of ordering.

[2] The proportion of lime to silica and alumina shall not be greater than the ratio (calculated in chemical equivalents) represented by $\frac{CaO}{SiO_2 + Al_2O_3} = 2{\cdot}85$. The molecular weight of lime = 56; silica = 60; alumina = 102.

which the test briquettes are submerged being renewed every seven days and the temperature thereof maintained between 58 and 64 degrees Fahrenheit.

• "*Briquettes* of neat cement of the shape and dimensions shown in fig. 88, having a minimum section of 1 inch square, shall be gauged for breaking at 7 and 28 days respectively, six briquettes for each period. The average tensile strength of the six briquettes shall be taken as the accepted tensile strength for each period. For breaking, the briquettes shall be held in strong, metal jaws of the shape shown in fig. 89, the briquettes being slightly greased where gripped by the jaws. The load must then be steadily

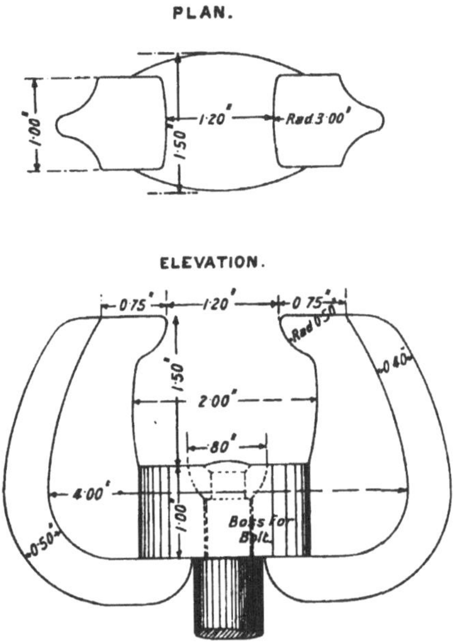


FIG. 89.—Standard Jaws for Briquette.

and uniformly applied, starting from zero and increasing at the rate of 100 lbs. in twelve seconds. The briquettes shall bear, on the average, not less than the following tensile stresses before breaking:—

7 days from gauging, 400 lbs.
28 ,, ,, 500 ,,

The increase from 7 to 28 days shall not be less than:—

25 per cent. when the 7 days' test falls between 400 to 450 lbs.
20 ,, ,, ,, ,, 450 to 500 ,,
15 ,, 500 to 550 ,,
10 ,, ,, 550 to 600 ,,
5 ,, is 600 lbs. or upwards.

"The cement shall also be tested by means of briquettes prepared from one part of cement to three parts by weight of dry standard sand, the briquettes being of the shape described for the neat cement tests; the mode of gauging, the filling of the moulds, and the breaking of the briquettes shall also be similar. The proportion of water used shall be such that the mixture is thoroughly wetted, and there shall be no superfluous water when the briquettes are formed. The cement and sand briquettes shall bear the following tensile stresses:—

7 days from gauging, 150 lbs.
28 ,, ,, 250 ,,

The increase from 7 to 28 days shall not be less than 20 per cent.

"The *standard sand* referred to is to be obtained from Leighton Buzzard. It must be thoroughly washed, dried, and passed through a sieve of 20 by 20 meshes per square inch, and must be retained on a sieve of 30 by 30 meshes per square inch, the wires of the sieves being ·0164 inch and ·0108 inch in diameter respectively.

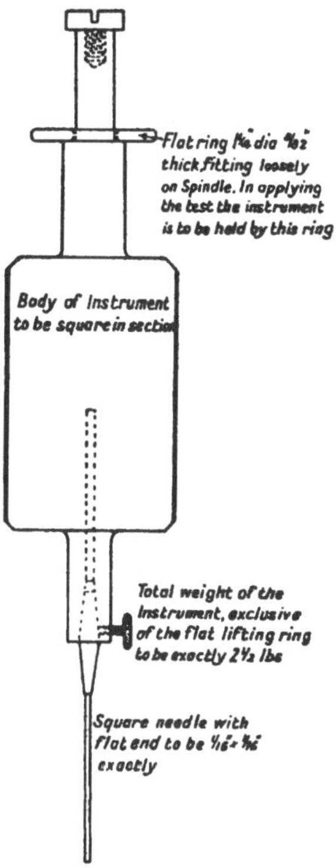


FIG. 90.—Needle for Cement Testing.

"There shall be three distinct gradations of *setting time*, which shall be designated as Quick, Medium, and Slow.[1]

"Quick: The final setting time shall be not less than ten minutes nor more than thirty minutes.

"Medium: The final setting time shall be not less than half an hour nor more than two hours.

"Slow: The final setting time shall be not less than two hours nor more than seven hours.

"The temperature of the air in the test-room at the time of gauging, and of the water used, shall be between 58 and 64 degrees Fahrenheit.[2]

"The cement shall be considered as finally set when a needle of the form shown in fig. 90, having a flat end $\frac{1}{16}$ inch square, weighing in all $2\frac{1}{2}$ lbs., fails to make an impression when its point is applied gently to the surface.

"The pats shall be mixed as previously described.

"The cement shall be tested by the *Le Chatelier method*, and shall in

[1] When a specially slow setting cement is required the minimum time of final setting shall be specified.

[2] The limits of temperature are applicable to the British Isles.

no case show a greater expansion than 10 millimetres after twenty-four hours' aeration and 5 millimetres after seven days' aeration.

"The apparatus for conducting the Le Chatelier test (fig. 91) consists of a small split cylinder of spring brass or other suitable metal of 0·5 millimetre (·0197 inch) in thickness, forming a mould 30 millimetres ($1\frac{3}{16}$ inches) internal diameter and 30 millimetres high. On either side of the split are attached two indicators with pointed ends AA, the distance from these ends to the centre of the cylinder being 165 millimetres ($6\frac{1}{2}$ inches).

"In conducting the test, the mould is to be placed upon a small piece of glass and filled with cement gauged in the usual way, care being taken to

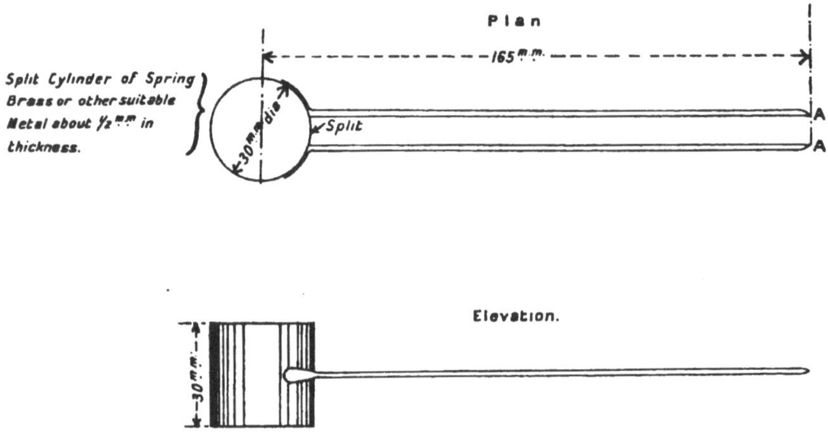


FIG. 91.—Apparatus for Le Chatelier Test.

keep the edges of the mould gently together while this operation is being performed. The mould is then covered with another glass plate; a small weight is to be placed on this, and the mould is then to be immediately placed in water at a temperature of 58 to 64 degrees Fahrenheit and left there for twenty-four hours.

"The distance separating the indicator points is then to be measured, and the mould placed in cold water, which is to be brought to boiling point in fifteen to thirty minutes and kept boiling for six hours. After cooling, the distance separating the points is again to be measured. The difference between the two measurements represents the expansion of the cement, which must not exceed the limits laid down in this specification."

The foregoing tests shall, as far as possible, be made within fourteen days from full delivery of each consignment of cement, and any consignment, the samples of which do not prove satisfactory in testing, shall be rejected.

Some of the test briquettes made as described above will be kept intact for a period of six weeks, and these will be examined from time to time. Should any of them show signs of cracking or disintegration within six weeks from date of gauging, the whole of the cement represented by the defective briquettes will be condemned.

The whole of any individual consignment of cement must be delivered on the site of the works sufficiently in advance of its intended use to allow of the foregoing tests being completed and adjudicated upon.

Contractors, or others using the cement, must provide on the site a suitable, water-tight, dry, wooden-floored store, capable of accommodating one-hundred ton lots so as to be kept distinct from each other. The cement is to be spread out over the floor to a depth not exceeding 3 feet, for a period of three weeks, immediately prior to using, and is to be turned over twice a week.[1] In humid weather the cement is to be protected from an excessive absorption of moisture. The stock must never be allowed to fall below six weeks' requirements.

Cement which has failed to come up to the requirements of the specification must be removed forthwith, and no cement must be used until a certificate of its efficiency has been issued.

For the purpose of *gauging* the ingredients for concrete, strongly constructed measuring boxes are to be provided.

The mixing of the concrete may be done either by hand or by a machine of approved make.

For hand-mixing the following procedure is to be adopted. A wooden platform must first be laid down, and upon this a cubic yard of the aggregate will be deposited and spread out into a uniform layer 12 inches in depth. This layer is to be evenly covered by the proper proportion of cement, and the whole turned over three times dry. Then water is to be added, applied through a rose-ended sprinkler, while the concrete is again turned over three times wet. On no account is the concrete to be deposited until perfect incorporation has been effected.

The quantity of water used must be adequate to bring the concrete to a viscous condition, of the consistency of slime. There must not be any excess, however, such as would wash the cement from the aggregate.[2]

In windy weather suitable screens are to be provided to prevent loss of cement.

In frosty weather no concrete is to be prepared without definite sanction and under appropriate conditions.[3]

No concrete is to be allowed to stand after being mixed, but must be used forthwith. It may not be thrown into foundations from a greater height than 6 feet, and it must be deposited in such a manner as to secure homogeneity and compactness.

Concrete work shall, as far as practicable, be carried on continuously in a series of layers not exceeding 3 feet in thickness, extending over the whole of the

[1] In view of modern methods of treating the cement prior to delivery, this clause is not now generally necessary.

[2] This applies to concrete used in situations free from water. For deposition under water, special measures are required which cannot be covered by a general specification.

[3] Such as the use of very salt water and, possibly, of sugar. The work will also require covering.

site. Where this is impossible, the higher part of it must be racked or stepped downwards to meet the lower, in steps about 5 feet high by 2 feet broad, and vertical joints must be avoided unless necessitated by particular requirements.

The scum arising from the concrete is to be allowed to drain away, and any that settles on the surface of a layer is to be carefully removed. After a layer has set, for which purpose two days must be allowed, an additional layer may be deposited, but not before the surface of the former layer has been well picked, washed with clean water, and brushed.

The surfaces of all brickwork, masonry, or concrete, on or against which concrete is to be laid, must be thoroughly cleaned and wetted immediately before the concrete is applied.

All wooden moulding boards are to have their surfaces paid over with oil, or a suitable composition, to prevent the concrete from adhering to them.

The exposed surfaces of all concrete work must present a fair and smooth appearance where such is desired, and any superficial irregularities must be made good with mortar composed of 1 part Portland cement to 2 parts of clean, sharp sand.

Where a facing of higher quality concrete is to be worked on to a lower quality backing, the division between the two portions is to be formed by a movable hand shutter. The two qualities (as 8 to 1 and 6 to 1) are to be deposited simultaneously and the shutter gradually raised, so that there may be thorough incorporation and the absence of any break or joint.

*Inspection.*—Finally, it may be observed that owing to the dependence of sound concrete upon perfect manipulation, both in mixing and in depositing, too much stress cannot be laid upon the desirability of appointing a trustworthy and competent man to personally supervise all concreting operations. In the case of work done by contract, it is a most essential step; in this way alone can the character of the workmanship be guaranteed, and without that, the best materials may prove practically worthless.

As a matter of interesting comparison, the conditions laid down in a modern Japanese specification are appended.

### Concrete in Blocks at Osaka Harbour Works, Japan.[1]

The proportions of the concrete were as follows:—

| | |
|---|---|
| Portland cement, . | 25 lbs. to 1 cubic foot of sand. |
| Sand, . . . | 2 } by volume. |
| Gravel, . . . | 3 } by volume. |

Since each block contained 120 cubic feet, the corresponding ingredients were:—

| | |
|---|---|
| Cement, . . . . . | 1500 lbs. |
| Sand, . . . . . | 60 cubic feet. |
| Gravel, . . . . . | 90 cubic feet. |

[1] Shima on Osaka Harbour Works, *Trans. Am. Soc. C.E.*, vol. liv.; Int. Eng. Conf., 1904.

Samples of the cement were taken from 2 per cent. to 5 per cent. of the barrels of every cargo. Extracts from the principal clauses of the specification are as follows:—

*Chemical Analysis.*—"If a sample of cement shows by chemical analysis that it contains either more than 1 per cent. of anhydrous sulphuric acid, or a trace of calcium sulphide, or more than 3 per cent. of magnesia, or more than 4 per cent. of ferric oxide, or that the hydraulic index is less than 42, the cement shall be rejected."

*Setting.*—"Cements which begin to set in less than one hour or finish setting in less than three hours or later than twelve hours, shall be rejected."

*Tensile Strength.*—"The tensile strength of the neat cement briquettes after seven days shall be not less than 285 lbs. per square inch, and after twenty-eight days not less than 500 lbs. per square inch; that of the standard sand mortar briquettes after seven days shall be not less than 110 lbs. per square inch, and after twenty-eight days not less than 215 lbs. per square inch. (The standard sand mortar consists of 1 part of cement to 3 parts of standard sand.)"

The sand was obtained at the mouth of the River Yamato; its grains were clean, sharp, and angular. The sand was screened before being used, on a sieve of $\frac{3}{16}$-inch square meshes.

No special variety of gravel was specified, but it was limited to that from sea beaches. It was obtained mostly from the north-western coast of Osaka Bay. The particles were hard and clean, but not very sharp. They were screened between 2-inch and $\frac{3}{8}$-inch sieves.

# CHAPTER VI.

## BREAKWATER DESIGN.

Importance of Breakwaters—Régime—The Sea Wave—Form, Height, and Length—Breaking Waves—Dynamical Value—Measurement of Wave Stroke—Dynamometers—Recorded Pressures—Instances of Wave Action—Classification of Breakwaters—Comparison in Cost of Construction and Maintenance and in Efficiency—Conditions of Stability—Stresses in Wall Breakwaters—Summation of Type Characteristics—Examples of Breakwater Design at Genoa, Marseilles, Algiers, Sandy Bay, and Tynemouth.

THE most important work, as also the most prominent and fundamental feature, in connection with artificially sheltered harbours and roadsteads, is the **Breakwater**. As the name implies, its function is to break up and disperse heavy seas, preventing them from exerting their destructive influence upon the area inclosed for the reception of shipping. Manifestly, then, a breakwater must be characterised by great strength and stability. The safety of helpless vessels and the efficiency of the harbour as a place of refuge are bound up in the essential permanence and immobility of the breakwater.

Before proceeding to an investigation of the principles which underlie the design of breakwaters and by which these objects may be attained, we have to pass in review the conditions and environment to which such structures must conform and the general circumstances attending their construction and maintenance.

**Régime of Breakwaters.**—Structures erected within the domain of the sea and submerged for the greater part of their bulk, if not altogether, are subjected to physical experiences of a nature very different from those which are characteristic of structures on land. The fact of immersion materially modifies the effect of gravity upon a body, reducing its apparent weight to a very considerable extent. That this condition must be applicable to maritime structures is obvious, unless, indeed, the foundation be absolutely impervious and there be an entire absence of ducts for the penetration of water—conditions which, in many cases, are quite unrealisable, and in most are so imperfectly guaranteed as to render them inacceptable as working hypotheses. The solvent properties of water combined with the extreme mobility of its particles, cause it to act in a most prejudicial and injurious manner upon much of the material used in breakwaters, as well as upon the foundation itself, and these merely mechanical effects are supplemented and aggravated by physical and molecular changes, resulting in deterioration in strength and durability. The intensity of the external forces which make for disruption is enormous, exceeding beyond all comparison the power of the wind on land structures. Wave agency is a thousandfold more potent than the most intense atmospheric movement. There are, moreover, insidious denizens of the sea, infesting it by millions, which, by their concerted action,

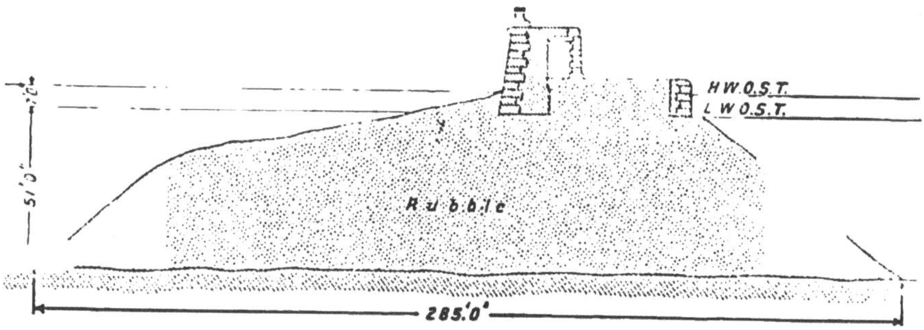


FIG. 92.—Section of Portland Breakwater.

are capable of undermining the hardest and soundest building materials, and that in the most secret and surreptitious manner, the damage being as unsuspected as it is irremediable.

Such inimical natural phenomena constitute the normal and characteristic environment of all maritime structures. They are bonded together, as it were, in an offensive alliance to urge incessant and unrelenting war upon man's handiwork—sapping, wearing, battering, making subtle inroads and open breaches, working now by patient effort, long sustained, and now by sudden, prodigious feats, month after month, year in and year out, knowing neither truce nor armistice.

**The Sea Wave.**—But by far the mightiest of the forces arrayed against the harbour barrier is the sea wave. This mysterious product of wind and water is endowed with tremendous disruptive power. It acts with all the magnipotent impulse of a huge battering ram, while, at the same time, it is equipped with the point of the pick and the edge of the wedge. It is, in fact, one of the most complex, the most volatile, the most pertinacious, and the most incomprehensible of natural forces.

From an engineering point of view, we have little to do with abstract theories of wave formation. Mathematically, the subject is too abstruse for any but very accomplished and capable mathematicians, and the intricacies of calculation are interesting only as academical exercises. Many of the theories advanced are merely tentative and lack substantial corroboration; others, while generally accepted, are still the subjects of speculation and inquiry. Thus, no useful purpose would be served by pursuing an investigation into the laws and phenomena of water undulation. Students who wish to do so, however, may consult the articles on Wave and Tide in the

*Encyclopedia Britannica* or the *Encyclopedia Metropolitana*. The late Sir George Airy, the distinguished Astronomer-Royal, also wrote a treatise on *Tides and Waves*, and this, with the works of Scott Russell and Weber, and later, of Wheeler, afford sufficient scope for reference.

Yet, while disclaiming any intention of probing into the depths of abstruse speculation, we cannot abstain from alluding in general terms to those principles of wave action which have reference to their physical effects upon engineering structures. Such information is essential to an appreciation of the problems of breakwater design.

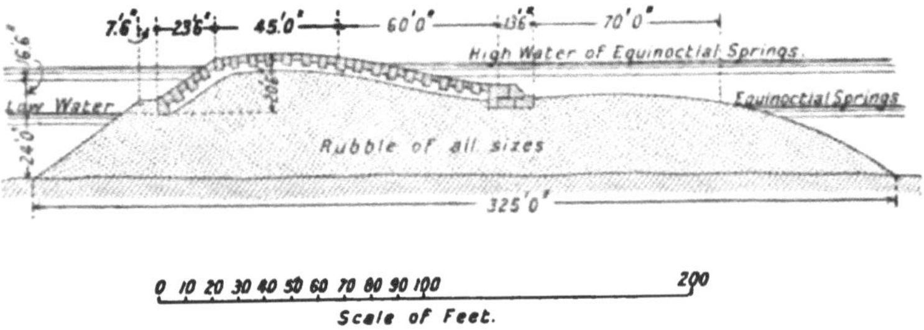


FIG. 93.—Section of Plymouth Breakwater.

For the present purpose, it suffices to state that water waves have conveniently been divided into two classes, viz., waves of oscillation, without forward motion, and waves of translation, possessing it. Yet, in spite of this distinction—a purely artificial one,—it seems probable that all waves are more or less waves of translation, causing the particles of which they are composed to advance permanently to some slight extent, at least. So far as sea waves are concerned, those which possess the power of exerting any appreciable effect on the stability of maritime works are undoubtedly waves of the second division.

**Form of Waves.**—The formation of storm waves takes place in the open sea, and their inception is, of course, due to the wind. The outline assumed is extremely variable, depending both upon the length of the undulation and its period, the lastnamed being the interval of time in which the wave traverses a distance equal to its length. The crest, or summit, of a wave is sometimes rounded, sometimes acute, and, in either case, it attains a height above mean sea level greater than the depth of the trough below it. In a swell in the open sea, the profile of a wave perhaps most nearly resembles a sinoidal curve, the slope directly exposed to wind action being more gradual and less steep than the leeward slope.

Under the conditions of modern investigation, however, as exemplified in the researches of Weber, Scott Russell, Enry, and Aimé, the hypothesis has been advanced that there is an orbital movement in waves, each particle of which they are composed pursuing a regular geometrical path. The precise nature of the path depends upon local conditions. Where the depth of water is sufficiently great, that is, where the depth is at least equal to the length of the wave from crest to crest, the motion of the particles of water is rotary along the circumference of a circle, as shown in fig. 94. The wave is one of

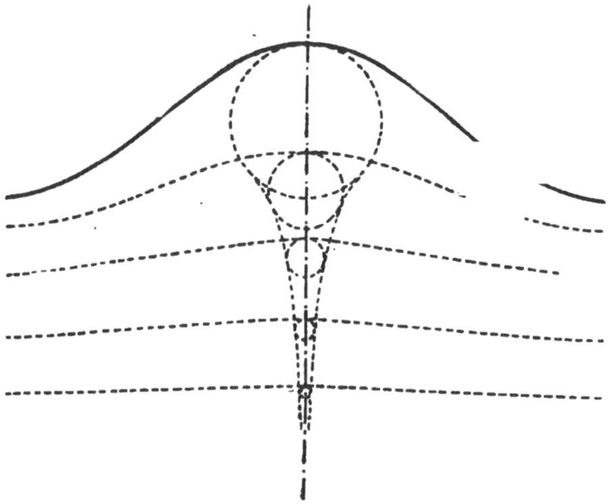

FIG. 94.—Wave in Deep Water.

oscillation, and each particle completes a revolution, returning approximately to its initial position. The profile of the wave, then, is a cycloidal curve traced out by a generating circle, which constitutes the orbit of the surface particles.

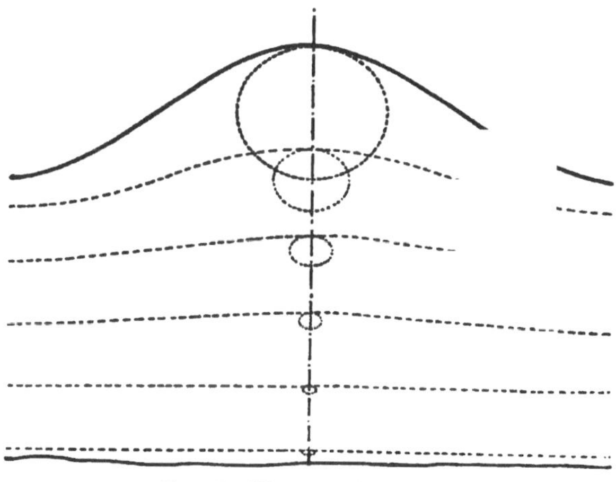

FIG. 95.—Wave in Shallow Water.

Accordingly, the actual momentary direction of motion of each of the particles is independent and variable. Thus, at the crest, the motion is horizontally forward; in the trough, it is horizontally backward; whilst at the midpoint between these extremes, it is purely vertical. Below the surface level the circular paths diminish rapidly to an insensible minuteness. At a depth equal to the length of the wave, the displacement of the water particles is $\frac{1}{535}$ of that of the surface particles, and at double the depth the ratio is reduced to $\frac{1}{286,500}$.

In shallow water of uniform depth, that is, in water the depth of which is less than the length of the wave, the orbit of the water particles is approximately elliptical with the major axis horizontal, as shown in fig. 95. The centre of the orbit lies slightly above the position of rest. With this exception, the same dispositions hold good as in the previous case as regards the movement of the particles. The ellipses of movement become flatter as the distance below the surface increases, until finally at the bottom there is horizontal motion only. In water which has a depth of only one-tenth of the length of the wave, the ratio of the elliptical axes at the surface is about $\frac{7}{12}$, and at nine-tenths of the depth it is $\frac{1}{16}$.

When, instead of remaining uniform in depth, the water in which a wave is travelling becomes increasingly shallow (fig. 96), the orbits of the particles

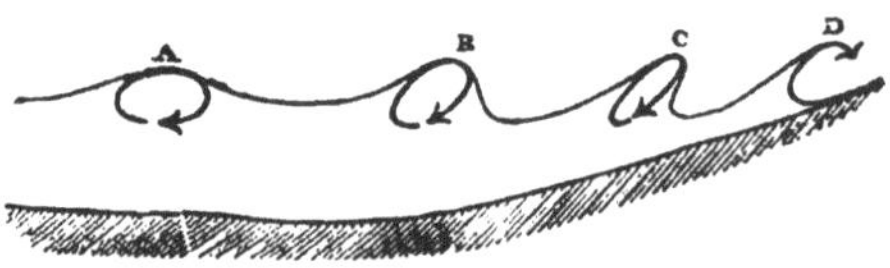


FIG. 96.

of water become correspondingly distorted. Owing to the friction exerted along the bottom, the major axis of revolution acquires an inclination to the horizontal, which is continually augmented. The wave ceases to be purely oscillatory: it undergoes a gradual transformation. The front of it becomes steeper than the back, the crest gaining more and more upon the trough until it actually overhangs. Then it falls forward and breaks into surf. At this point the wave is altogether a wave of translation, and the forward motion of the particles is exactly equal to the velocity of the wave. It is in this phase that waves possess their most formidable potency.

Any sudden change in the level of the ground over which a wave is travelling is capable of producing the disruption of the wave. This effect is not confined to shallow reaches, but extends to depths as great as 16 to 20 fathoms, or even more, in the open sea. Thus, on the Herreca reef, seven miles from land, breakers are apparent in tempestuous weather in a depth of 90 feet of water.

**Height of Waves.**—The inception of waves being due to the wind, their development manifestly depends upon the extent of surface acted upon. Waves generated without restriction are capable, under propitious circumstances, of attaining a very high degree of development, both as regards height and length. On the Lake of Geneva, for instance, storm waves are stated to reach a height of 10 feet; in the German Ocean, from 12 to 15 feet; in the

Mediterranean Sea, from 15 to 20 feet; in the Bay of Biscay, from 25 to 30 feet; in the open Atlantic, from 30 to 40 feet; and in the Pacific (off Cape Horn and the Cape of Good Hope) from 50 to 60 feet. Other estimates of a much higher nature have been made, but it is open to question whether they have not been influenced by an unconscious tendency to exaggeration on the part of the observer, due to the inspiring nature of the spectacle, or been founded upon mistaken and erroneous data. There is, in such cases, a strong and an acknowledged inducement to use picturesque language and to speak of waves as "mountains high," to which, of course, numerical values must as far as possible correspond. Indeed, viewed at close quarters, a formidable wall of water towering suddenly above the spectator's line of sight, even when on the upper deck of a vessel, can hardly fail to produce an illusory sense of enormous magnitude and overwhelming menace. So far, however, as unquestionable records go, it may safely be asserted that 50 feet is about the maximum height attainable by unbroken waves, and this view is supported by the opinions and testimony of Sir George Airy, Captain Scoresby, and other observers.

The heights of waves breaking against the cliffs and headlands of a rocky coast do not, of course, come within this category. The summits of columns of water thrown up by the force of impact attain, as might be expected, to much greater altitudes. The effect is particularly noticeable in the case of lighthouses and prominences with vertical or nearly vertical faces. Thus, at The Hague heights of 75 feet, at Bell Rock 100 feet, and at Eddystone 150 feet, have frequently been recorded, and Lord Dunraven has observed heights of 150 feet on the precipitous South West Coast of Ireland.

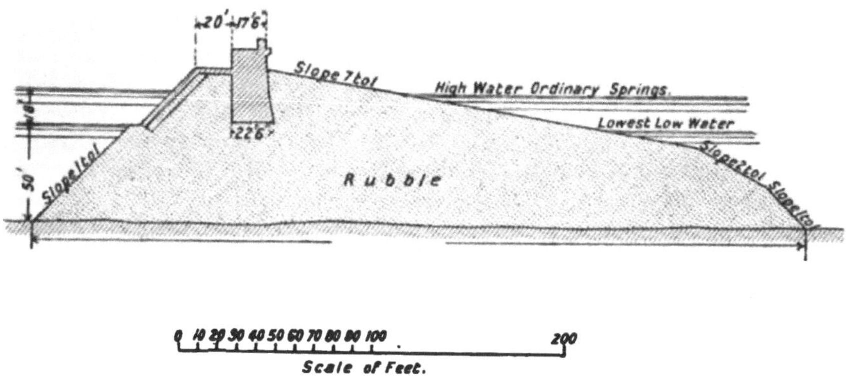


FIG. 97.—Section of Holyhead Breakwater.

On the basis of the evident connection existing between the development of waves and the generating distance, Stevenson, the eminent harbour engineer, devised an empirical formula to determine the height of waves from the *Fetch*, or extent of sea available for the purpose of generation. Taking H as the height, in feet, of the wave, and F as the length, in miles, of the fetch, he found that approximately

$$H = 1{\cdot}5\sqrt{F} \quad . \quad . \quad . \quad . \quad . \quad . \quad (\alpha)$$

or, more closely, for short fetches of less than 30 miles,

$$H = 1{\cdot}5\sqrt{F} + (2{\cdot}5 - \sqrt[4]{F}) \quad . \quad . \quad . \quad . \quad . \quad (\beta)$$

The following instances afford a comparison of the results of calculation based upon these formulæ, with the values obtained from actual observation.

HEIGHT OF WAVES.

| Place of Observation. | Length of Fetch, Nautical Miles. | Observed Height of Waves in Feet. | Calculated Height in Feet. | |
|---|---|---|---|---|
| | | | Formula (α). | Formula (β). |
| Scalpa Flow . . | 1·0 | 4·0 | 1·5 | 3·0 |
| Firth of Forth . . | 1·3 | 1·8 | 1·8 | 3·2 |
| Lough Foyle . . | 7·5 | 4·0 | 4·1 | 4·96 |
| Clyde. . . . | 9·0 | 4·0 | 4·5 | 5·25 |
| Colonsay . . . | 9·0 | 5·0 | 4·5 | 5·25 |
| Lough Foyle . . | 11·0 | 5·0 | 5·0 | 5·7 |
| Anstruther . . | 24·0 | 6·5 | 7·5 | 7·7 |
| Lake of Geneva . . | 30·0 | 8·2 | 8·2 | 8·37 |
| Buckie . . . | 40·0 | 8·0 | 9·5 | ... |
| Douglas, I.O.M. . . | 65·0 | 10·1 | 12·0 | ... |
| Kingstown . . . | 114·0 | 15·0 | 16·0 | ... |
| Sunderland . . | 165·0 | 15·0 | 19·3 | ... |
| Peterhead . . . | 400·0 | 22·6 | 30·0 | ... |

It is manifest, however, that waves cannot attain their full development where there is inadequate depth. No wave can have a height greater than the depth of water through which it passes. Consequently, the intervention of shoals in the path of a wave serves to limit its size. Reefs and sandbanks, even though entirely submerged, materially reduce the range of undulation, and the length of fetch must be gauged accordingly. Thus the effective length of open sea may be much less than the apparent length.

But, even with this restriction, the fetch is far from being an exact or reliable indication of wave height in every locality. It is true that the maximum height attainable can be calculated therefrom with some approach to accuracy; but the fact must not be overlooked that winds may not always, or often, or indeed ever, blow along the line of maximum fetch. There is a stretch of 500 miles of open sea leading to the harbour of Kurrachee with unrestricted depth, yet the highest waves—those from the south west—are said not to exceed 15 feet. Other instances might be adduced—such as Peterhead in the preceding table—to show that the maximum fetch alone is by no means an infallible criterion of wave height. There is also another point of no slight importance. Not only is it possible for the severest gales to blow from some other quarter of the compass than that lying in the direction of the greatest fetch, but it is also a matter of experience that heavy rollers are frequently deflected so as to reach a point on the coast which does not lie upon their direct path. This is exemplified in the case of a headland and bay in fig. 98, and the same effect is noticeable at the pierheads of artificially sheltered harbours. Islands also act as pivots in many cases, causing the waves to wheel round and break upon their leeward shores. The result is due, no doubt, to the retardation produced by shallowing ground upon the nearer portion of waves approaching a coast obliquely, or running parallel thereto.

Furthermore, the convergency due to narrowing inlets tends to accentuate, to a greater degree even, the eccentricities of wave development. Pent

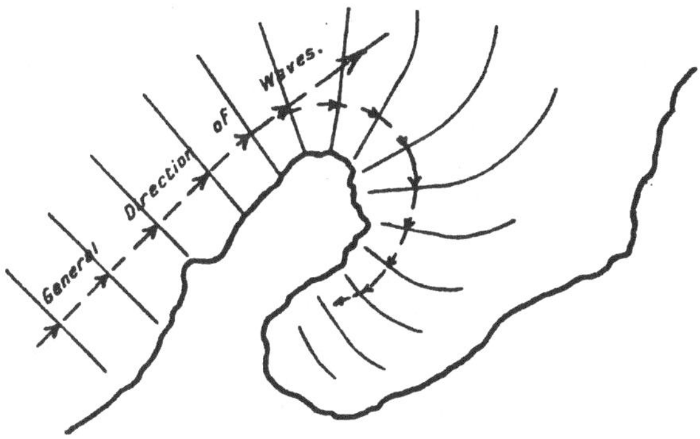


FIG. 98.—Deflection of Waves by Headland.

between lateral arms drawing gradually closer together, the volume of water is raised above the level it would normally assume, and so gives rise to breakers of a character equally marked.

As a corollary to what has been said, it is evident that waves of great height cannot reach any coast-line, and, for that matter, any artificial barrier, unless there be an unbroken extent of deep water penetrating close into them.

**Length of Waves.**—The length of waves is a feature which seems to be independent of the height, though it is connected in some way with the amount of exposure to wind action, and it influences the force of the wave. In the Atlantic Ocean waves of from 500 to 600 feet between crests have been observed, while in the Pacific they are stated to reach anything from 600 to 1000 feet.

The length of waves, however, in the open sea is a difficult matter to determine satisfactorily, owing to the absence of any reliable linear standard. Alongside jetties and piers, the obstacles in the way of exact measurement are not so great, and serviceable computations may be made with the aid of Bertin's formula. Observing the length of time in seconds which elapses between the passage of the same point by two successive crests—in other words, the period of the wave, calling this period P and the length of the wave in feet L,—we have

$$L = \frac{P^2 g}{2\pi};$$

or, fairly approximately,

$$L = 5\ P^2.$$

The length of the wave in conjunction with the depth of water determines the speed of movement of the wave and, conjointly, the velocity of the particles of which it is composed. The relationship existing between these elements will be discussed a little later.

**Breaking Waves.**—We have now to consider the manner in which a wave acts upon any fixed obstacle in its path, whether it be the beach upon which it is spent or an artificial barrier which causes its abrupt collapse.

Dealing first with the oscillatory wave, and assuming that it reaches a wall or other obstruction having an abrupt, vertical face, we find that it is reflected in the manner indicated by fig. 99. The particles of water in contact with the wall (A) move up and down through a height which is twice the height of the original wave, as also do the particles in the trough (C) half a wave-length distant. At a point (B) midway between the trough and the wall, that is, one quarter of a wave-length from either, the particles move horizontally backwards and forwards, while at intermediate points the path of the particles is inclined at various angles. The whole motion, in fact, is the inverse of that which occurs in the unobstructed wave.

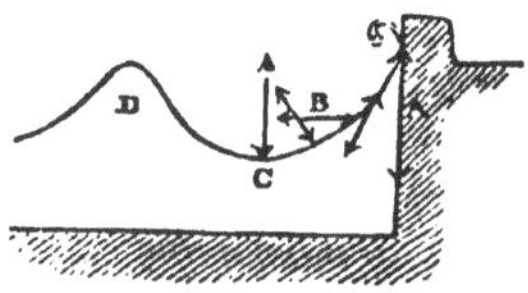

FIG. 99.

When, on the other hand, without meeting with any abrupt obstacle, the wave advances into rapidly shoaling water, its energy is communicated to smaller and successively decreasing masses. Consequently there is a tendency to produce in those masses an agitation of increasing violence. But this effect is generally diminished, and sometimes entirely counteracted, by the loss of energy due to friction along the bottom, and to surging. On the other hand, the influence of concentration arising from funnel-shaped inlets is clearly to intensify the agitation, and the same effect is producible by submerged rocks with deep narrow gorges between, in passing through which the water is heaped up into masses of considerable volume.

When, however, the bottom friction has produced the necessary retardation, the crest of the wave falls forward, as has already been explained, and impact takes place at the precise stage at which the forward motion of the particles has become equal to the velocity of the wave, so that the stroke of the latter is delivered with maximum effect.

Taking all these diverse phenomena into consideration, it is evident that breaking waves result in the generation of four separate and distinct forces acting individually and collectively upon all obstacles and structures in their path.

(1) A direct horizontal force, exerting compression.

(2) A deflected vertical force, acting upwards and tending to shear off any projections beyond the face line of the obstacle, whether cliff or wall.

(3) A vertical downward force due to the collapse of the wave and exercising a particularly disturbing effect on mounds in shallow water and beaches.

(4) The suction due to back-draught or after-tow. This also produces its most noticeable results on foundation beds, whether natural or artificial.

Applying these unmistakable and fundamental phenomena to the question of breakwater design, it will be recognised that the forces to which they give rise are as follows:—

(1) A powerful momentary impact, combined with

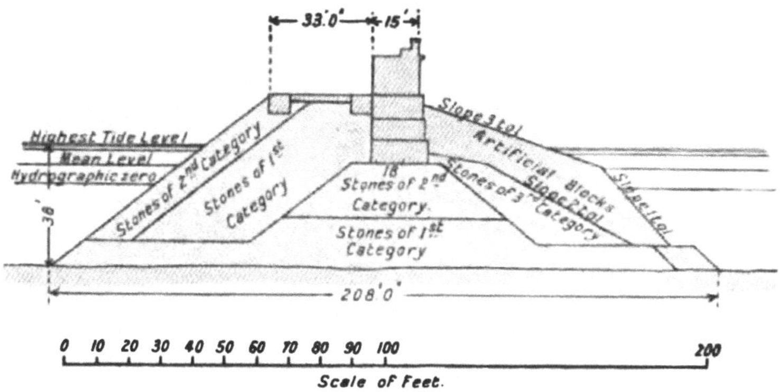


FIG. 100.—Section of Leixoes Breakwater.

(2) Hydrostatic pressure continuous for some short period, however minute, after the first shock.

Attending these principal forces there will be several subsidiary results, such as:—

(1) A vibration of the whole structure, tending to weaken the connection of the various parts.

(2) A series of impulses imparted to the water contained in the pores, joints, and interstices of the structure, producing internal pressures in various directions.

(3) The alternate condensation and expansion of the volumes of air which are confined in cavities and which may be unable to escape freely or not at all.

The exact determination of these stresses is practically impossible. Something, however, may be done towards estimating their scope and extent. How far this lies within the range of definite and effective calculation is our next concern.

**The Dynamical Value of Wave Action.**—The difficulties attending a determination of the precise effort of a wave are due to several causes. In the first place, there is the incompressibility of water combined with the extreme mobility of its particles. Arrested suddenly in the course of motion, it produces all the percussive effects of a solid body in an infinite number of directions. No clearer evidence of this could be produced than the phenomenon known as water-hammer. If the outlet valve of a hydraulic service-main be shut down abruptly, a blow is administered to the pipe which may be, and often is, sufficient to produce rupture, even at a considerable distance from the outlet, unless, as is generally the case, a relief valve is provided to prevent such a disastrous effect.

In the second place, the wave-stroke is both abrupt and continuous. Its first action is a blow, sharp and decisive and of high momentary intensity.

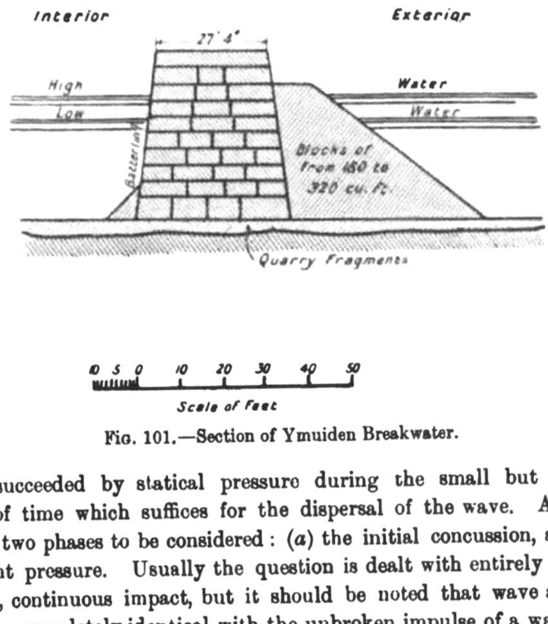


Fig. 101.—Section of Ymuiden Breakwater.

This is succeeded by statical pressure during the small but perceptible interval of time which suffices for the dispersal of the wave. Accordingly, there are two phases to be considered: (*a*) the initial concussion, and (*b*) the subsequent pressure. Usually the question is dealt with entirely as a matter of simple, continuous impact, but it should be noted that wave action is far from being completely identical with the unbroken impulse of a water-jet.

Now, according to the principles of dynamics, the reaction of a surface subjected to continuous impact is measured by the rate at which momentum is destroyed. If, therefore, $w$ be the weight of a unit volume of water, $\frac{wv}{g}$ is the mass which impinges on unit surface in unit time, and $\frac{wv^2}{g}$ is the rate at which momentum is consumed. Hence, if $p$ be the pressure on unit surface, we have

$$p=\frac{wv^2}{g}. \qquad . \quad . \quad . \quad . \quad . \quad . \qquad (a)$$

Accordingly, if it be possible to determine the velocity of impact, the pressure momentarily exerted may be deduced from the above relationship.

Now, the velocity of a wave in shallow water has been found to be nearly the same as that which would be acquired by a heavy body falling freely from rest under the influence of gravitation through a height equal to the semi-depth of the water plus three-fourths of the height of the wave. In symbols,

$$v_1 = \sqrt{2g\left(\frac{d}{2} + \frac{3h}{4}\right)} \quad . \quad . \quad . \quad . \quad . \quad (\beta)$$

When the depth of water exceeds the length of the wave, the speed of the latter is practically independent of the depth, and is almost exactly equal to the velocity acquired by a body falling freely through a height equal to one-half the radius of the circle, the circumference of which constitutes the length of the wave. That is:—

$$v_2 = \sqrt{2g \cdot \frac{l}{4\pi}} \quad . \quad . \quad . \quad . \quad . \quad . \quad (\gamma)$$

Reverting to the first, ($\alpha$), of the foregoing equations, we may recall the fact that if the velocity of a falling body be expressed in terms of the height ($h$), or perhaps, more correctly, the depth below the position of rest, we have

$$v^2 = 2gh\,;$$

and, substituting this value for $v^2$ in ($\alpha$),

$$p = 2wh \quad . \quad . \quad . \quad . \quad . \quad . \quad (\delta)$$

which leads us to the conclusion that the pressure intensity of a water-jet is equal to the weight of a column of water *twice* as high as the amount of free fall required to generate the specified velocity in a heavy particle.

We see, however, that the numerical coefficient in this last expression is not applicable to wave pressures, owing to the modifications in the value of $v$ given above in ($\beta$) and ($\gamma$). As a matter of fact, the value of the coefficient is very variously stated by different authorities, so that all that is justifiable is to write in general terms,

$$p = kwh \quad . \quad . \quad . \quad . \quad . \quad . \quad (\epsilon)$$

where $k$ has a fairly wide range of values.

| | | | |
|---|---|---|---|
| Thus, | Dubuat | gives | $k = 1{\cdot}85$ |
| | Rayleigh | ,, | $k = 1{\cdot}96$ |
| | Mariotte | ,, | $k = 1{\cdot}25$ |
| | Gaillard | ,, | $k = 1{\cdot}31$ |
| | Jœssel | ,, | $k = 1{\cdot}62$ |
| | Thibault | ,, | $k = 1{\cdot}85$, |

the average value being 1·6.

It is to be noted, however, that Gaillard considers the velocity of the crest of the wave, while breaking, to exceed the velocity of the body of the wave by 30 per cent., so that, if we give effect to this modification, his coefficient becomes raised to 2·33. This value, though much higher than any of the values of the other experimentalists quoted above, is not without support from the observations of Bidone, who obtained coefficients ranging from 1·5 to 2·3 for the pressure of water-jets.

The values assigned to $k$ in the foregoing equation are all based on the assumption that the line of action of the wave is perpendicular to the surface on which it impinges. When the line of incidence makes an angle $a$ with the surface, the coefficient undergoes further modification, and, according to Lord Rayleigh, $k$ becomes

$$\frac{2g\pi \sin a}{2 + g\pi \sin a}.$$

When $a = 90°$ it will be noticed that this expression becomes approximately 2, which, to a certain extent, coincides with the value of $k$ given previously.

One point of interest about the fundamental equation $p = kwh$ is that it may be written

$$p = \frac{w}{2g} \times 2kgh\,;$$

and since $w$, the weight of salt water in lbs. per cubic foot, differs imperceptibly from the value of $2g$, the equation becomes practically

$$p = 2kgh\,; \qquad (\zeta)$$

or, giving $k$ its mean value of say 1·6,

$$p = 3{\cdot}2gh \qquad (\eta)$$

which is also transformable into

$$p = 1{\cdot}6v^2 \qquad (\theta)$$

**Measurement of Wave-stroke.**—It is a matter for regret that few or no appliances are available for satisfactorily comparing the results of theoretical calculation with actual pressures. It is true that various kinds of apparatus have been contrived for the express purpose of registering the compressive force of the wave-stroke, but for certain reasons these records cannot be considered an absolutely reliable criterion. The recoil of a spring is far from being a satisfactory method of gauging the colliding force of incompressible bodies. The very elasticity of the spring robs it of one of the most characteristic features of the ideal breakwater, and the retreat of the surface plate before the impulse of the wave is not in accordance with actual conditions. The real intensity of the blow, in fact, lies in the absence of yielding in either body. Theoretically, the effect of such impact is infinite, and in practice it must often far transcend the imperfect records of a none too sensitive spring dynamometer.

Furthermore, the assumption of uniform distribution of pressure involved in such means of measurement is untenable. Wave power is at least as subtle and irregular as wind pressure. Waves strike hardest in isolated places at uncertain intervals. The location of the dynamometer may or may not coincide with these places; in any case, it is a matter of mere hazard and surmise.

Yet, imperfectly as they realise ideal conditions, instruments of this class are as yet the only available means of obtaining practical data in regard to the force of waves. Stevenson's apparatus is perhaps the best known. It is illustrated in fig. 102. Several improvements have been contrived since it was first designed, but in principle it consists of a flat disc, perpendicular to which, and behind it, are arranged four rods passing through a firmly fixed cylinder. The disc is set fronting the sea, and when it is struck by a wave the rods are forced back simultaneously through the cylinder, thereby extending a spring connected with the front of the latter. On each rod is a leathern ring, which, prior to movement, is in contact with the back plate of the cylinder. The passage of the rods through this plate is unrestrained; but the rings cannot pass, and so they are forced along the rods. When the latter resume their original position under the recoil of the spring, the distance travelled by the rings is a measure of the intensity of the blow.

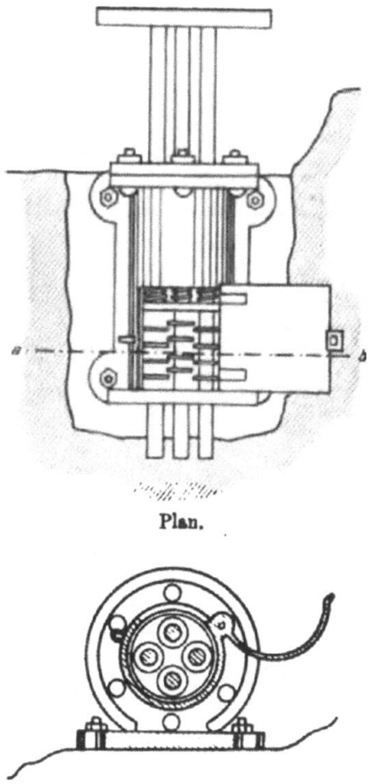

Plan.

Section.

FIG. 102.—Stevenson's Wave-stroke Dynamometer.

An instrument on these lines, but with special features, was constructed a short time back by Messrs W. H. Bailey and Co., Ltd., of Manchester, for use on the coast of Japan. It is illustrated in fig. 103. The principal modification consists in placing the instrument on trunnions with a swivel base-plate, so that it may be adjusted both horizontally and vertically to any desired angle. A pencil attached to the index-rod and a revolving drum, enable the record to be kept graphically over a continuous period.

The calibration of these instruments is effected in the same way as ordinary spring balances; that is, by the imposition of dead loads. This method is open to the objection already stated, that statical pressure is quite a different thing from dynamical force, and a more appropriate system would be to calibrate by means of falling weights in units of kinetic energy. Yet, even then, there would be the difficulty of the conversion of these last into their statical equivalents. No satisfactory solution has yet been put forward.

A dynamometer, lately devised by Major Gaillard of the Corps of Engineers of the U.S. Army, possesses a cylinder fitted with an elastic diaphragm and filled with liquid, which is in communication with a gauge. The mobility of the fluid particles enables the wave-stroke to be administered with less loss of energy than in the case of the solid plate, where the inertia

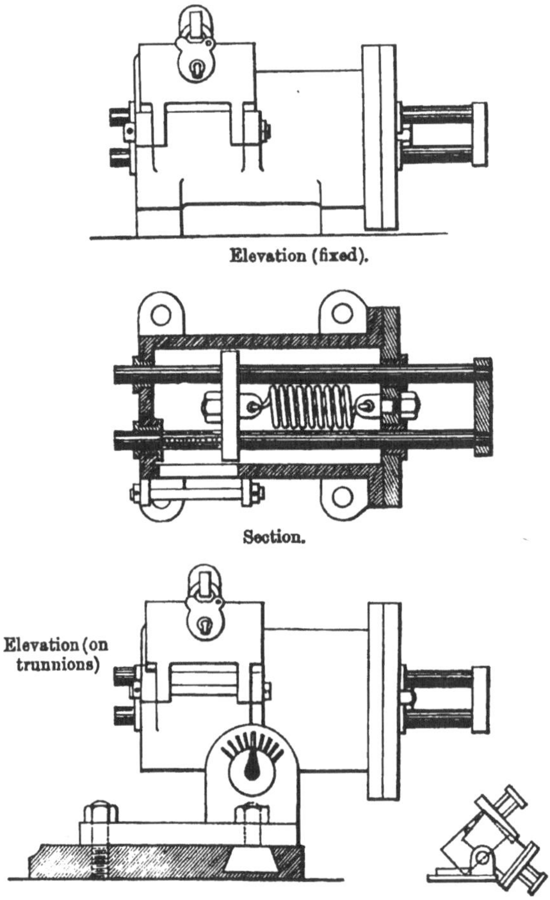


FIG. 103.—Bailey's Wave-stroke Dynamometer.

of the moving parts has to be overcome. But the appliance is characterised by the same absence of conformity with actual conditions, to which attention has already been drawn.

The maximum pressure actually recorded by the marine dynamometer does not appear to have exceeded 3½ tons per square foot. At Skerryvore (in the Atlantic) a pressure of from 2½ to 2¾ tons per square foot has been observed; at Bell Rock (German Ocean) 1½ tons; at Dunbar (East Lothian)

3½ tons; and at Buckie (Banffshire) 3 tons. Stevenson noted that the force of impact on a rising slope is six times as great as the pressure on a steep wall.

Some inferential records may be adduced in confirmation of the above. From experiments made with concrete blocks sliding upon a well-wetted concrete floor, Mr Shield determined a frictional coefficient of ·7· In 1891, a section of the breakwater at Peterhead, weighing 3300 tons in a single mass, was slewed bodily to the extent of 2 inches, without any dislocation of the substructure. The waves, therefore, must have exerted a pressure of

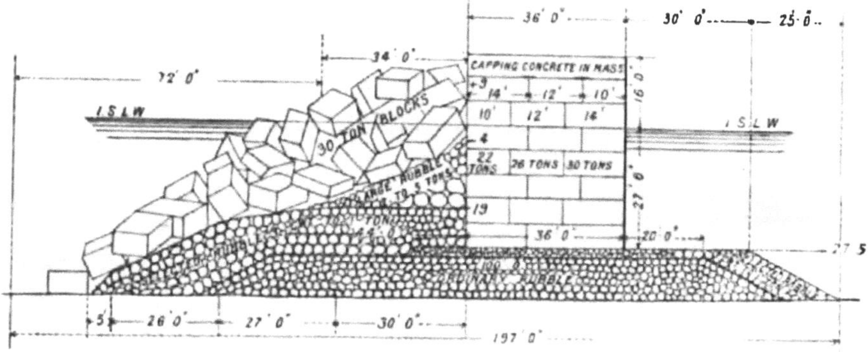


FIG. 104.—Section of Madras Breakwater.

over 2310 tons, which, upon the surface exposed, works out to an average pressure of rather more than 2 tons per square foot.[1]

At Penzance, Mr Frank Latham has noted pressures of from 18 to 20 cwts. per square foot. At Cherbourg, wave pressure is stated to vary from 5 to 7 cwts. per square foot.

**Examples of Wave Action.**—Noteworthy instances of wave action are numerous, and many of them remarkable to a degree verging on the incredible. At Genoa, in 1898, blocks of artificial stone weighing 40 tons each are said to have been driven a distance of over 160 feet.[2] At Wick, in 1872, a huge monolith weighing 1350 tons is recorded as having been removed bodily from its seating, and deposited intact some considerable distance away. Another enormous mass of 2600 tons at the same place was broken into two pieces and similarly displaced.[3] Yet another instance is afforded by the movement of the 3300-ton mass at Peterhead, already alluded to.

*Storm at Genoa.*—It will not be without interest to consider, in some detail, the description of a great storm which damaged the breakwaters at Genoa in 1898. The following account is condensed from the report of M. Bernardini to the International Maritime Congress at Milan in 1905.

The maximum fetch at Genoa is about 600 nautical miles, and the sector of exposure has an angle of 30° open to the south west.

[1] *Min. Proc. Inst. C.E.*, vol. cxxxviii. p. 400.
[2] Bernardini on the Galliera Mole, *Proc. Inst. Nav. Cong. Milan*, 1905.
[3] *Min. Proc. Inst. C.E.* vol. xliii.

On the evening of 26th February 1898, after several days of rainy weather and rough sea, both wind and waves became suddenly intensified. The gale augmented in violence as night advanced, reaching its maximum shortly after midnight. The wave crests, mounting higher and higher, finally leapt over the parapet of the Galliera mole and fell upon the inner quay. The light at the pierhead was visible until 3 A.M., when it suddenly went out. Although, as M. Bernardini admits, it was easy to confuse the spray with the wave itself, and the grandeur of the scene was a temptation

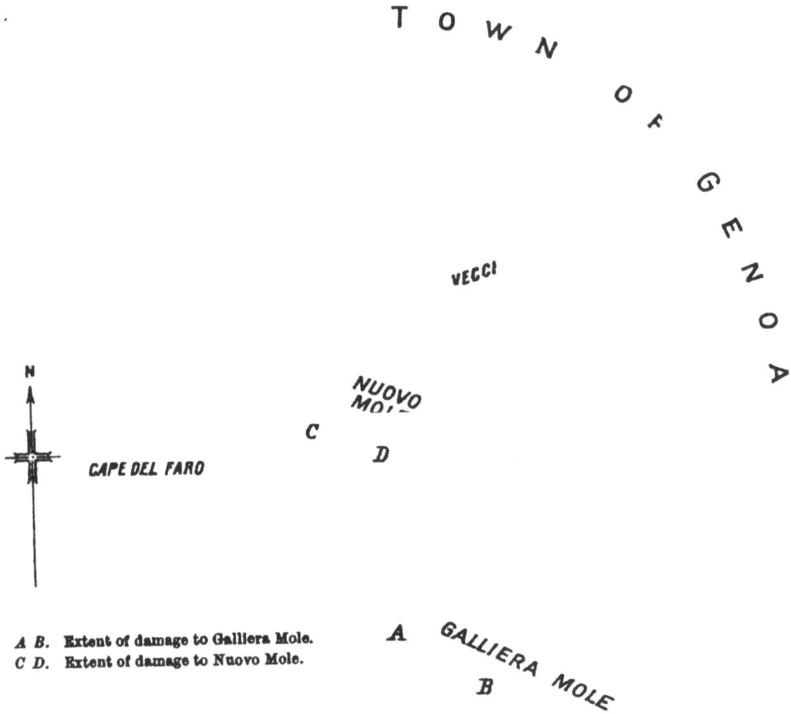


FIG. 105.—Harbour of Genoa.

towards exaggeration, it can certainly be said that the columns of water thrown up by the force of the waves, as they broke against the mole, attained a height of 65 feet during the early hours of the morning of 27th November. This is without taking into account a few small columns here and there, which rose to much greater heights—at least 100 feet. The height of the waves themselves is estimated from careful observation to have been about 25 feet.

It is curious to note that, contrary to what might have been expected, the atmospheric pressure did not fall in proportion to the unusual violence of the gale. In fact, the diagram recorded by the local hydrographic bureau indicates that the minimum pressure was attained on the night preceding this great storm, which exceeded all previous memorable storms in intensity.

The breakwater at Genoa is somewhat of the shape of a slightly distorted Z, and is divided into two sections, known as the Nuovo mole, adjacent to the shore, and the Galliera mole, further out. The Nuovo mole is 2950 feet long,

FIG. 106.

while the Galliera mole comprises two arms 2155 feet and 2765 feet long respectively. In all, the breakwater is a mile and a half long. The damage wrought by the storm was as follows.

Of the Nuovo mole a length of rather over 800 feet flanking the deepest portion of the sea had its foundation laid bare, both the natural and artificial protection blocks being swept away and the front apron demolished, so that the foot of the wall lay exposed to the full effect of the waves. The nature

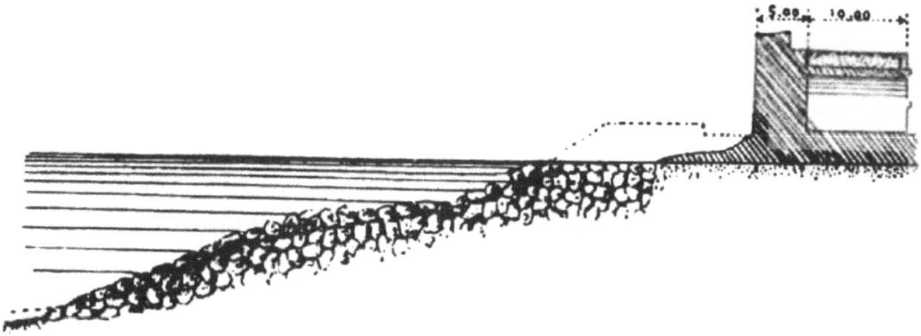

FIG. 107.—Part Section of Nuovo Mole.

of this transformation is indicated in fig. 107, which shows the section of the mole both prior and subsequent to the storm.

The Galliera mole had never suffered in any way from the frequent southwesterly gales to which it had been exposed during the ten years of its existence preceding the storm of 1898. All that had been necessary was to renew from time to time the cement coating which had been applied as a preservative to the artificial blocks lying above water. When, therefore, the storm broke, every part of the mole was in perfect repair.

During the early hours of the morning of 27th November, despite the fact that the outline of the mole was almost completely obscured by foam and spray, the shelter wall of the outer arm was observed to be split into several sections for a distance of 500 feet from its junction with the inner arm. Some of these sections had merely shifted in position, but others had been completely overturned on to the inside quay. As the day advanced the breach was extended, until eventually it was 65 feet wide, the wall continuing to break away and small portions of it to be swept into the harbour. Fig. 109 is a plan of the damaged portion, and figs. 110, 111, and 112 are cross sections at various points.

In the first length, A.B., for a distance of about 230 feet from the commencement of the bend, the pitching or covering of artificial stone blocks was torn off to depths varying from 6 feet 6 inches to nearly 20 feet, the blocks in some cases being deposited along the outer slope, and in other cases projected a long way out on each side. The protecting apron was completely swept away, but the foundation of the mole structure was not damaged beyond a few cracks near the base, which were neither large nor deep.

In the length B.C., the topmost course of artificial blocks of stone was overturned, and the lower courses were dislocated and partially damaged. The shelter wall was broken into five enormous blocks, of which four were shifted parallel to their original position and one (number IV. on plan) was overturned on to the quay. Detailed particulars of the blocks are as follows:—

| No. of Block on plan. | Volume in cub. ft. | Weight in tons. |
|---|---|---|
| I. | 15,510 | 1,012 |
| II. | 6,980 | 455 |
| III. | 4,935 | 322 |
| IV. | 13,395 | 894 |
| V. | 14,030 | 915 |

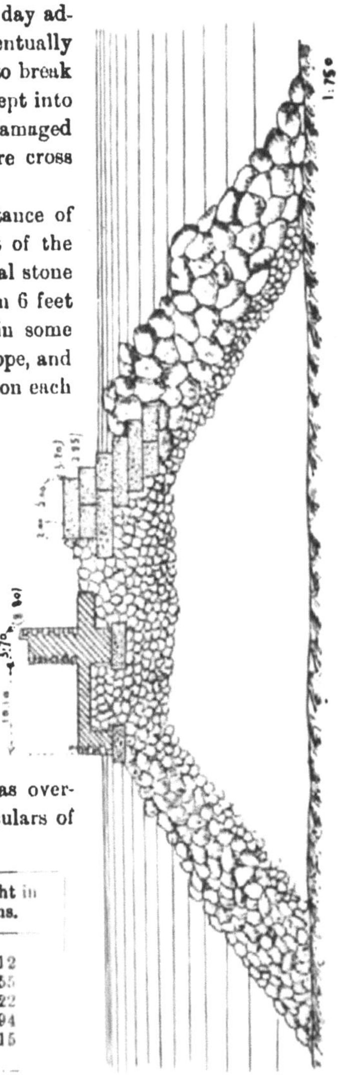

FIG. 108.—Section of Galliera Mole as constructed.

In the length C.D. the shelter wall was entirely demolished and the artificial blocks disturbed, though not to the extent experienced in the

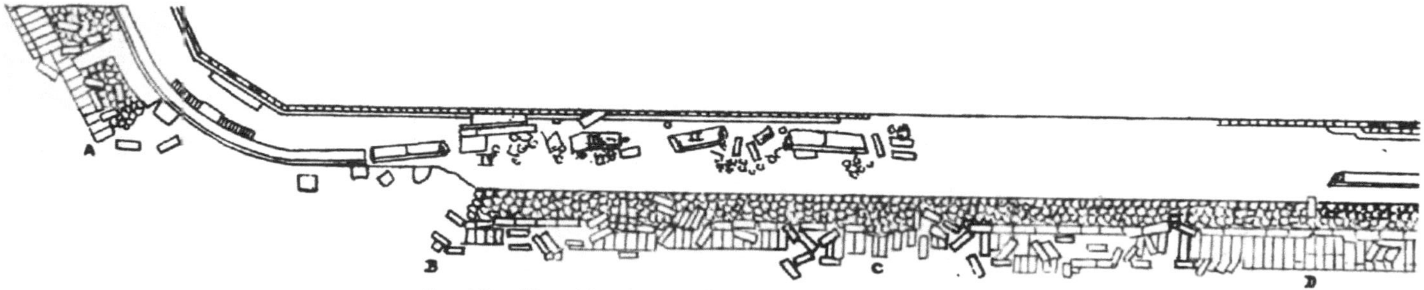


Fig. 109.—Plan of Portion of Galliera Mole showing extent of damage.

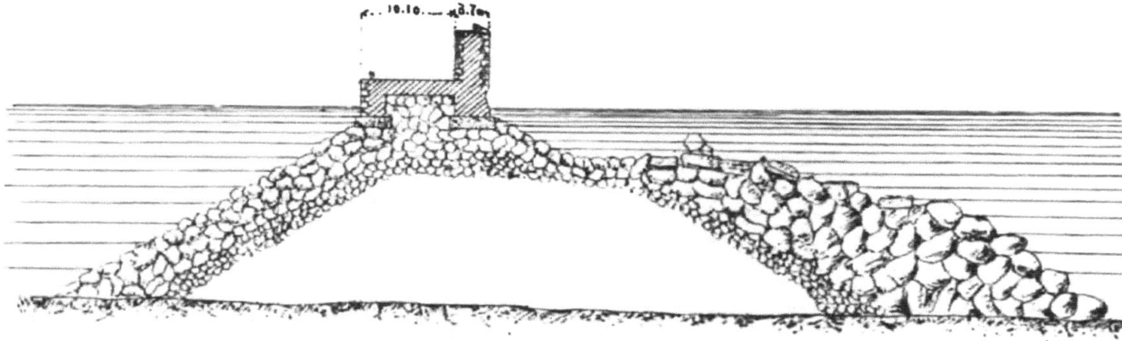

Fig. 110.—Section of Damaged Portion of Galliera Mole from A to B, fig. 109.

adjoining sections. One characteristic feature of the damage was the dis-

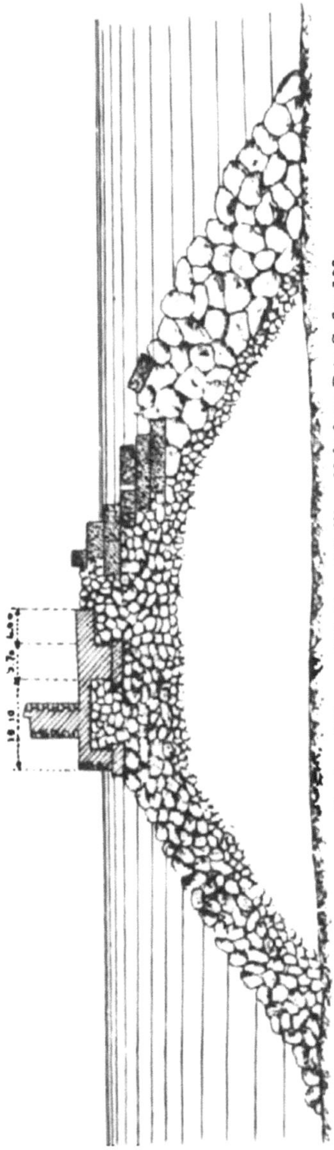

FIG. 111.—Section of Damaged Portion of Galliera Mole from B to C, fig. 109.

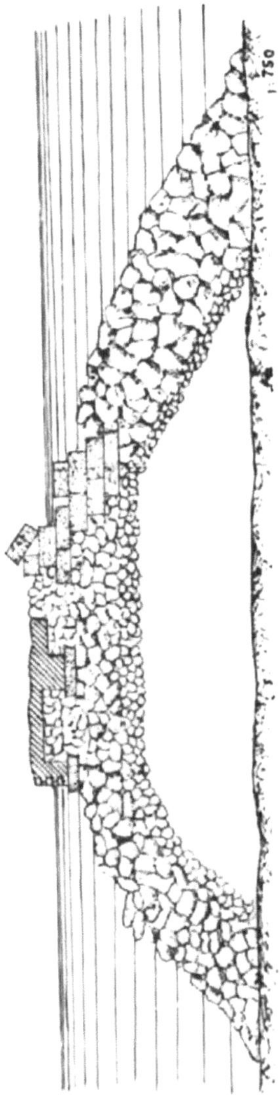


FIG. 112.—Section of Damaged Portion of Galliera Mole from C to D, fig. 109.

ruption effected by the air compressed within and below the masonry, which caused the latter to be projected upwards as if by explosion.

Several of the artificial blocks, laid as headers in the upper course, were displaced in a manner which deserves attention. They were found leaning against adjoining blocks, as if they had been acted upon simultaneously by two forces, one vertical and the other lateral.

Throughout the undamaged section of mole, the protection blocks were set back about a yard. One bollard, struck by a portion of falling shelter wall, was sheared flush with the quay level. Several blocks of concrete, each weighing about 40 tons and having a volume of over 600 cubic feet, were driven a distance of 165 feet. This movement, of course, could not have been accomplished by any single stroke, but must have been the cumulative effect of repeated blows.

*Storm at Bilbao.*—Equally remarkable is the account of the damage wrought by a storm on the last day of the year 1894 at the port of Bilbao.

On that evening the action of the waves became so violent that the whole mass of protecting blocks covering the breakwater was completely carried away. These blocks had each a volume of 39¼ cubic yards and a weight of over 60 tons; they had been laid with the greatest care in contact with one another, forming an apron to the superstructure 26 feet by 16 feet, and consisting of two rows in width and depth alike. The toe of the superstructure being then unprotected, the latter work was soon undermined and demolished. The most striking feat of the storm, however, was the removal of a large monolithic mass of 1046 cubic yards volume and 1700 tons weight placed at the extremity of the breakwater: it was carried a distance of 105 feet into the interior of the harbour.

These instances suffice to exhibit the vagaries which attend a demonstration of wave power by nature in her more violent moods. We pass on now to an application of these facts to breakwater design.

**Classification of Breakwaters.**—Practically all breakwaters fall within the limits of two types, the respective characteristics of which are

(1) the heap, or mound, and

(2) the wall.

The former of these is a heterogeneous assemblage of natural rubble, or undressed stone, in pieces of varying size, supplemented in many cases by artificial blocks of bulk larger than can be conveniently quarried in the natural state, the whole being deposited pell-mell, without any regard to bond or bedding.

The latter involves in whole, or mainly, the construction, in a regular and systematic manner, of a masonry or concrete wall, with vertical, or nearly vertical, faces.

Subsidiary classes form a series of gradations between these two distinctive types, so that strict lines of demarcation are not always easy to draw. The combination of wall and mound in varying proportions constitutes indeed, by far, the bulk of instances in modern practice. Sometimes the mound predominates and is simply capped by a slight superstructure of regular masonry, as at Algiers and Oran; in other cases, it is reduced to a minimum,

Fig. 113.—View of Genoa Breakwater after Storm of 1898.

[To face p. 132.

becoming a mere foundation layer for a wall of massive and substantial proportions, such as is exemplified at Ymuiden and Zeebrugge.

The advantages and disadvantages attaching to each of the two principal

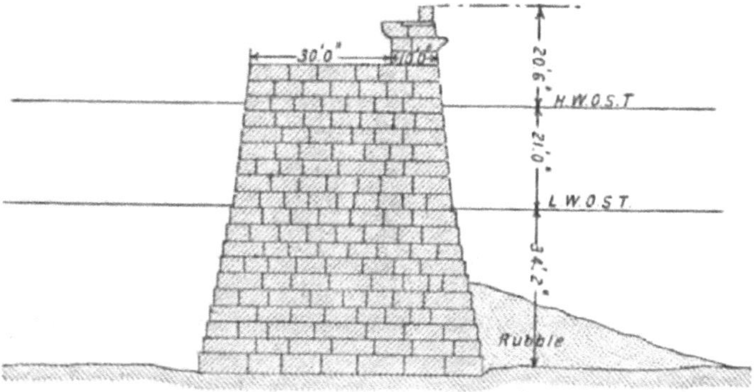


FIG. 114.—Section of Battery Pier, Douglas, I.O.M.

types may be considered under the heads of (1) Cost of Construction, (2) Cost of Maintenance, and (3) Efficiency.

**Cost of Construction.**—As regards the first point, much, of course, depends upon the locality of the breakwater and its coastal environment.

Where stone is plentiful and quarries lie conveniently adjacent to the site, the rubble mound will commend itself on account of the facility with which it can be formed, and the comparative economy resulting from the use of undressed stone, with its attendant unskilled labour. Such was the case at Portland, where there was not only an abundance of stone, but also a practically unlimited supply of convict labour.

In the absence, however, of these essential conditions, and provided the depth of water be not great and the foundation be sufficiently firm, the wall, involving, as it does, a much less quantity of material, will be found preferable. Especially will this be the case where skilled labour happens to be plentiful and cheap. Even the difficulty of a defective foundation may be overcome by one or other of several expedients without perceptibly altering the relative positions. But where the sea bottom lies at a great depth, the superior economy of the pure wall cannot be maintained.

Taking (merely for comparative purposes) the cost of rubble stone at the quarry at say eighteen pence per ton, and allowing $1\frac{1}{3}$ to $1\frac{1}{2}$ tons to the cubic yard of stone, *in situ* (after deducting 20 to 30 per cent. for interstices), the total cost of obtaining and depositing a rubble mound under favourable conditions may be stated at from 3s. to 3s. 6d. per cubic yard of volume.[1] This, of course, applies to natural rubble deposited at random in the body of,

[1] The cost of the rubble mound at Holyhead ranged from 2s. 3d. to 2s. 7d. per ton, deposited in place. The cost of quarrying was 9d. per ton.

At Sandy Bay, U.S.A., 1s. 9d. per ton was paid for ordinary rubble deposited *in situ*; larger blocks, averaging 5 tons, were rated at 4s. 10d.

and forming the bulk of, the breakwater. Larger and special material for protecting the surface slopes will run to 6s. and 7s. per cubic yard. Possibly 4s. or 5s. per cubic yard might be taken as an average all-round cost for the whole.

For dressed masonry or concrete work, either in the form of blocks or in bulk, set mainly below the water level, it is difficult to assess a rate without a full knowledge of the circumstances and resources at disposal; yet it would be unjustifiable to imagine the work as capable of being carried out at a lower rate than 2s. a cubic foot, and it might easily attain a very much higher figure.[1] Even this minimum rate is from eleven to thirteen times that of rubble work, so that, *ceteris paribus*, the bulk of the mound would have to exceed the bulk of the wall in something like the same ratio before it ceased to be the more economical method.

The cost of composite breakwaters combining a foundation mound with an upper wall will, of course, lie between both extremes, and probably, in the majority of cases, it will prove to be rather more than half the cost of an equivalent upright wall.

Actual examples affording any degree of serviceable comparison are difficult to quote, as so much depends upon the particular circumstances and conditions of each case. It would, in fact, be necessary to go very minutely into detail in order to estimate the relative value of each variation from its fundamental type, and, apart from this, no effective comparison could be made. All that can be said is that breakwaters have cost anything from £50 to £400 per lineal foot. The lower limit appertains to minor structures only. Among those of greater importance may be cited the following. Portland breakwater cost approximately £130 per foot run; Holyhead, £160; Colombo, £170; Alderney, £235; Plymouth, £300; Peterhead, £300[2]; and Dover, £370.[2] Other instances will be found in connection with their detailed descriptions.

**Cost of Maintenance.**—A comparison of the expenditure upon upkeep of the wall and the mound admits of only one conclusion.

The wall, provided it be carefully and properly constructed in the first instance, calls for no further attention save for such rare and occasional damage as results from some storm of exceptional severity.

The mound, on the other hand, is peculiarly susceptible to the constant fretting and attritional action of waves. Concussion and back-draught, or suction, constitute two alternating forces continuously and incessantly at work, even in times of moderate and calm weather. Rough rubble is smoothed and rounded by repeated movement, until it is easily sucked out of position and rolled away. The surface slopes thus become gradually less steep, while the flattening correspondingly increases the power of the waves, converting them more and more from the oscillatory into the translatory variety. The

[1] Particulars of ashlar work at Holyhead breakwater: Runcorn sandstone below zero, 2s. 11d. per cubic foot. Anglesea limestone below zero, 3s. 5½d. per cubic foot. Runcorn limestone above zero, 1s. 9d. per cubic foot. Anglesea limestone above zero, 2s. 3½d. per cubic foot.

[2] Incomplete; estimates only.

ultimate dispersal of a rubble mound left entirely to itself is only a matter of time. The preservation of a mound breakwater necessitates, therefore, a constant replenishment of material.[1]

The pitching of seaward slopes with ashlar work, or with massive concrete blocks, goes far to neutralise the destructive action; but the protection afforded is not always complete, and in cases where it has proved effectual, the result has only been attained by a much greater outlay than could justifiably be assigned to the formation of a simple rubble mound.

**Efficiency.**—The efficiency of a type is, after all, the consideration of greatest importance. Cheap construction and maintenance, though points to be carefully weighed, must inevitably be subservient to the attainment of the object in view.

The wall, rising up sheer from a sea bottom below the zone of disturbance with its exposed face vertical, or practically so, receives the wave before any conversion of oscillation into translation can take place. The wave is deflected upwards, and it falls back and down upon a bed of water too deep to permit of any deleterious influence upon the foundation.

On the rubble mound, with face slopes of 1, 2, 3, 4, and 5 to 1, the stroke of the converted wave is delivered with powerful and inimical effect—not only as regards the breakwater structure, but also the area which it incloses. The mass of water rushing up the seaward slope eventually falls over the crest, beating down upon the inner face and tending to effect a breach which must ultimately lead to serious results. Furthermore, even if the wave do not surmount the crest of the mound, the undulations of the sea are transmitted through the interstices of the stone mass and the harbourage area is kept more or less in a state of agitation. This action will be the more evident as the stones or blocks are of greater size, involving vacuities of corresponding magnitude. Amid large-sized artificial blocks deposited irregularly, the voids will amount to at least 25 or 30 per cent. of the whole; and as these blocks are employed to crown the majority of mound breakwaters, the protective value of the type falls considerably below that of a wall.[2]

In order that a composite breakwater may possess the efficiency of the wall, it is necessary that its superstructure should commence at a depth of at least 5 fathoms, otherwise the back-draught of the waves will exercise an undermining influence upon the rubble foundation. The peculiar drawback attaching to this class of breakwaters is that due to irregular settlement, whereby the superimposed wall is liable to be cracked and fissured. This point will receive further notice in the next chapter.

From the foregoing remarks, it will be seen that no absolute preference

[1] There is, as might be expected, considerable variation in cost at different localities. The maintenance of the mound at Holyhead is stated to be 1s. 3d. per linear foot per annum; at Genoa it is 7s.; at Naples 13s.; while, during a certain period, the Alderney breakwater involved an expenditure of from 25s. to 45s. This, however, was quite an abnormal experience. The renewal of large artificial blocks on the seaward slope of Cette forms an annual charge of 29s. per foot run.

[2] At Marseilles it has been found that external waves 3 feet high give rise to fluctuations of from 4 to 6 inches within the area sheltered by the breakwater.

can be attached to any specific type of breakwater for general adoption. Questions of cost and maintenance, the degree of efficiency desired, the nature of the sea bottom, and the extent of exposure—these are all matters which have to be individually weighed before any definite decision can be arrived at. At the present time, there are breakwaters, either in course of construction or recently completed, of the pure wall type at Dover and Tynemouth, of the pure mound type at Brest and Marseilles, and of the composite type at Zeebrugge, Bilbao, and Peterhead.

We now enter upon a discussion of the conditions affecting the stability of breakwaters.

**The Stability of Mounds.**—It has already been pointed out that mounds are lacking in the quality of permanence. This applies more particularly to their upper portions which are under the constant influence of hydrodynamic action. The equilibrium of the lower portion is simply a question of quiescent hydrostatic pressure. Wave influence does not extend to an indefinite depth. Below the level at which its effects are felt, it has been found that rubble mounds will stand at slopes of 45 or 50 degrees.

The limiting depth of wave influence, however, is a matter of some uncertainty. It has generally been assumed, until recently, that a depth of 30 feet below the surface level marks the extreme boundary of the zone of appreciable disturbance; but there are on record instances of serious wave action at greater depths. Thus at Peterhead Harbour, in October 1898, blocks weighing upwards of 41 tons each were displaced by waves at a depth of 36½ feet below low water of ordinary spring tides. Instances of this nature, however, are very rare, and in the majority of cases the standard limit may still be counted upon as generally reliable.

The disturbing influence of waves is most keenly felt between the levels of high and low water, and it is in this region that the most trying ordeals of a breakwater are experienced. A difficulty underlying the situation is that in proportion as the slope is flattened to maintain its equilibrium, the disruptive effort of the wave is fostered and increased. Hence the introduction of huge blocks and monoliths to withstand impact. These blocks, which rarely weigh less than 25 or 30 tons a-piece, and often considerably more, may be deposited either in courses or at random. In the former case, they may be stepped so as to form a general inclination of 1 to 1; but if deposited at random, a flatter slope will be necessary.

The blocks, when artificial, are generally made in the form of rectangular solids: parallelopipeds in preference to cubes; and they should be laid as headers—that is, with their ends facing the line of wave action. In this way the minimum face area is exposed to the stroke, and there is the maximum resistance to overturning.

Natural blocks are heavier per unit volume than the majority of artificial blocks, and, for this reason, have claims to preference. They are also less liable to disintegration, but they are difficult to procure economically to large dimensions, and their irregular shapes render it impossible to bed them systematically. They have a tendency, also, towards becoming rounded like boulders, and this does not improve their steadiness *in situ*.

Mounds are most commonly formed in assorted layers, with the smaller material at the base and the largest at the top and on the flanks. Apart from the additional expense involved in selecting the material and of laying it in proper order, there is this further consideration, that such mounds are less compact and less solid than mounds which are formed by an indiscriminate

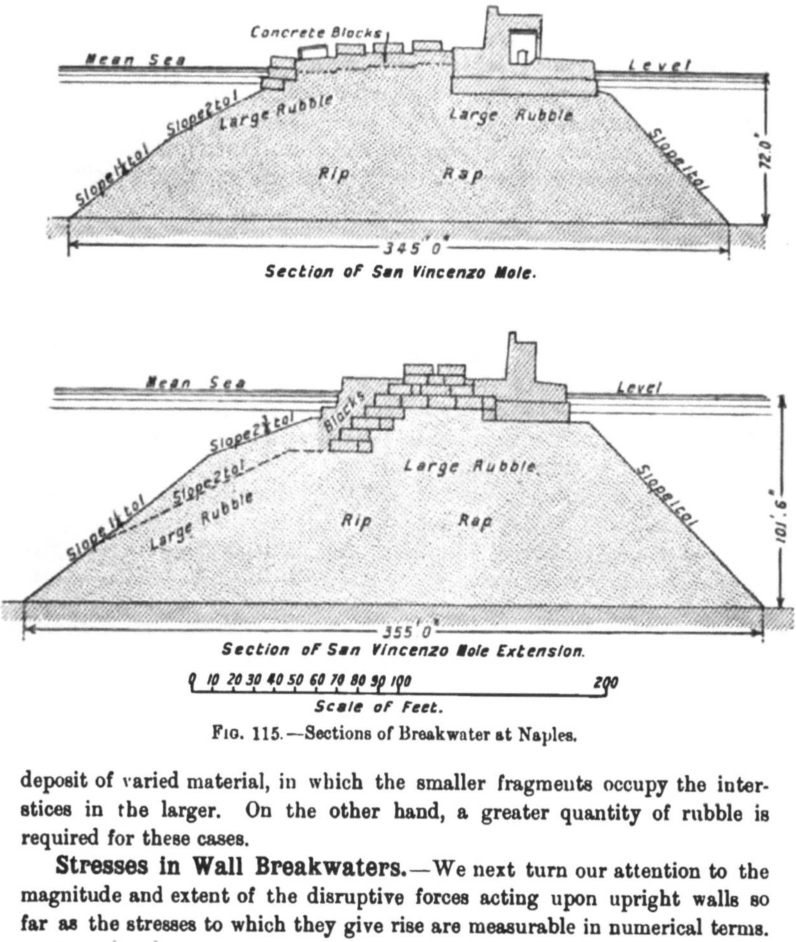


FIG. 115.—Sections of Breakwater at Naples.

deposit of varied material, in which the smaller fragments occupy the interstices in the larger. On the other hand, a greater quantity of rubble is required for these cases.

**Stresses in Wall Breakwaters.**—We next turn our attention to the magnitude and extent of the disruptive forces acting upon upright walls so far as the stresses to which they give rise are measurable in numerical terms.

Considered as a structure exposed to the effects of wave action, a wall breakwater may fail partly, or wholly, in one or other of the following ways:—

(1) By the shearing of some bed-joint, or by the sliding of one component block upon another;

(2) By overturning as a solid mass in sections of variable size;

(3) By the uplifting and dislocation of a horizontal course, or layer; and

(4) By fracture and shattering.

Assuming, for the sake of convenience and simplicity, that we are dealing with a single block or monolith of assigned dimensions, we may express the value of the adhesiveness of the bed-joint or base in its resistance to shear as from 3 to 5 tons per square foot if the cementing material be hydraulic mortar of good quality, and from 6 to 9 tons per square foot if Portland cement mortar, further assuming that, in each case, the proportion of sand to the matrix does not rise above 3 to 1.

Accordingly, the condition for equilibrium is that the horizontal component of wave pressure measured in tons shall not exceed the area of the bed-joint in square feet multiplied by some coefficient ranging from 5 to 9 in accordance with the nature of the cementing material of the joint.

This is in regard to shearing action. If the joint be already fractured, or if the adhesiveness be neglected, then resistance to movement can only be forthcoming through the agency of friction. The coefficient of friction has already been stated at ·7 (*vide* p. 126) for smooth concrete blocks, and a value of ·65 to ·7 will hold for all surfaces of masonry and brickwork in contact. For stone on rock the same value will suffice, but for brick or stone on moist, unctuous clay, the coefficient must be reduced as low as ·3· If the weight of a given block be W, then something like ·7W is the force required to move it over a masonry or rocky surface, and from ·3W to ·5W over an earthen one.

A distinction must, however, be made as regards the weight of the block, whether it be submerged entirely, partially, or not at all. Substances immersed in water lose a part of their weight equivalent to the weight of the volume of water which they displace. Consequently, the effective weight of a completely immersed block is less than its weight in air by the weight of an equal volume of water, which, in the case of sea-water, it is customary to estimate at the rate of 64 lbs. per cubic foot.

The fact may be expressed in another form by stating that the weight of the block is equal to $(d-1)$ times the weight of an equal volume of water, $d$ being the density of the block compared with that of water as unity. This relationship has an important bearing on our next consideration.

Parenthetically, it may be pointed out that sliding action is very materially assisted by small smooth stones and pebbles more or less spherical in shape, which not infrequently intrude themselves between the detached blocks protecting the outer slopes of certain breakwaters.

The resistance of breakwaters or their component parts to overturning arises from their (effective) weight and from the tensional strength of the joints. This latter source should, however, not be counted upon. Beyond affording some slight additional margin of security, its assistance is so slight as to be negligible, especially when compared with the inertia of the mass.

The **overturning effort** is due to the horizontal pressure of the wave, which exerts a moment about any point of the base measurable as $Fx$, where $x$ is the height above the base at which the effect of impact is assumed to be concentrated and F is the force of impact in statical units of pressure. If the block be small, and if its entire vertical surface encounter the full stroke of the wave, it is not unjustifiable to assume that the value of $x$ is $\frac{h}{2}$ or the semi-height of the block. It is, of course, a matter of conjecture, but evidently it represents the extreme condition of things in an unfavourable sense, and therefore is a reliable basis of calculation.

But over surfaces of considerable extent the hypothesis of uniform intensity of pressure is not strictly tenable, and indeed, in certain cases, is very far from representing the actual effect of wave impact. The equivalent pressures at various points of an extensive surface are equally variable. The maximum occurs approximately at mean water level, and the force decreases above and below this point, probably in the ratio corresponding to ordinates of a parabolic curve.

Now, the stability of a block is a function of $(d-1)$ times the volume, for the moment of resistance to overturning is the product of the effective weight into a moiety of the width of the base. For critical equilibrium, therefore, we have :—

$$Fx = W\frac{b}{2}.$$

If, then, W varies as $(d-1)$ V, it is noteworthy that any increase in $d$ involves a much greater increase in W. Thus, if $d$ be increased, say, from 2 to 3, the value of W is increased from V to 2V, an increment, in the one case, of 50 per cent., and in the other of 100 per cent. Hence the great importance to be attached to the use, for sea work, of materials having a high specific gravity.

Although the influence of the bed-joint, in so far as it affords tensional resistance to the overturning action, is wisely neglected, on the other hand, it is not safe or desirable to ignore the effect of the corresponding compression upon the inner edge or line about which overturning may take place.

The resultant of the overturning force and the gravitation of the wall will often produce a very powerful and concentrated pressure upon a small area of the bed-joint, which may be beyond its capacity to resist. Thus, if the line of action of the resultant fall upon one or other of the two points which trisect the base, the intensity of pressure on the edge nearer the point is twice as great as the mean of the pressure over the whole area, and for any further eccentricity of the resultant, the ratio is greatly magnified. The following expression serves to convey a value for the intensity of pressure, $p$, on the nearer edge in terms of the eccentricity ($x$), the length ($l$) of the base-line, and the mean pressure ($a$) :—

$$p = a + \frac{6ax}{l}.^{*}$$

The maximum value of $p$ consistent with safety is about 10 to 12 tons per square foot on Portland cement concrete, 8 to 10 tons on hard rock, 4 to 5 tons on rubble masonry, and from 2 to 3 tons on gravel, sand, or clay.

* See *Dock Engineering*, p. 176.

**Uplifting** is an action which takes place through the application of wave force to the underside of a mass. Obviously the dead (effective) weight of the mass is the resisting element, and the problem is a simple case of the equilibrium of two opposing forces, each of which has already been defined and described.

The **fracture**, or shattering, of a homogeneous block rarely results from the direct impact of the wave. When it does take place, it is probably caused by a prior dislocation, resulting in collision with other parts of the structure. The fracture of joints has been considered under the heading of resistance to shear. Blocks may also be fractured by unequal subsidence in the wall. This possibility applies more particularly to composite breakwaters, where the rubble foundation mound is subject to irregular settlement. The results can only be guarded against by avoiding the use of bond in the building of the wall, or by the adoption of what is termed "sloping" bond, as exemplified at Kurrachee and Colombo (p. 169).

**Milan Conference; Report on Breakwaters.**—The subject of Breakwater Design formed one of the topics of discussion at the International Maritime Congress of 1905. Papers, some of which have already been noticed, were presented by eminent engineers of various countries, and a general report was submitted to the Congress. This report was drawn up by Professor lo Gatto, and a transcript of his conclusions cannot fail to be of interest. They were as follows:—

"Breakwaters built of rubble, although expensive in upkeep, are suitable for very sheltered sites in shallow water, provided good and cheap material is procurable. This type is not affected by the muddy or soft nature of the sea bottom.

"When the structure is exposed to very heavy seas, the rubble type of mole can still be adopted, under the conditions mentioned above, provided a revetment of concrete blocks is added outside down to a certain depth. The method of depositing these blocks at random appears the best as regards resistance and maintenance, on condition that the profile of the protected slope is so designed that it will shear the waves at sea-level. On the other hand, the method of setting the blocks in regular courses offers serious objections, as they are liable to be disturbed by the settlement of the rubble base and to be completely destroyed during gales, and, in any case, they cannot be maintained in good condition without abandoning the principle of the system itself.

"Breakwaters with a rubble hearting and a double revetment of protecting blocks, laid in regular courses, are not at all reliable in very heavy seas, but they can render very useful service in sheltered sites and in waters of moderate depth, especially if the works are not of very great importance.

"Breakwaters with vertical, or almost vertical sides, are very suitable for moderate depths and hard sea-beds, where there is no fear of the undermining effect of the backwash and currents. They are very expensive and consequently inapplicable to unimportant works.

"The composite type of breakwater, consisting of a base formed by a loose rubble mound, surmounted by a vertical superstructure, is peculiarly suitable for tidal seas and for seas with a slight tidal rise and fall, provided the water in that case be very deep. In the case of tidal seas, there is no objection to stopping the superstructure at low water level.

"The type of construction in which the superstructure is made entirely of blocks laid in regular courses, can be adopted for seas which have but a slight tidal rise and fall, provided the site be sheltered. This type is not sufficiently reliable for heavy seas, and in some cases the system of large monolithic caissons can be adopted instead with advantage, on condition that the width of the blocks be suitably proportioned to their length, that the loose rubble of the hearting is perfectly compact, and that the very dangerous effects of the backwash at the seaward base of the blocks, which are produced by the impact of the waves, be counteracted by using exceptionally good material for the upper part of the apron or outside road on the sea face, or by loading and preserving this apron by means of protective blocks deposited at the base of the caissons."

The recommendations of this report formed the subject of some discussion and not a little adverse criticism on the part of the Congress in regard to several of the opinions therein expressed. It was evident that unanimity could not be attained, and finally, the Congress limited the expression of its views to the following resolution:—

"The Congress refers to the information furnished by the written reports and oral observations; it considers that engineers will find there information of great value for the construction of breakwaters, especially in regard to the force of waves, but, by reason of the great diversity of cases, it does not think that it should formulate any absolute conclusions."

With this summation of the special advantages and disadvantages attaching to the various types of breakwater exemplified at the present day, we bring our remarks on breakwater design to a close, simply adding some detailed reference to a few selected cases, chosen in illustration of the principles laid down in the preceding pages.

**Breakwaters at Marseilles.**—The main undertaking, begun in 1845, has a length, at the present time, of 4530 yards, including an extension of 600 yards completed in 1904. The same principle of construction has been maintained throughout a period of sixty years with unvarying success.

A section of the breakwater is exhibited in fig. 116. The core is a bed of small rubble, having a depth or thickness of 10 feet, and lying upon the sea bottom at a depth of 55 feet below low water level. It is overlaid by layers of natural stone of increasing dimensions, ranging from 2 cwts. to nearly 4 tons a-piece. The quay shelter wall is a masonry structure founded upon the topmost layer of blocks.

The exterior slope is 4 to 3 for its lower portion, extending from the foundation to low water level. At this point it flattens abruptly to nearly 3 to 1. The effect of this sudden transition is to create a sharp ridge at the water-line, with the result that the waves are cut at the point where thei action is most potent. The upper part of a wave, therefore, falls dead upon the flat slope above, or, at the worst, upon the masonry apron in front of the shelter wall, in neither case capable of producing any deleterious results.

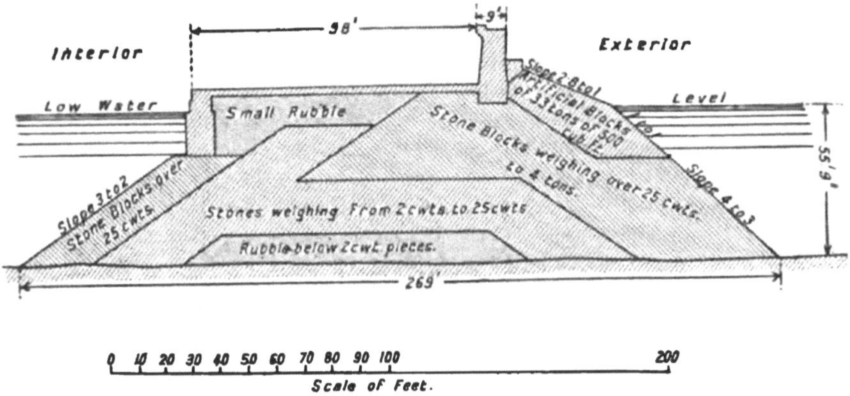


FIG. 116.—Section of Grand Jetée, Marseilles.

The parapet thus receives no appreciable shock, and spray alone passes, at times, over its crest to fall upon the interior quay.

The blocks, forming the flattened slope referred to, are huge monoliths, rectangular in shape, with a length twice as great as their width, having a volume of 500 cubic feet and a weight of about 33 tons a-piece. They are

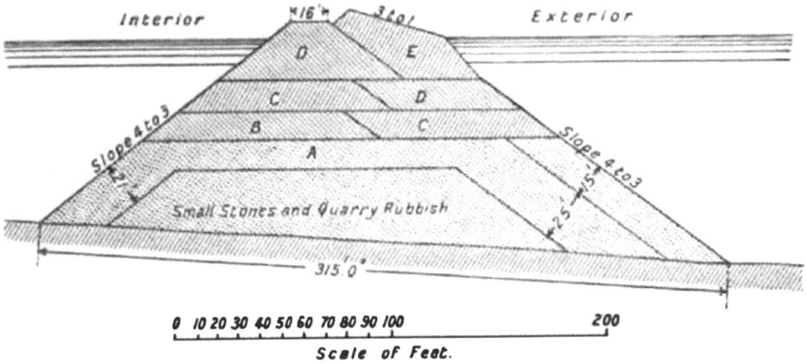


FIG. 117.—Section of Breakwater Extension, Grand Jetée, Marseilles.

A. Rubble stone up to 2 cwts. a-piece.
B. Natural blocks from 2 to 25 cwts. a-piece.
C. ,, 25 to 75 ,,
D. ,, over 75 ..
E. Artificial blocks of 33 tons.

deposited so as to lie longitudinally in the direction of the onset of the waves.

The external profile of the breakwater has proved to be extremely stable and is kept up at a very trifling expense in the way of repairs. For a length of 1200 yards, constructed prior to 1865, the annual cost of maintenance is just under 1s. 9d. per lineal yard per annum. The remaining and later portion of the breakwater costs practically nothing for upkeep.

Apart from the parapet wall and the quay, the cost of the breakwater is stated by Baron Quinette de Rochemont[1] to have been as follows, according to the depths of water in which it was founded:—

In depths of 33 feet, £39, 13s. per foot run.
„ 65 „ 71, 16s. „
„ 100 „ 118, 4s. „

According to M. de Joly,[2] however, the cost of the original breakwater, including the parapet and quay wall in a depth of 60 feet, was £127 per linear foot, a figure which is evidently somewhat in excess of those quoted above, even when allowance is made for the additional work covered. M. de Joly's cost for the extension, however, is in accordance with Baron de Rochemont's figure for the same depth, viz.: 100 feet.

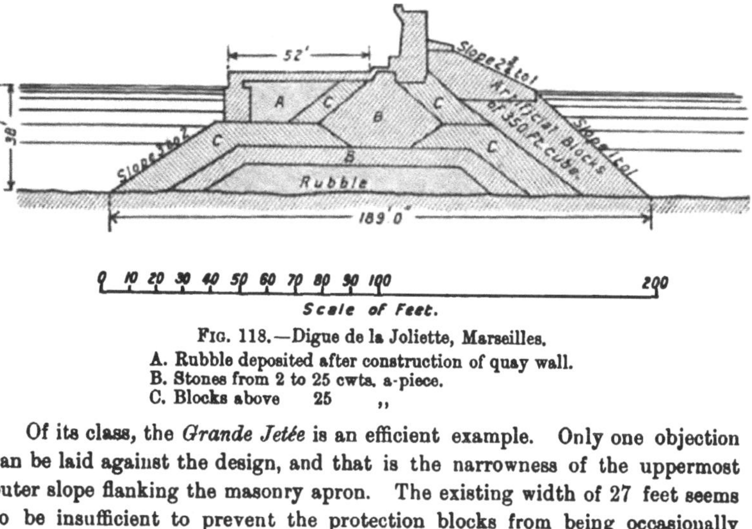


FIG. 118.—Digue de la Joliette, Marseilles.
A. Rubble deposited after construction of quay wall.
B. Stones from 2 to 25 cwts. a-piece.
C. Blocks above 25 „

Of its class, the *Grande Jetée* is an efficient example. Only one objection can be laid against the design, and that is the narrowness of the uppermost outer slope flanking the masonry apron. The existing width of 27 feet seems to be insufficient to prevent the protection blocks from being occasionally rolled off by the waves into deep water.

Settlements in the mass of the breakwater, though they have been by no means inconsiderable in themselves, appear not to have given rise to any serious dislocation of the parapet wall. Indeed, it is said that it is only possible to observe, on scrutiny, a few vertical cracks here and there, with widths of mere fractions of an inch. The shelter wall and its apron are not bonded together: they are simply in contiguity. Separation was inevitable, since they rest upon distinctly different foundations, the wall upon material of smaller size and greater compactness than the apron.

The *Joliette Digue* (fig. 118), constructed in about 38 feet of water, follows

[1] *Cours de Travaux Maritimes*, 1ère Partie, 1896.
[2] *Report on French Breakwaters* to Tenth Int. Nav. Cong., Milan, 1905.

somewhat generally the same lines of construction as the *Grande Jetée*, in so far as regards the disposition of the material and the external slopes. The quay, however, is only about half the width of that in the previous case. The cost of this breakwater, including the outer protection blocks, amounted to nearly £154 per yard. The parapet wall with its ashlar work came to £61, 10s. per yard, and the formation of the inner quay involved another £44, making the total cost, approximately, £260 per lineal yard.

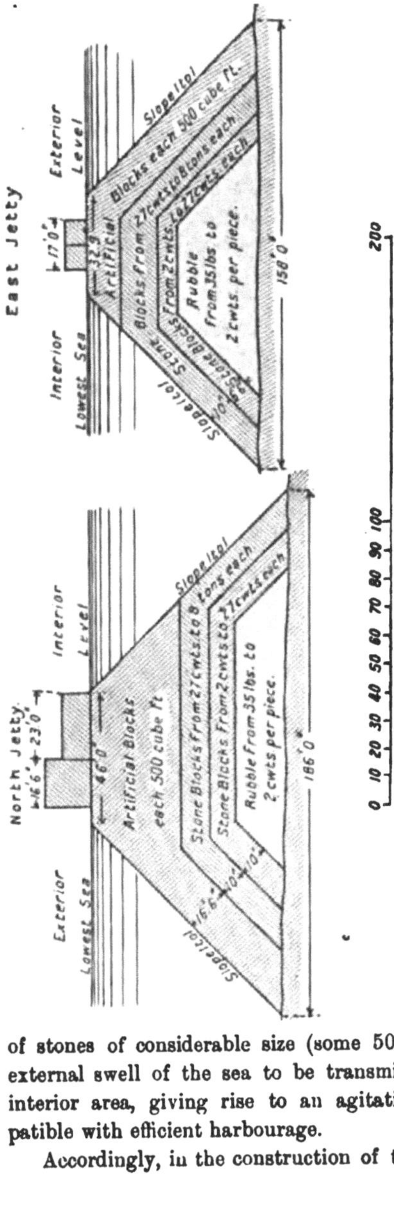


Fig. 119.—Sections of Breakwaters at Algiers.

**Breakwaters at Algiers.**—The principle underlying the design of the north and east breakwaters at Algiers, is that of approximating the rubble mound as closely as possible to the form and functions of an upright wall. The mound has been laid to the very steep slope of 45 degrees throughout, and the superstructure occupies the whole of the narrow summit. Such a design is open to very strong and grave objections. The impulse of the wave, abruptly checked by the face of the parapet wall, is converted into a powerful downward force directed against the outer slope, or, alternatively, the waves, rising to an abnormal height, clear the parapet wall and break with considerable impact upon the inner side. In each case the tendency to disturbance is very pronounced, and movements frequently take place. Furthermore, the voids and interstices in the uppermost layer of rubble, which consists of stones of considerable size (some 500 cubic feet in volume), causes the external swell of the sea to be transmitted through the breakwater to the interior area, giving rise to an agitation of the surface, which is incompatible with efficient harbourage.

Accordingly, in the construction of the inner port at the Agha, begun in

1899, the substructure of the jetty includes a solid wall of concrete blocks brought up from a depth of 18 feet below the water level. These blocks are not laid in bond, that is, breaking joint as at Genoa or Cette, on account of the highly compressible foundation, which is sand and mud. They are superimposed in such a manner as to form a series of piers, disconnected except for the masonry crown, which, though fairly continuous, is jointed every 25 or 30 feet. So far, the jetty has stood satisfactorily, but its construction is of too recent a date to admit of any definite pronouncement of its value. The cost is stated by M. de Joly to amount to £350 per lineal yard, or about the same as the Marseilles breakwater extension, which, however, is of much greater sectional area and in water of much greater depth.

**Breakwater at Sandy Bay, Mass., U.S.A.**[1]—"The subject of an extensive harbour of refuge at Sandy Bay has been under consideration since 1882. The project submitted to Congress was for the construction, at a cost of $4,000,000, of a breakwater 9000 feet long in the location shown on plan in fig. 7. The proposed breakwater was to be a rubble mound surmounted by a masonry superstructure founded 15 feet below low water. The mound was to be 40 feet wide at the top. The superstructure was to be trapezoidal in section, to rise 8 feet above high water, and to be 15 feet wide at the top. Below low water, it was to be laid 'dry'; above low water, in mortar."

No work was ever done upon the superstructure above described, and, in fact, no project for the construction of a superstructure was adopted until 1892. In 1884 the plan for the substructure was changed to that of a mound 40 feet wide at the top, rising to 22 feet instead of 15 feet below low water.

The depth of water at mean low water varies from 6 feet at Avery's Ledge, the extreme southerly end of the breakwater, to about 89 feet at the extreme westerly end, and averages about 45 feet along the southerly arm and about 65 feet along the westerly arm. The bottom along the line of the work is nearly all ledge, except at the westerly end, where it is sand and shells. In the anchorage area, the holding-ground is excellent, being sand mixed with mud.

The work done prior to 1892, up to which time $450,000 had been appropriated, consisted in the placing of about 500,000 tons of stone in the substructure.

In the early part of 1892, a board was appointed to recommend a project for the superstructure and any changes that might be desirable in the existing project for the substructure. The section adopted is shown in fig. 120.

By 1898, 600 feet at the northerly end of the southerly arm had been completed to full section, 1200 feet more had been carried up to low water, and 2800 feet more had been founded. The 600 feet of superstructure was formed of stones weighing not less than 4 tons each and averaging 6 tons,

[1] McKinstry on Breakwaters, *Trans. Am. Soc. C.E.*, vol. liv.; Int. Eng. Cong., 1904.

and the southerly 250 feet of it had settled some 2 feet. In the early part of the year, in a storm of exceptional severity, the 600 feet of completed superstructure was torn down to a height of about 5 feet above mean low water.

Modifications in design, recommended by a board of inquiry and adopted in September 1902, are shown in fig. 121.

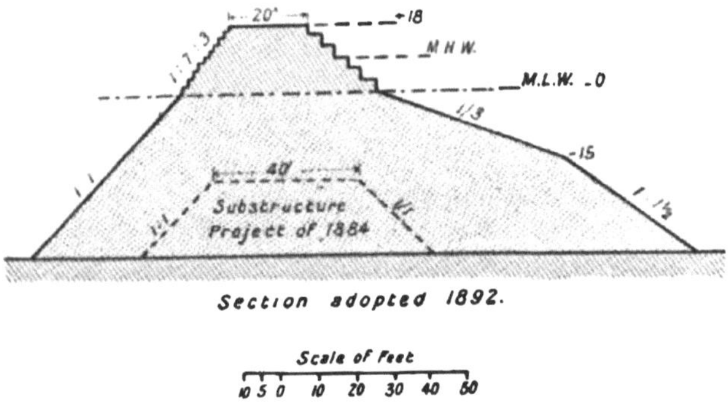


FIG. 120.—Section of Breakwater, Sandy Bay, U.S.A.

The capstones in the new plan are to weigh not less than 20 tons, to be 20 feet long by 3 feet by 5 feet in end-section, laid on edge, and in as close contact as possible. The course below the capstones is to contain two stones, each weighing about 10 tons, the outer stone to be at least 15 feet long and

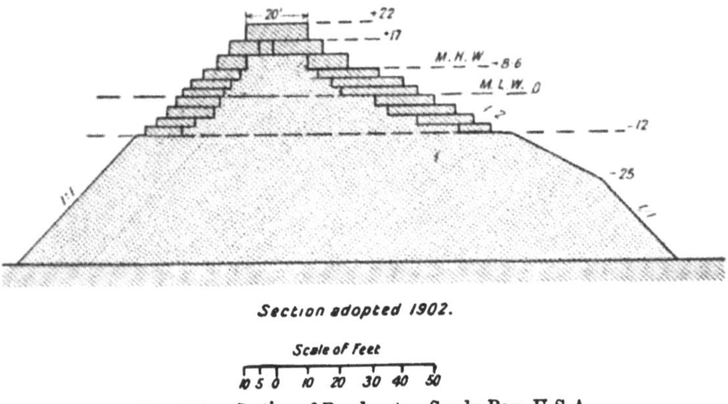


FIG. 121.—Section of Breakwater, Sandy Bay, U.S.A.

the inner one at least 10 feet. Below this course, to a depth of 12 feet below mean low water, the stones in the outer face weigh at least 8 tons, and in the inner face, at least 3 tons; and all are to be laid horizontal and as headers.

Including what had been spent up to 1902, the total estimate of cost for the work was $6,904,952. It is estimated that, when completed, the work will contain 6,301,407 short tons[1] of stone. To September 1904, 1,690,178 tons had been deposited.

**Piers at Tynemouth.**—The Tyne piers commenced in 1855 for the purpose of sheltering the mouth of the River Tyne, constitute an example of the composite type in which the wall predominates. The rubble mound, which acts as a foundation to the wall (see fig. 122), has since been discarded in connection with the reconstruction of a portion of the north pier, due to a breach in it, 100 yards wide, made by a storm in 1897.

The mural portion of the original structure consisted of two longitudinal masonry walls connected at frequent intervals by cross walls, the cavities or pockets between being filled near the shoreward end with quarry débris, and, further seaward, with mass concrete.

"The depth of the foundations of the superstructure varies from low-water level at the shoreward end to 27 feet lower at the pierhead. This

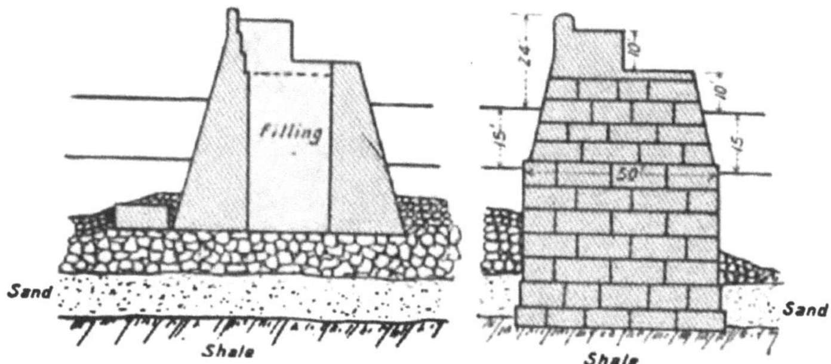


Typical Section of Old Work. Typical Section of New Work.

FIG. 122.—River Tyne Breakwater.

depth of foundation at the pierhead is much greater than was originally contemplated, it having been discovered, while the work was in progress, that wave action took place at much greater depths than had previously been supposed. The depth of the foundations would probably have been carried still lower had it not been that the rubble mound was deposited very much in advance of the superstructure, in order to ensure its being sufficiently consolidated before being built upon. The whole work seems to have stood well until the winter 1893–94, after which it was found that some of the foreshore blocks had been moved and the foundations of a short length of pier exposed."

After the more serious breach of 1897, the question of reconstruction was considered, and, upon careful deliberation, the Tyne Commissioners decided to form a length of new work within the line of the old work, as shown in fig. 123.

"In the new work the rubble mound is being dispensed with, and the foundations are being taken down to a hard shale, the depth averaging about

[1] Tons of 2000 lbs.

20 feet more than that of the original structure. Above the low-water level, the sectional outline of the work is identical with that of the old" (*vide* fig. 122).

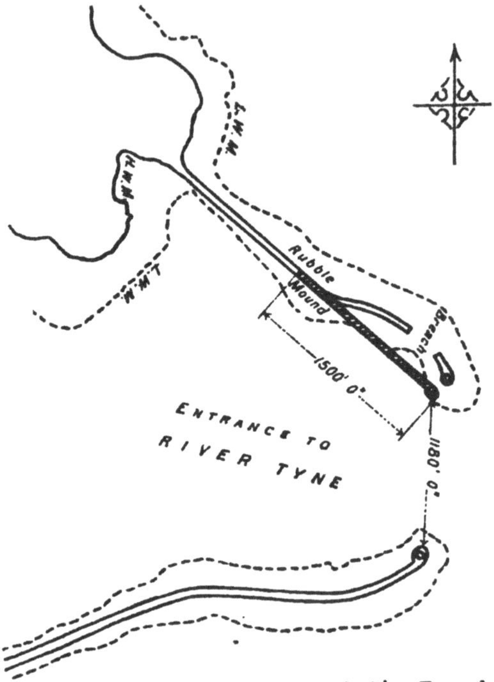


FIG. 123.—Breakwater Piers at the mouth of the River Tyne, showing Reconstruction of North Pier.

The quotations are from a notice on the Tyne North Pier Reconstruction by Mr Ivan C. Barling, the resident engineer, contributed to the summer meeting of the Institution of Mechanical Engineers, July 1902.

# CHAPTER VII.

## BREAKWATER CONSTRUCTION.

Mound Construction—Barge System: Loading, Aligning, Discharging—Staging System: Erection and Maintenance, Trackways—Low Level System—Wall Construction: Staging and End-on Systems—Functions of Titans, Mammoths, and Goliaths—Caisson System—Foundations: Nature and Characteristics—Settlement—Wall Foundations—Piers—Piling—Limiting Loads—Surface Treatment—Levelling—Benching—Deposition of Concrete under Water—Bagwork—Block Making—Bond—Sloping Bond—Grouting—Minor Breakwaters—Crib and Box Work—Fascine Work—Examples of Breakwater Construction from Tynemouth, Alderney, Zeebrugge, Cette, Bilbao, Bizerta, and Dover.

METHODS of breakwater construction are naturally as diverse as the local conditions which govern them, yet they fall, without undue constraint, under the same heads as those enumerated in our classification of the principles of breakwater design. Thus, we have the special methods appertaining to the formation of the mound and to the building of the wall. We will subdivide our observations accordingly.

**Mound Construction.**—For the purposes of a mound, no preliminary dredging operations are necessary. The material for the mound may be deposited upon the sea-bottom direct, for, from the very nature of things, it will spread itself sufficiently to distribute its weight within the limits of support, or it will sink until it reaches some firmer substratum by which the settlement becomes arrested. Nevertheless, it should be pointed out that dredging has not infrequently been resorted to when the surface of the sea floor is mud of a particularly impalpable character, and likely to prove treacherous. At Trieste, for instance, in consequence of certain mishaps, it was found necessary to remove a proportion of the softer mud. The rubble work did not subsequently sink so deeply as before, yet the settlement continued still to be considerable, amounting to 9 or 10 feet in depth. We shall have occasion later on to discuss more fully the question of settlement in foundations. Meanwhile, we are concerned solely with methods of construction.

Rubble may be deposited in one or other of three ways. These are:—

(1) By tipping or discharging from barges, scows, or other vessels afloat.

(2) By discharging from travelling gantries or from cranes, running on temporary overhead staging.

(3) By discharging from wagons passing over roads laid at or about the level of the top of the mound. The wagons are tipped in advance of the

finished mound, upon which the roads are continuously extended as the work proceeds.

**The Barge System.**—The first method is best adapted to sheltered situations—the difficulties of discharging from vessels in a rough sea must be sufficiently obvious; but it necessitates the existence of sufficient depth of water for the loaded barge or scow, together with the additional clearance required, when, as is generally the case, the latter is fitted with hopper doors and the material is dropped through the bottom of the vessel. Certain barges permit of lateral discharge, but the type is not common; and in cases where there are no doors for the purpose, it is attended by some risk of capsizal, with attendant danger to the men engaged upon the work.

A corollary to the foregoing restriction is that hopper barges depositing

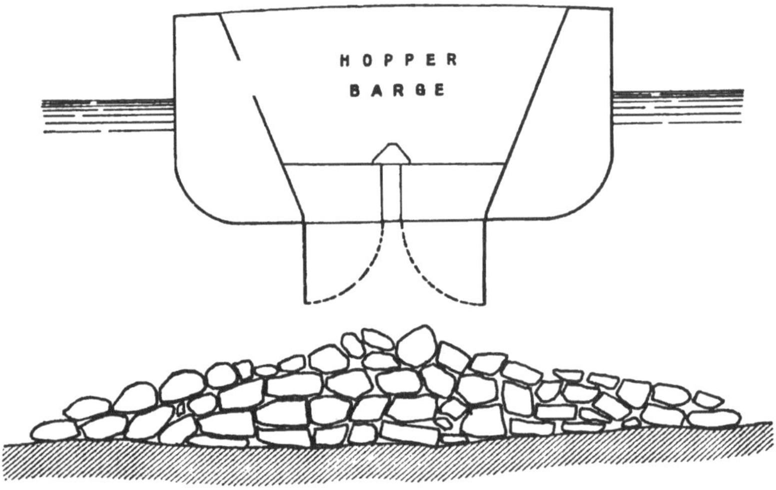


FIG. 124.—Hopper Barge discharging load.

material through bottom doors cannot be employed for the entire construction of a breakwater consisting of a rubble mound only. Even though there be considerable tidal fluctuation, admitting of the higher stages of the work being carried out during periods of high water, yet it is evident that the mound cannot be brought to surface level, and must, indeed, cease at depths below it, which may be anything from 10 to 15 feet, or even more.

One advantage attaching to the employment of barges is the opportunity afforded for depositing rubble uniformly and simultaneously over the whole site of a breakwater. This advantage is shared, but not to the same degree of freedom, by the second method. With floating plant there is no restriction whatever, and the work may be prosecuted over a very extensive area without incurring any higher expenditure or any greater risk of misadventure. As will be seen when we come to deal with the question of settlement, this consideration has a very important bearing on the permanence of breakwaters.

On the other hand, it must be pointed out that the progress of work carried out by floating plant is very much at the mercy of the wind and waves; so that long spells of unfavourable weather result in an almost complete cessation of operations, the period of which is therefore materially protracted. During the winter months, in particular, these interruptions are sure to be frequent and prolonged. Accordingly, in many cases, the method of staging is to be preferred, especially when it is necessary to complete the undertaking with despatch.

The loading of the barges is usually performed at the pier of an adjacent quarry by the ordinary means of tipping through a shoot, the stone being conveyed to the quay edge in wagons running on rails, the gauge of which is generally small. In the case of large blocks, cranes are necessary, both for loading and unloading. The loading crane is situated on the quay; the other is usually mounted on an attendant barge. A pair of sheer legs may take the place of a crane.

On arriving at its destination, each hopper barge, containing random rubble, is adjusted in position with the aid of suitable sight-lines fixed on the shore, or of any convenient landmarks. It is difficult to make satisfactory use of floating objects for this purpose, as they are necessarily moored in a flexible manner, and changes of tide and current may make sensible alterations in their positions, the exact extent of which depends, of course, upon the length of the moorings.

Satisfactory adjustment having been achieved, the hopper doors are released and the material falls through the bottom of the hopper. It may then be necessary to trim it, especially if the deposit forms part of the upper layers. In tidal situations this may be done at periods of low water; otherwise, the services of divers are required. Care should be taken both by accurate alignment and judicious deposit to reduce the labour of trimming to a minimum, as it adds considerably to the cost of the undertaking. In shallow water the trimming and levelling of a rubble bed may be not unsatisfactorily achieved by supplementary hand-tipping, the inequalities in level being indicated by a sounding-lead.

Rubble should be evenly and systematically distributed over the entire width of base which the breakwater is intended to occupy, as also, where possible, over the entire length. Broken ridges and isolated heaps of stone give rise to currents and so to scouring; and although any effects of this action may be rectified by subsequent deposits, yet an additional supply of material is entailed, as well as loss of time and labour. At Cette, excavations ranging from 3 to 5 feet in depth were found to have been produced by scour alongside rubble deposits which had been irregularly made.

**The Staging System.**—The use of staging, though primarily more expensive than any other method of procedure, is attended by many direct and indirect benefits. It promotes, to a very great extent, the unbroken sequence of operations, which is, perhaps, the highest desideratum from every point of view, and it affords greater protection to those engaged upon those operations than, at anyrate, can be guaranteed by the barge system. Interruptions of more than a few hours' duration during tempestuous weather occur but rarely, and there is little time lost in waiting for the subsidence of the sea after a storm has spent its force. Of course, this assumes that the staging itself suffers no appreciable damage. It cannot be denied that temporary structures of slender build exposed to the full force of a gale run some risk of destruction—partial, if not complete. Collapses of greater or less extent have proved this beyond question, yet the instances are not so numerous as to warrant the attachment of very serious importance to the objection, and the particulars are not infrequently exaggerated. Thus, writing in 1904, Sir William Matthews, K.C.M.G., says:—

"Notwithstanding the alarming reports which have appeared in the press, from time to time, with regard to the works at Dover, it is satisfactory to state that practically no damage whatever, from the first, has been occasioned

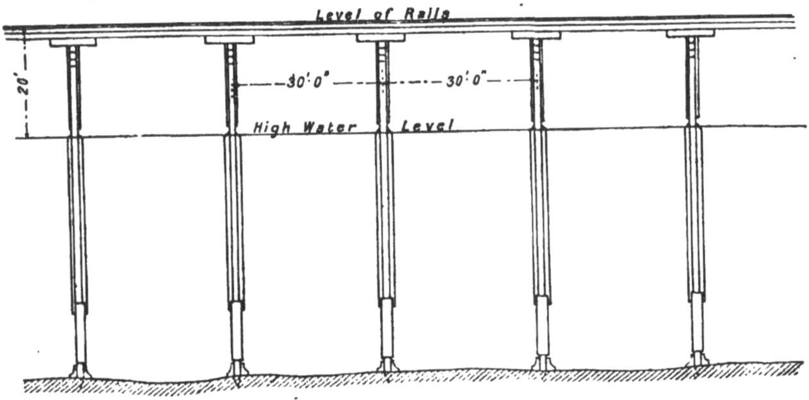


Fig. 125.—Longitudinal Elevation of Staging, Holyhead Breakwater.

to the permanent works, and only comparatively insignificant damage, having regard to the magnitude of the undertaking, has been caused to the temporary structures. Although it was alleged that during the great gale in September 1903 one thousand feet of breakwater works and staging had been carried away, the only loss which was occasioned was the turning over of one span of temporary staging of 50 feet in length with the plant thereon, which, at that time, occupied an isolated position." [1]

A more serious source of danger to sea staging is insect attack, and it is the more to be feared in that the depredations of sea-worms may remain undetected for some time. Constant inspection, therefore, is absolutely essential, and there can be no feeling of security. We are dealing, however, with this matter more at length in another section.

Apart from these drawbacks, staging forms a steadier base for working purposes than a barge or vessel. Platforms may be affixed to it, or suspended

[1] Matthews on Harbours of Great Britain, *Trans. Am. Soc. C.E.*, vol. liv.

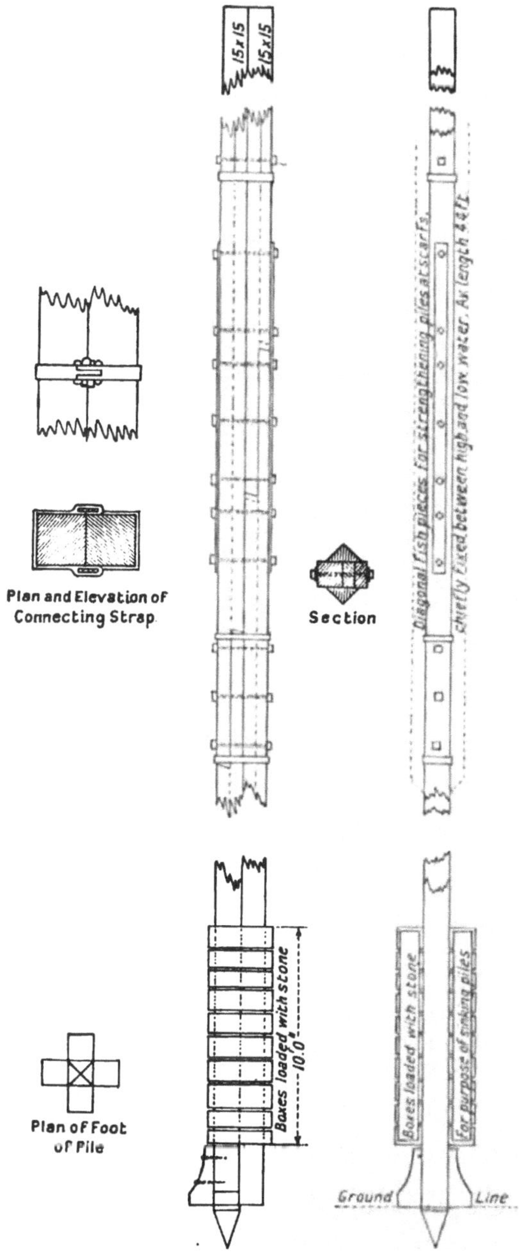


Fig. 126.—Staging Piles, Holyhead Breakwater.

from it at any desired level. It acts as an aid to the alignment of the work, and it allows appliances of a more powerful and efficient character than floating plant to be employed. Some of these features are equally characteristic of the third or low level system, but we shall see that there are also corresponding defects in the latter which are not applicable to staging.

Staging, as usually practised, takes the form of a series of piles in one or more rows of double line driven at regular intervals (say from 15 to 50 feet apart) and connected by longitudinal runners (or, for the longer spans, by wrought iron or steel girders), and bracing, with side strutting, cross bearers, etc. It thus forms a track, or a number of parallel tracks, for wagons, travelling gantries, and cranes; and, in order that as little surface as possible may be presented to wave action, these roads or tracks should be located well above the highest sea level, say not less than 20 feet, and preferably 5 or 10

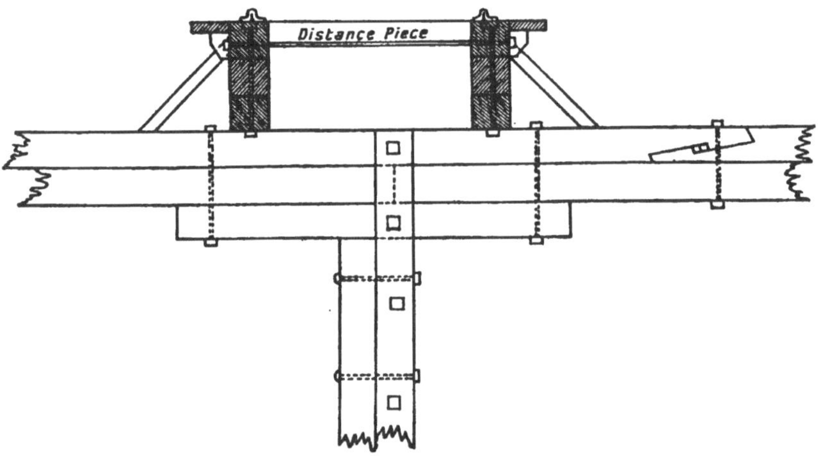


FIG. 127.—Rail Track, Holyhead Breakwater Staging.

feet more, though, of course, any increase in the height is made at the expense of stability. Furthermore, the solid strutting, which characterises much land staging, is best replaced by slender tension members—chains and wire rope stays attached to secure moorings; or, if the surging of these under wave action be deemed undesirable, second-hand railway metals will be found eminently useful for the purpose.

The piles, wherever possible, are driven into the ground by a pile-driver rigged up on a barge or floating platform, or supported on a carriage which projects over from the land or the staging previously completed. The floating pile-driver (or rather, a number of such appliances) can, in still water, construct the road at a much quicker rate than the stage pile-driver, which is limited in the scope of its operations.

In ordinary firm ground, the above is the usual course. If the ground, however, be of a very soft and yielding character, it will be desirable to substitute screw piles with a broad-bladed screw at the foot to afford the necessary surface bearing, or the pile may simply be set upright upon a large iron base-plate in the form of a shoe. A very broad and fairly thick stone slab, carefully set upon the surface of the ground, will often afford a sufficiently substantial base. This method, however, entails much cross strutting between the piles. Finally, if the sea bottom be rocky, the lower ends may be let into sockets drilled in the rock and steadied by concrete filling, or the pile may be shod with a stout iron spike, capable of being driven several inches, at least, into the solid.

For rubble work, the tracks are such as will suit the wagons in which the material is conveyed. The number of these tracks and their distances apart will depend upon the actual extent of the breakwater, but 25 feet or so seems to constitute a fairly average distance between track centres, and there are few breakwaters where the number of such tracks need exceed six, affording a width of 175 feet over all.

The stone, having been conveyed directly on to the stage in wagons, is tipped either by hand or by automatic arrangement, the wagons being tilted at the ends or at the sides. The staging method is particularly convenient on account of its adaptation to an organised continuous supply of stone, and the ease with which wagons may be marshalled and discharged. But it does not command the same extent of area as the barge system, unless the staging be erected from end to end in the first instance, which is unlikely, owing to the delay, risk, and cost.

Under general circumstances, staging may be utilised several times over in different positions; in other words, it is not necessary to provide for a length of staging equivalent to that of the breakwater. As the work is completed, the rear staging may be moved forward, connection with the ground level being maintained by sloping ways. There is inevitably some interruption while the change is being

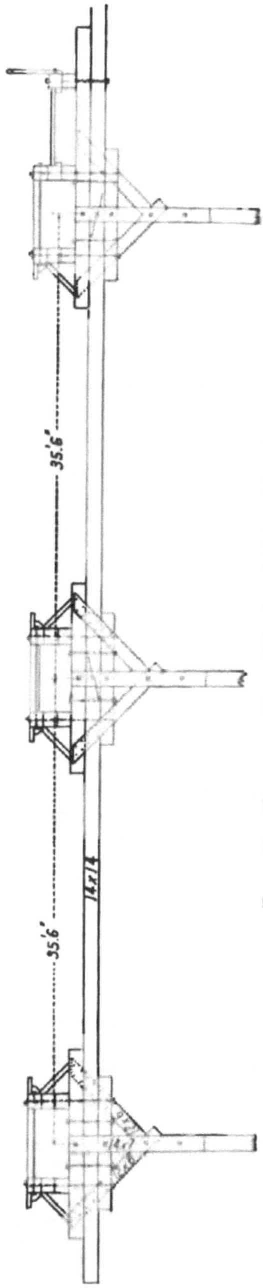


FIG. 128.—Transverse View of Staging, Holyhead Breakwater.

made, and any marked restriction of the working length tends to cripple progress and increase the cost.

**Low Level System.**—The low level system of tipping, by means of wagons running along tracks laid on the solid breakwater structure as it advances, saves the cost of staging and reduces the amount of work done during the process of discharge; but, at the same time, it is a method which is greatly restricted in scope, in that operations are limited to the immediate vicinity of the completed mound; for, until the bank be raised at least above sea level, any extension of the railway track is impracticable. And allowing, under the most favourable circumstances, that the lines can be laid close to the water level, there is the added risk, the probability, the certainty even, that rough weather will cause frequent irruptions and settlements, so that there will be more or less constant relaying of the tracks—all entailing delay and expense. On the other hand, it may be urged that the use of the completed section of the mound as a roadway for the transport of materials assists to consolidate it, and to reveal any sources of weakness which it may possess. This is no doubt true; but whether it affords sufficient justification for the adoption of a system, which is otherwise slow and restricted, is a point which must be determined by the special circumstances of the undertaking. It is a form of construction which is not generally suitable for works on a large scale. For small embankments, however, it may be considered convenient and economical if time be not a matter of importance, and it produces substantial and reliable work.

Leaving the mound type of breakwater at this point, we pass on to methods of wall construction.

**Wall Construction.**—The masonry wall, built with prepared blocks of ashlar or concrete, carefully bedded and laid in accurate alignment, manifestly calls for more elaborate and less rudimentary appliances than are available for the formation of mounds. Other kinds of wall, such as those consisting of concrete deposited in mass in a fluid condition,[1] or built up of sacks and bags laid in courses, also demand special apparatus. The methods of construction generally adopted may be ranged under the headings of:—

(1) The Staging System.

(2) The End-on, or Over-end, Low Level System.

(3) The Caisson or Buoyant Monolith System.

Floating plant, while useful enough as an adjunct, cannot be relied upon alone to carry out operations with sufficient exactitude.

**The Staging System.**—As regards the staging system, there is little to add to what has already been written in connection with mounds. The same lines of formation are followed and the process of depositing is the same, with the exception that, instead of being tipped in bulk, each block of stone is laid individually in position. Cranes or gantries are therefore an integral part of the system, and the tracks will be arranged to suit their requirements. For concrete work, platforms may either be erected on the staging itself, where

[1] The deposition of concrete in a plastic or partially-set condition is a practice to be deprecated.

the materials can be incorporated and discharged into shoots conveying it to its destined situation, or mixing may take place in the yard ashore, and the concrete be conveyed in skips to its appointed place. The former method has the advantage of greater convenience of output, the concrete machines being allocable in various parts, so as to command an extensive range, and there being no tendency to block the service-lines.

**The End-on System.**—The low-level system practised with a single powerful crane running upon a track laid over the finished portion of the work is open to the objection, already stated, of limited scope. The work proceeds outward from the land, and it cannot be attacked from several points as in the case of staging. Yet the method is one which has been adopted in a very great number of modern instances. A strong point in its favour, particularly when dealing with huge blocks of 30 to 50 tons and more, is the greater stability of the working base. On the other hand, there are many occasions when its full lifting power is not in request, and when a much less powerful machine could do the work required at the moment.

The machine employed in connection with this system of construction goes by the generic name of a "Titan" (fig. 129). In principle it con-

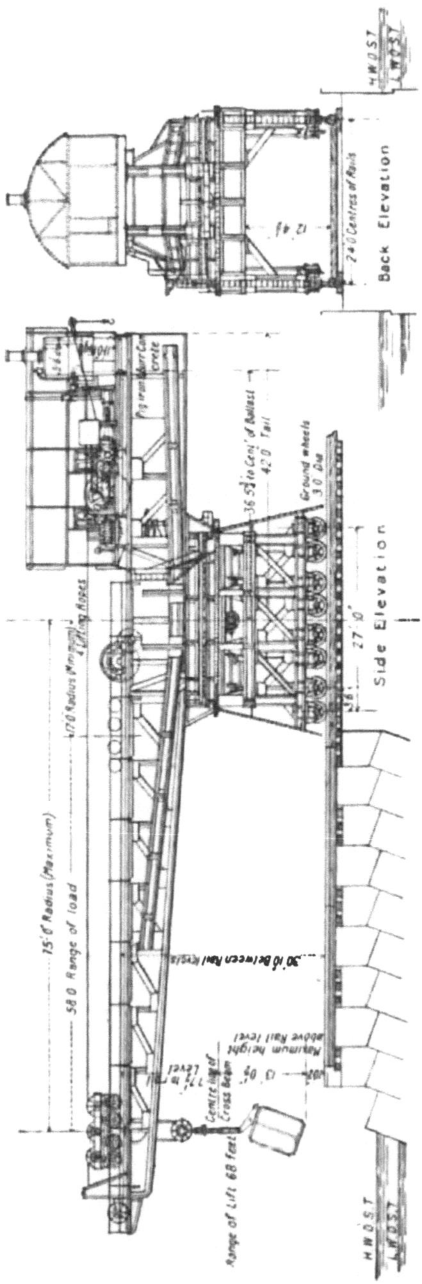


FIG. 129.—Titan.

sists of a huge cantilever crane with a substantial wheel base. There are two variants in design. In one case the cantilever arm is a girder trussed within its flanges; in the other it is supported by means of tension rods from above. The former obviously lends itself to greater stiffness and steadiness, while the latter is lighter and carries the arm at a lower level for the same over-all height.

Apart from their systems of trussing, Titans differ in that some have a fixed base, while others are pivoted upon their carriages. The former class, generally differentiated by the term "Mammoth," are provided with a carrier having longitudinal and transverse motions; the action of the latter class is radial. The radial machines can command a wider lateral range than the rectilinear machines, but they are not so conveniently adaptable to setting out work, a diagonal movement being less easily regulated to alignment in dual directions than a direct one. However, radial machines are capable of depositing wave-breakers along each flank of a breakwater to some distance outside, and this is a feature in which they decidedly excel the alternative type. Moreover, with Mammoths, the block has to be run under the machine before it can be picked up, but with Titans this is not the case. This is not unimportant, owing to the moorings.

The Titan is served with monoliths by a "Goliath" (fig. 130)—the generic name for an overhead traveller, the carrier of which runs on tracks transversely to the road of a wheel base of considerable span. The blocks are loaded on to trollies by the Goliath, and so conveyed from the block-yard to the breakwater, there to be set in position by the Titan. There is, however, nothing rigorous about the practice. The yard machine may be, and is, in some cases, a Titan.

Examples of both these machines are shown in the accompanying figures.

The **Caisson System** is an adaptation of the power of natural buoyancy to transportation purposes. Gigantic boxes of iron framework incased in concrete are formed in a sheltered recess or inlet on the coast or in an inner dock. When built to the required size—which is such that when sunk in position their topmost edges will project slightly above the surface of the sea at low water,—they are temporarily strutted in the interior, launched, and towed out to the site they are intended to occupy. Great care has to be exercised in aligning these huge boxes and in maintaining their perpendicularity while foundering. When this delicate operation has been successfully performed by admitting water to the interior of the caissons, they are filled with fluid concrete, stone rubble, and small blocks, so as to form ultimately a solid monolith.

The method involves some risk, especially on an exposed coast. The caissons are very unwieldy: they call for powerful towing and directing appliances; but once in position and rendered solid throughout, they constitute a most potent defence against breaking seas. As regards cost, it is not apparent that they are more expensive than other forms of break-

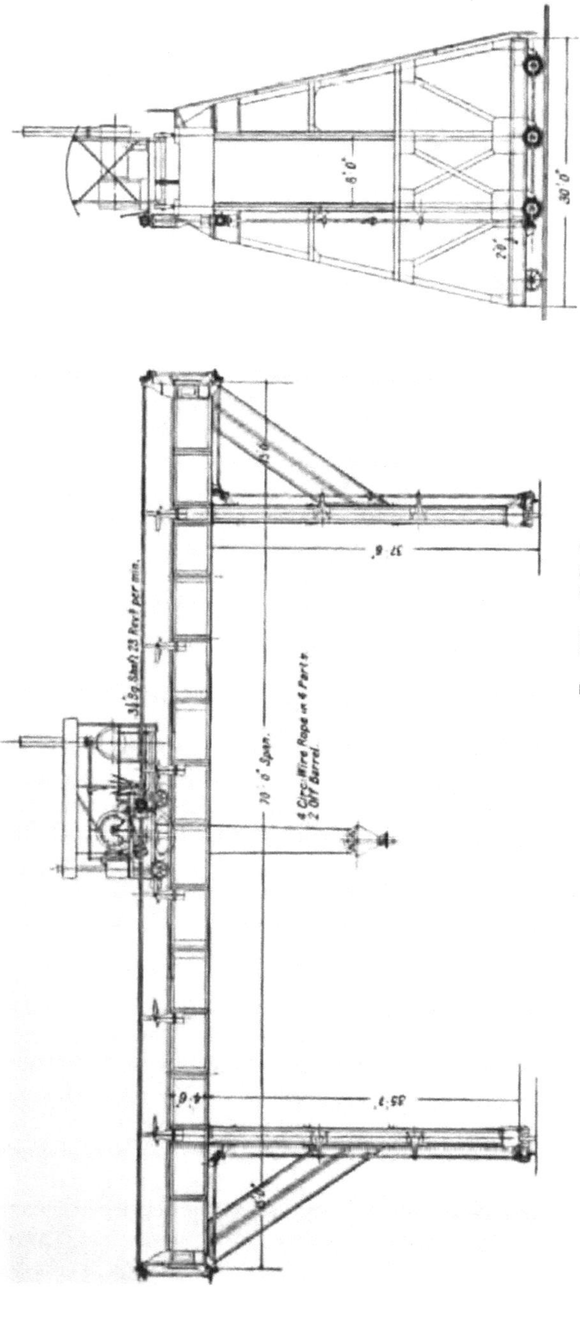


Fig. 130.—Goliath

water, but works on which they are adopted are liable to stoppages and delays arising from tempestuous weather.

In certain instances, as at Dublin, solid blocks of masonry have been built on an adjacent quay and transported by a floating crane, in an almost wholly immersed condition, whereby their effective weight has been very materially reduced.

Apart from and independent of any particular system of excavation, there are general features of breakwater construction which call for careful consideration.

The first and most important of these is the foundation.

## Foundations.

It would be impossible almost to devote to this subject more attention than it merits. Very great and serious harm may accrue to a breakwater founded upon a base insufficiently firm and secure. Even if the damage be remediable, there is the expense of repairs, which will probably become a matter of periodic recurrence. These repairs will naturally be of a more pronounced character in the case of regularly bonded structures, such as walls of ashlar work, which, when disturbed or deranged in any way, involve the provision of special appliances and skilled labour to reinstate them.

Accordingly, it will be well to consider the characteristics and qualifications of a good foundation. These may be classed under two heads: Incompressibility and Permanence.

**Incompressibility.**—A theoretically ideal foundation is incompressible: it does not yield in any way to the load imposed upon it. Such a foundation, however, except in the harder varieties of rock, is almost impossible of realisation.[1] The greater part of the material constituting the sea bottom is more or less of a compressible nature, though in some cases the compression may be but slight. Thus, in addition to the softer kinds of rock, sand and gravel and some varieties of marl are very little, if at all, affected by heavy loads, provided precautions be taken to prevent lateral escape. All other materials are compressible to a marked degree: mud, silt, the softer kinds of marl, clay (particularly when moist and plastic), peat, etc.

While an incompressible foundation is undoubtedly desirable, some slight yielding is no insuperable objection, provided the settlement be uniform. It is of no great moment if the whole superstructure sink a little; but if a portion only gives way, fracture between the stationary and yielding parts is bound to occur. Hence, a foundation should be as far as possible homogeneous. A building is safer on an all-clay foundation than on one of rock and clay. Where the foundation is varied in character, therefore, special precautions are necessary to ensure equal bearing power. The pressure on the weaker material should be distributed over a larger area; the dividing line between the two strata should be distinguished by augmented bond, such as is afforded by

[1] When obtained, it is not an unmixed blessing, as the levelling of an indurated surface is troublesome.

tie-rods or bars; and great care should be exercised in construction. The better course, wherever practicable, is to excavate to the lower level, at which the harder stratum is found.

It must be borne in mind that some settlement is inevitable. It will take place, if not in the foundation, at anyrate in the structure itself, especially in mounds formed of rubble work. The numerous vacuities in the mass and their proportionately great volume, combined with inequalities of bedding and support, lead to a shrinkage of the entire mass, which is very considerable in the earlier stages of its existence, and is more or less a constant characteristic. The diminution arising from this cause, however, is readily made good in ordinary cases by the simple deposition of additional material; but it is manifest that where the mound is acting as a substratum or base for a wall, the effects of shrinkage cannot be so easily effaced, nor can the wall itself escape a share in untoward consequences. Hence the obvious necessity of allowing such mounds adequate time to take a firm bearing.

Moreover, it must not be overlooked that in addition to that arising from its own inherent tendencies, some further subsidence must occur when the weight of the wall is imposed upon a mound. Allowance must be made, in the first instance, for this and for other contingencies.

Settlement, therefore, in some form or other, must be looked upon as inevitable, and the essential point is to ensure its uniformity. Well-constructed breakwaters have sunk to the extent of 10 or 12 per cent. of their total height without appreciably affecting the appearance or the stability of the superstructure; but this has only been so because the process was gradual and regular. Sudden and irregular changes cannot fail to produce fracture, especially in bonded work, concerning which we must speak later.

**Permanence.**—The second point of a good foundation is permanence, or unalterability. Certain mineral substances, when exposed to external influences, undergo physical and chemical changes which naturally modify their characteristics. The hardest rocks, such as granite, are known to disintegrate and decay under atmospheric agencies alone. Marine, and particularly submarine, agencies are much more drastic in action. The attacks of sea-worms, the erosive power of currents, the dissolving properties of water, and the percussive action of waves—all these are sources of change and deterioration.

As far as possible, therefore, a foundation should be guarded from destructive influences. Even when the ground is naturally firm and durable, it is very desirable to protect the surface in the immediate neighbourhood of the breakwater from scour. To this end, in the case of upright walls, rubble and riprap are deposited alongside, so as to form an apron covering the toe, and, in more exposed cases, large blocks and monoliths are similarly utilised.

**Wall Foundations.**—Before dismissing the subject of foundations, we must make a few remarks on the manner in which they are prepared for breakwater piers of the upright wall type.

11

In all cases, it is essential to remove the surface coating of mud, ooze, and weed, which covers the sea floor. This may be done by dredging or with the aid of divers.

If the stratum thus exposed be sufficiently firm for the purpose, the breakwater pier may be laid upon it forthwith. Otherwise, it will be necessary to excavate further until a satisfactory base is obtained. In the event of the desirable stratum lying at a great depth, as revealed by borings, shafts may be sunk and the work built up in the form of piers inside them (fig. 131).

These shafts may consist of steel plating with walings of steel tees and angles, strutted with similar sections or with timber balks. They are usually so built, in lengths of convenient dimensions, such as can be handled by a crane or other lifting appliance. The lowermost length is fitted with a V-shaped cutting edge of hard steel. A sufficient number of lengths are bolted together to bring the shaft above the water level when resting on the sea bottom. Excavation is then carried on in the interior by grab buckets with

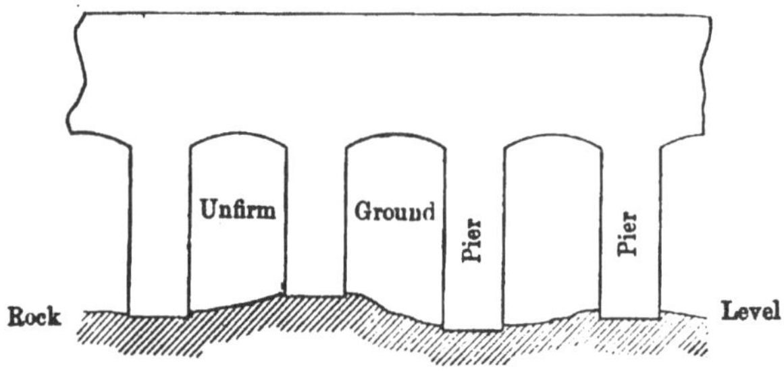


FIG. 131.—Pier Foundations.

frequent inspection by divers. As the shaft sinks under its own weight, combined with that of kentledge, additional lengths are added at the top. When the solid stratum is reached, the interior of the shaft is filled with concrete. The spaces lying between successive piers are arched over at or about the level of the sea floor (fig. 131).

Another method of transmitting the weight of a breakwater to a lower stratum, is by means of timber-piling driven at short intervals over the whole area of the site. The required depth must, of course, lie within the range of ordinary logs, say from 40 to 50 feet. Piles of greater length are expensive and difficult to obtain. When driven to their utmost extent, the heads of the piles are cut off by divers and cased in rich concrete (say 3 to 1) to a depth of at least 2 feet below the mud level in order to secure immunity from vermicular attack. A foundation layer of concrete may then be distributed over the whole area.

In certain circumstances it may suffice to inclose the site within sheet-piling; remove an upper layer of material, a foot or two in thickness, and deposit concrete.

**Limiting Loads.**—In all these cases it is necessary to bear in mind the limiting resistance to compression of the stratum founded upon. The following values may be adopted for use in ordinary cases.

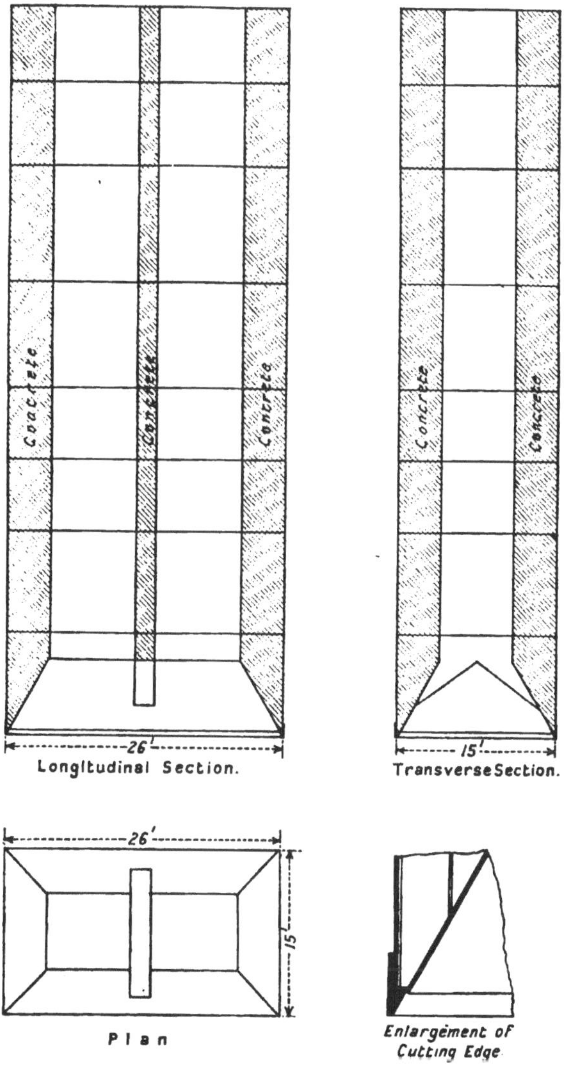


FIG. 132.—Foundation Caisson for Piers.

Concrete will safely stand from 10 to 15 tons' compression per square foot of area; hard rock from 9 to 10 tons; soft rock and stiff clay from 2 to 3 tons; and sand and gravel from 1½ to 2 tons.

Timber piles driven to a hard bottom will support a load of 10 cwts. per square inch of cross-sectional area; if dependent entirely upon the frictional resistance of the ground against its sides, and not upon basal support, the bearing power will vary with the perimeter of the pile; but in any ease not more than 2 cwts. per square inch of sectional area should be imposed.

The loads actually due to the substance of a breakwater may be computed from the following table, which gives the weight in lbs. per cubic foot of various minerals. When completely immersed in salt water, they lose 64 lbs. of the weight given; but there are circumstances under which the deduction is not justifiable, at anyrate for purposes of calculation.

APPROXIMATE WEIGHT PER CUBIC FOOT OF MINERAL SUBSTANCES.

| | lbs. | | lbs. | | lbs. |
|---|---|---|---|---|---|
| Basalt | 187 | Limestone—*contd.* | | Sandstone—*contd.* | |
| Brick | 115 to 135 | Purbeck | 150 | Talacre | 150 |
| Granite— | | Chilmark | 155 | York | 157 |
| Cornish | 164 | Kentish rag | 166 | Dundee | 159 |
| Aberdeen | 166 | Marble | 170 | Monmouth | 168 |
| Guernsey | 187 | Magnesian | 175 | Slate— | |
| Limestone— | | Masonry | 116 to 144 | Cornwall | 157 |
| Bath | 120 | Sandstone— | | Westmoreland | 173 |
| Portland | 130 | Red | 130 | Welsh | 180 |
| Chalk | 145 | Craigleith | 141 | Trap rock | 170 |

**Surface Treatment.**—The surface of a reliable natural foundation generally requires some treatment before it is ready to receive the first course of wall structure.

In rock there are always numerous cracks, crevices, and fissures, and a general unevenness of surface. Cavities and pockets containing soft material should be cleaned out and filled with concrete prior to extending the concrete over the entire site. To prevent lateral escape of the concrete while soft, it should be flanked on each side, temporarily, with bags of sand, planking on edge secured to iron pins driven into the rock, or by slabs of stone. Small irregular apertures may be staunched by packing with clay, or by covering with strips of jute or canvas.

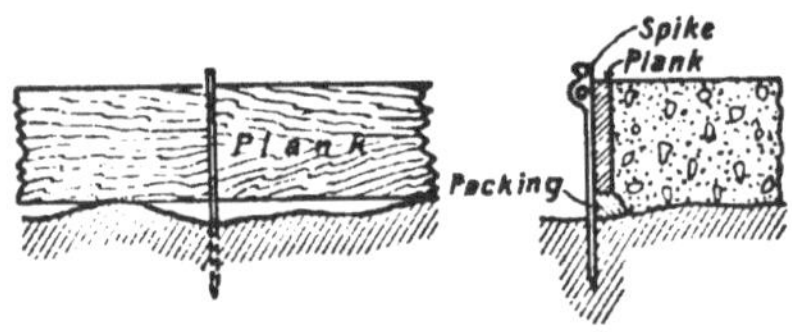

FIG. 133.—Moulds for Concrete Foundation.

Where the top of the rock, however, is not very hard, it may be found preferable to dress it down to a level surface, or to a series of benched beds of sufficient area to receive one or more blocks. Dips should likewise, where possible, be benched out to prevent any tendency of the wall to slide over the sloping surface (fig. 135).

The levelling of the surface is, of course, only absolutely essential for blockwork. For walls built of concrete in mass, though benching is desirable, all that is strictly needful is to erect the side-moulds within which the concrete is to be deposited. At first sight this would appear to be a simple operation, but the difficulties of setting temporary wooden moulds under water are anything but negligible. Where piling is practicable, a series of uprights may be driven at regular intervals, fitted with grooves within which panels of sheeting may be slid down, and raised again as the work proceeds. At the junction of the planking with the ground, a broad strip of canvas can be laid, forming a lining to the adjacent surfaces of each, for a width of 18 inches or

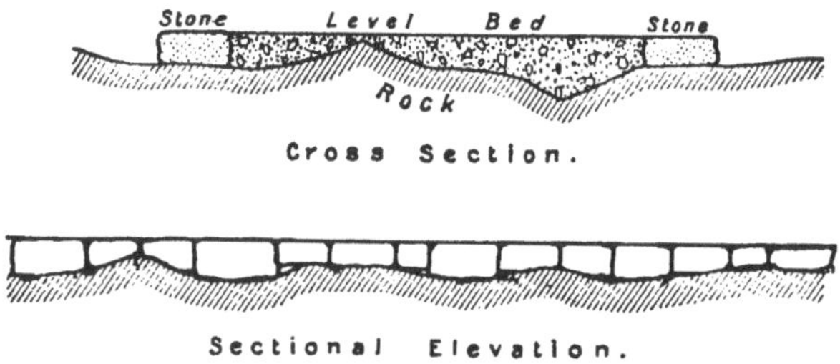


FIG. 134.—Concrete and Stone Foundation Work.

2 feet. The vertical portion will be backed to the planking, and the horizontal portion weighted to the ground with stone.

For mass concrete on a rocky bed, where guide piles are impracticable, the plan to adopt would be to lay external facing blocks and to deposit mass concrete in the space inclosed.

The deposition of concrete under water is an operation requiring the utmost care for its satisfactory accomplishment, the danger being that the cement may be washed out of the aggregate. It is useless, therefore, to

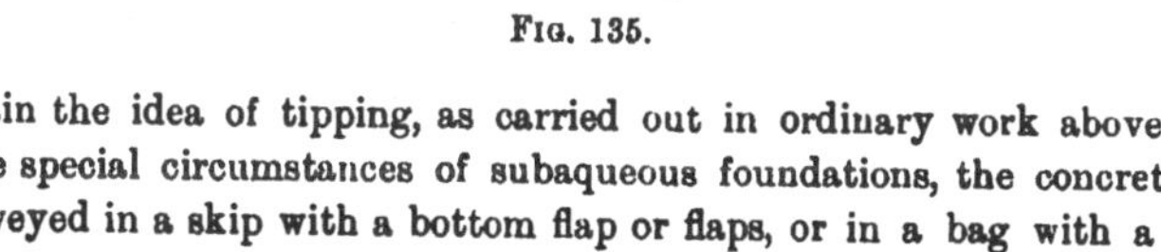


FIG. 135.

entertain the idea of tipping, as carried out in ordinary work above water. For the special circumstances of subaqueous foundations, the concrete must be conveyed in a skip with a bottom flap or flaps, or in a bag with a double mouth, that at the lower end being temporarily bound with a looped rope, capable of being released by a tripping rope. The skip, or bag, is lowered right to the bottom, or as near thereto as is consistent with discharge, and the contents are allowed to flow quietly into place, with as little manipulation as possible. It will be evident that concrete for submarine work should be rich in cement, say 4 to 1.

One caution to observe in tidal situations is that of so arranging the periods of deposit that a mass of freshly placed concrete may not be subjected, while setting, to the disturbing action of a choppy sea surface. Less than 18 inches or 2 feet of water is insufficient to prevent even small waves from exercising a deleterious influence, chafing the concrete, and robbing it of its cement. Therefore, wherever possible, advantage should be taken of the variation in the tidal level, during springs and neaps, to suspend concreting from time to time, either at some little depth below the surface or altogether out of range. A quick-setting cement will prove of considerable value in tidal situations.

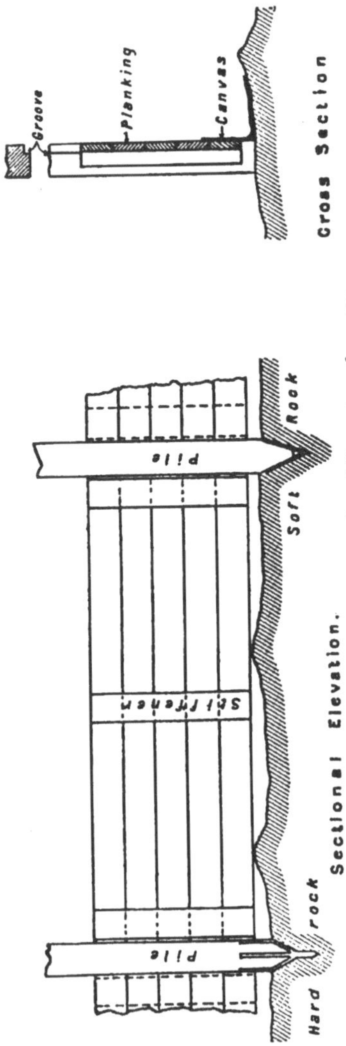

FIG. 18. —Moulds for reception of Concrete under water.

It is needless to remark that when operations are carried on within the shelter of a diving-bell, the same restrictions do not apply, although it must be admitted that sudden outbursts of air may do more damage than the fretting action of waves. Yet, with care, these may be avoided. The work is then only limited by the convenience of arrangements in regard to shifts.

**Bagwork.**—The dispersive action of waves, and, indeed, the solvent action of comparatively still water, has led to the introduction of a system of concrete laying in sacks or bags. These bags have, in certain cases, attained a very considerable size and weight, the latter reaching 100 tons and over. But small bags of 5 or 10 tons, or thereabouts, are most common. They are often employed for regularising the surface of an uneven bed destined to receive blocks. The bags are of jute or canvas, strongly made. After being filled with concrete, they must be deposited immediately, while the material is plastic, so that each bag may adapt itself to the inequalities of its environment. This adaptation of bulk to various positions is one of the chief advantages claimed for bagwork. The system is not without certain drawbacks; the bags are liable to burst. This, of course, could be remedied by strengthening and improving the sacking. Moreover, the bags cannot be brought to a perfectly level surface; neither can they be compacted very closely in successive rows; and further, they are liable to work loose and be sucked out by the sea, or, failing that, the ends may be broken off by waves. These defects, however, are not vital; careful setting will go far to minimise them; and many breakwaters in existence have been partially, or wholly, constructed of bagwork.

The jute sacking generally used for the purpose weighs from 25 to 30 ozs. per superficial yard.

**Block-making.**—The making of blocks for breakwater building calls for little explanation. The blocks are built of concrete in moulds at a block-yard adjacent to the site. The weights range from 5 tons upwards, according to the capacity of the setting machine. There is, however, practically no limit to size, since huge monoliths may be deposited by special means. At Dublin a wall has been built with foundation blocks weighing 350 tons each, while the caisson blocks at Zeebrugge weighed no less than 4500 tons each.

Blocks should be allowed to mature for at least a couple of months before depositing in position, though they may be removed from the moulds at the end of a fortnight. The season of the year and the temperature produce variations in the time of maturing.

In order to facilitate the placing of blocks, they are usually constructed with two vertical or slightly inclined perforations, through which are passed iron bars with T or angle ends, capable of engaging against the underside of the block when turned through a right angle. When the blocks are very heavy, the T heads should be provided with hard wood or iron-bearing surfaces to prevent the concrete from suffering damage.

**Bond.**—The problem of bond in breakwater construction is a difficult one. Theoretically, the effect of introducing a system of interlocking is to materially strengthen the breakwater by binding together, in close association, the separate elements of which it is composed. Practically, there are the consequences of unequal settlement to be considered, whereby the sinking of any part of the breakwater will probably fracture the blocks connecting that part with the portions adjoining. The evils attaching to such a contingency can only be averted by discarding the idea of bonding horizontally. Vertical or sloping joints then become inevitable. The breakwater can be connected longitudinally, wherever this is done, by means of dowel- or joggle-joints. Such connections offer no resistance to settlement. A *dowel*-joint consists of a square-shaped aperture, set diagonally, one-half or a V-shaped portion being cut out of each of two stones. When these are joined together, the aperture is filled with a piece of stone of diamond section or with concrete. A *joggle*-joint differs only from a dowel-joint in that the connection is formed by a projection on one piece fitting into the aperture in another. It is the stronger method of the two, but more expensive, because a considerable quantity of one stone has to be cut away in order to form the joggle.[1]

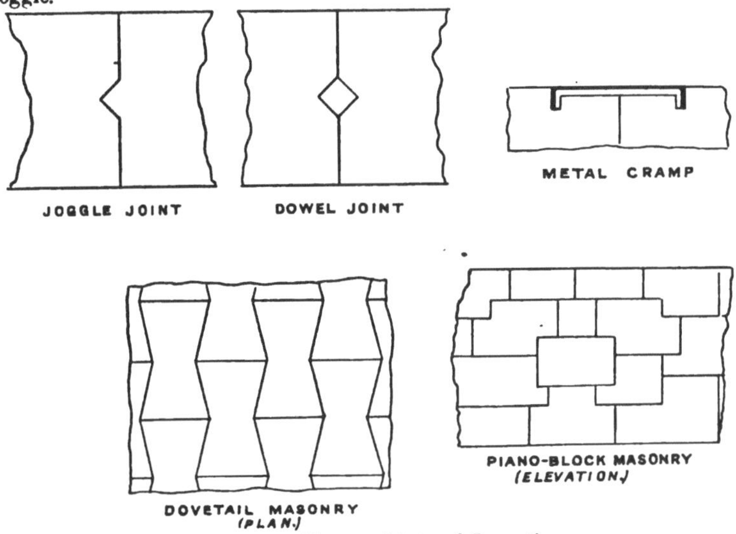


FIGS. 137-141.—Masonry Joints and Connections.

*Dovetailing* is another method of bonding masonry, but it necessitates much elaboration in fitting and jointing, and therefore is costly. It has been very often used for lighthouse work.

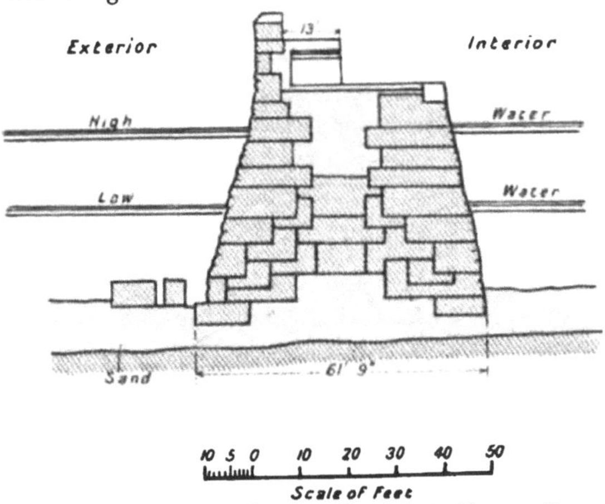


FIG. 142.—Piano-blockwork at North Jetty, Tynemouth.

Somewhat similar, though perhaps less intricate, is the system of "*piano*" *blocks* used by Mr Messent in the construction of the Tyne breakwaters. The

[1] The distinction between a dowel and a joggle is, however, not always observed, and the terms are often treated as synonymous for either class of joint

object of these was to prevent the sliding of the blocks transversely when struck broadside by a heavy sea. The system, however, did not prove a success, and is unlikely to be repeated.

Both the foregoing arrangements are costly and tedious. The object in view may be achieved to some extent by means of *bed-plugs*. These are projections of stone or iron standing up above the level of one course of blocks and fitting into apertures in the underside of the next course.

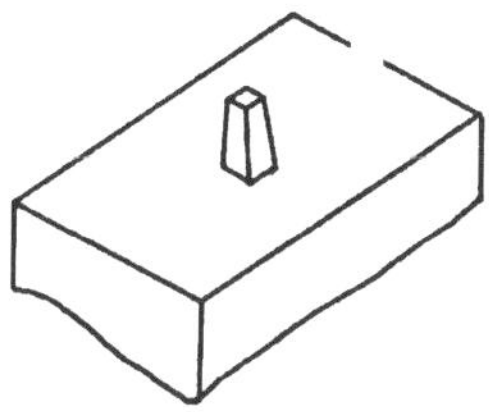

FIG. 143.—Bed-plug.

*Metal cramps* form an effective connection, provided they are protected from the possibility of rusting and corrosion. This can only be satisfactorily realised by bedding them below the surface of the stone and completely inclosing them in Portland cement.

**Sloping Bond.**—The term sloping bond has been applied to an arrangement of blocks whereby they lie tilted on end a little out of the vertical—the angle of inclination varying from 80 to 60 degrees, or rather less. By this system the blocks are fairly free to slide, in case of settlement, without disturbing the adjoining courses. When, however, as in many cases, dowelling and bed-plugs are also introduced, this freedom of action exists only to a restricted extent, the frictional resistance to movement being considerable.

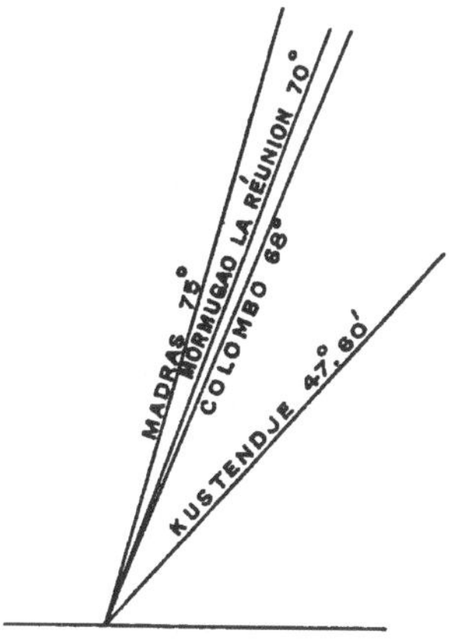


FIG. 144.—Sloping Bond of Breakwaters. See also fig. 129, p. 157.

The horizontal bonding of blocks — of dubious advantage, as it is in the upper part of a breakwater where the blocks can be accurately adjusted, and the bed-joints well flushed with cement—is a matter of almost positive harm in the courses which lie below water level, where, in most cases, blocks have to be laid without bedding, and where the joints are left open. It is manifest that, under such circumstances, the blocks are not bearing equally on their beds, and it is readily conceivable that a long block extending over three others in a lower course might only be supported at each end. The risk of fracture would then be very great.

During the progress of the work, and especially at the commencement of a winter season, or other period when operations are intermittent or entirely suspended, care should be taken to see that the end blocks of the work actually executed are amply secured.

**Grouting under Water.**—The joints of work under water may be filled, to a certain extent, by means of grouting from the surface. A pipe or tube is arranged so as to communicate with the part proposed to be dealt with, and through this tube, under a considerable head or under direct pressure from a ram or piston, fluid concrete is forced into all the adjacent cavities. To prevent the escape of the concrete, however, all the face joints must necessarily be closely caulked. This has been done in certain cases by forming slightly dovetailed grooves near the outer edges of the blocks and ramming them with rolls of canvas containing neat cement, as shown in sketch (fig. 145). These were allowed to harden, and the joints packed with shingle before grouting.

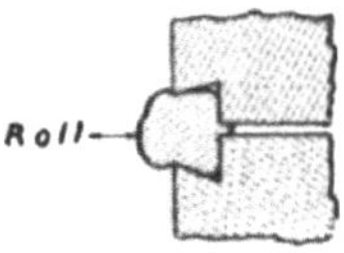


FIG. 145.—Joint Packing.

In the case of wide joints, the apertures may be faced with brickwork in cement, or with small bags, containing neat cement, stacked compactly.

The concrete for grouting purposes should not be too fluid. Other materials are also used, such as clay worked up with hydraulic lime, and sand mixed with iron filings and sal ammoniac; but Portland cement concrete, on the whole, is preferable.

**Minor Breakwaters.**—All breakwaters are not planned on the same scale. Massive construction may be both necessary and possible in the case of leading harbours and ports. But there are also small harbours, where any great outlay on protection works is out of the question, and where some expedient must be contrived for affording reasonable protection at moderate cost. It is both interesting and instructive, therefore, to consider the steps which have been and may be taken to meet these cases.

**Crib and Box Breakwaters.**—The submersible caisson of steel and concrete has had its prototype in a long, wooden box, floated out into position and filled with rubble. Such was the form of breakwater adopted in many early instances, and still practised at some ports on the Baltic seaboard. These boxes or cribs were braced, at intervals of 8 to 10 feet, by transverse partitions, and, in wide boxes, there were often one or two longitudinal partitions as well. Floors of planking were arranged in several—about three—tiers, with a charge of stone incased in each. Both the outer casing and the inner partitions were constructed in solid timber.

"Stone cribs of this type of construction constantly required repairs, since, apart from scouring, the terminal boxes were damaged by every strong sea; the planks and the upper balks were torn off, although secured by means of strong iron bolts, and the stones were hurled about."[1]

The dams were strengthened, as far as possible, by driving piles through the inclosures, in two rows with cross ties, and by depositing a mound of huge stones in front of the seaward face. But no measures proved completely satisfactory. Breakwaters such as these could only be employed in comparatively shallow and but moderately exposed positions. Depths of 15 feet of water probably mark the limit to which they may be advantageously applied

Somewhat similar to the foregoing are the timber cribs used on the North American Lakes. They are box-shaped frames of timber constructed in open

[1] Anderson on Breakwaters, *Proc. Int. Nav. Cong. Milan*, 1905.

work, with numerous compartments formed by means of transverse and longitudinal ties. The compartments form receptacles for stone rubble. From the crudeness of their build, these cribs can only be looked upon as of the nature of temporary structures. They are referred to in somewhat greater detail in *Dock Engineering*.[1]

**Fascine Work.**—Another form of construction adopted in certain localities for moles and breakwaters is known as fascine work, and consists of bundles of brushwood arranged as mattresses, which are sunk in position in successive layers and weighted with stone. Piles are then driven through the mattresses into the sandy bottom to prevent displacement. In process of time the interstices of the mattresses become filled with sand and drift, forming a solid mass. The system is more particularly characteristic of the low-lying coasts of Holland and Denmark, though it is also to be found on Prussian shores. Fascine mattresses are also described at some length in *Dock Engineering*,[2] and they are alluded to in Chapter IX. of the present volume in connection with channel training-works; but, as in the previous instances, they have exhibited no great resisting powers to rough seas.

The following is a description of some early breakwaters on the Baltic littoral:—

"The fascine dams consisted, according to the depths of water, of one or several layers of fascines 3 feet to 4 feet thick, which were floated down from the inner harbour where they had been made, and sunk on the spot. The upper layer was afterwards covered with a packing and with a stratum of small stones and rubble, about 3 feet in height and rounded on the top. This cover was paved with large, approximately cubical, stones of granite. The capping, which had a width of about 13 feet and was slightly arched, was scarcely 6 feet above mean water level. The slope from the capping down to the outer edge was 3 to 1 on the sea side and 2 to 1 on the harbour side. The thickness of the stone paving was 3 feet in the capping and 2 feet elsewhere. The stone layer was further secured by strong oak piles from 7 to 10 feet long and 6 inches square, called caisson piles; they were driven at distances apart of 6 feet along the edge of the capping, and of 18 inches along the water-line.

"These fascines were exposed to heavy damage, for every storm from the sea lifted the paving stones of the slope, especially at the head and on the sea side, from their seats, and carried them inland, or hurled them up the slope and over the mole into the inner harbour. The reason for this was mainly that the stones did not lie sufficiently close upon the flat slopes, and that they could be loosened separately and disturbed by the waves, lacking sufficient weight in themselves to resist this action. In order to make the surface of the slope as plane as possible, with a view to avoiding points of attack for the impinging waves, the stones had been placed with their roughly hewn, approximately square, heavy portions—that is to say with their bases—upwards, and with the tapering parts downwards. In this

[1] pp. 286, 287. [2] pp. 282 *et seq.*

position they were only secured by being packed underneath with smaller stones and by leaning against one another. The stones were not further fixed, therefore, and although they were as closely packed on the surface as was possible, the remaining gaps, especially between the lower portions of the stones, afforded the waves sufficient front for attack. It was for this reason that the joints of the stones were at a later period closed with concrete, when a quiet sea and low water permitted such operations. These measures diminished the destruction, but did not by any means prevent it; for the chief trouble arose from the insufficient loading of the stones, which could not be altered, and the part of the structure which was most exposed, *i.e.*, the toe on the sea side, could not be strengthened and properly secured. The toe of the slope on the harbour side could be strengthened by forcing in several layers of balks behind the pile walls, and it was against these balks that the cubical stones of the slope were resting. Nothing of the kind, however, was possible on the open water side. Rubble mounds were useless; for the stones were driven inland or raised up the flat slopes of the moles into the harbour. There was nothing left, therefore, but to place further fascines in front of the slope to fill up the hollows formed alongside the mole, and to restrict further injury to the pavement."[1]

Another simple method of breakwater formation is to drive a double row of piles and fill the intermediate space with rubble, the piles being retained in position by longitudinal walings and transverse ties. The piles may be either of whole timber or of iron. In one case, at Touapsé on the Black Sea, railway metals were used for the purpose. On account of the disruptive tendency of the hearting, it is necessary to have good stout piling with strong transverse pieces capable of offering ample resistance to the lateral pressure imposed upon them. The piles are sometimes driven in an inclined direction, pointing inwards from the bottom towards the top so as to increase their stability.

## Examples of Breakwater Construction.

**North Pier at Tynemouth.**[2]—"The new length of pier (1500 feet long) is being made of Portland cement concrete blocks bonded from side to side of the pier, no mass work being used except above high water level. The heaviest blocks weigh (in air) from 30 to 40 tons, and those exposed to the sea are faced with Aberdeen granite. The blocks below low water are built without mortar joints, but they are interlocked by round joggles and other means to such an extent as to render relative movement among them impossible. Above low water the blocks are built with mortar beds, the joints being also grouted up. The material overlying the new foundations is excavated by means of grabs, and as soon as the grab has worked down to the shale, a diving-bell is used to level the bed for the blocks.

[1] Anderson on Breakwaters, *Proc. Int. Nav. Cong. Milan*, 1905.

[2] Barling on Tyne North Pier Reconstruction, *Min. Proc. Inst. Mech. E.*, Newcastle Meeting, July 1902.

"The diving-bell now in use is 12 feet long by 9 feet wide by 6 feet high, and four men work in it at a time. The pressure, which, of course, varies with the depth of water, is about 20 lbs. on the square inch above that of the atmosphere, and up to now (1902) there has been no case of sickness due to working under air-pressure. When the bed is prepared, the blocks are set by helmet divers, and as great care is taken to get them level and true as if they formed part of an architectural structure above water. The reason for commencing the work at some distance seaward of the junction, was to admit of the work being carried on at two faces, and thus extended seaward and shoreward simultaneously."

**Breakwater Construction at Alderney.**—The following detailed description of the operations in connection with the construction of the wall or superstructure of Alderney breakwater will be found extremely instructive. It is quoted from an account[1] by Mr (now Sir) John Jackson, who acted as contractor's agent on the works for a period of nine years. The breakwater unfortunately subsequently acquired an unenviable reputation on account of the large annual expenditure incurred in its maintenance.

"In building the walls, as no machinery of any kind could remain out during the winter, the works had to be recommenced every year. The first operation was taking down a machine called the 'Samson,' invented in Alderney. This was like a railway turn-table on wheels, with balks of timber 76 feet long, placed across and over-trussed; one end, which projected past the side of the table farther than the other, was called the jib, and at the other end was the balance-weight. A double-purchase crab was fixed in the centre, which worked a chain over a travelling sheave near the end of the jib; and the whole revolved on the under frame. The gauge was 15 feet: just the space left between the travellers or gantries spanning the sea and harbour walls. This machine was capable of lifting a weight of 4 tons in the water at a distance of 30 feet from the outside edge of the turn-table. The first operation was to stretch the jib outside the end of the previous season's work, the foreman labourer standing on the outer end. This man held a copper wire attached to a large cast-iron plumb-bob, which, at slack tide, he let down to the place where the first pile or upright of the stage was to stand, and at this spot a helmeted diver excavated a hole in the bank to receive the pile, to the end of which was fastened a stone weighing 15 cwts. When the hole was excavated to the required depth, the divers retired, and the pile was lowered by the Samson into its place. Four piles were set in a row 30 feet apart, and longitudinal-trussed beams, 2 feet 4 inches by 1 foot 2 inches, were placed from pile to pile and formed a bay of staging, which consisted of 1050 cubic feet of timber and 3 tons 8 cwts. of wrought iron in trussed rods, knees, bolts, straps, etc. The carpenters erected a bay of staging in a week. When the stage had advanced seawards three lengths or bays, six travellers or gantries, each capable of lifting 20 tons, were taken down the wall—an operation performed by the carpenters generally in one day. The gantries spanned the sea

[1] *Min. Proc. Inst. C.E.*, vol. xxxvii. p. 87 *et seq.*

and harbour walls, and the space between was occupied by two lines of railway for conveying men and materials; and in this way the whole width of the top of the stage, viz., 70 feet, was occupied.

"About the middle of May in every year the first block was lowered by the helmet divers' gantry to its place. The helmet divers' stage was suspended by iron rods from the beams of the main stage, and hung about 10 feet below it. To this stage wrought iron ladders were attached for the convenience of divers descending to their work. Six divers were under water together—four on the seaside and two on the harbour side. They remained down four hours at a time, when there was a shift; and there were three shifts in the day. The life-line men and pumpers remained on the work all day, but the pumpers were relieved every half-hour. The divers' apparatus and the stage were removed every night, so treacherous was the sea, for even in summer it was not safe to leave anything at the level of the divers' stage; but at the height of the main stage, 10 feet higher, or 20 feet above high water, the sea seldom disturbed anything. The mode of bringing the work up was by taking advantage of the spring tides; thus it was expected of the divers that, in a fortnight, they would bring the diving work up to the level of low water, for a distance seaward of 60 feet, ready for the masons. Whatever excavation was required for the lowest course, it being a great deal more at some times than others, or however rough the sea had been, the divers never failed to prepare a length of 60 feet; but they frequently went down a second time for an extra shift to accomplish this. The average day's work of a diver in the year 1860 was 8½ cubic yards of building; in 1861 it was 11 cubic yards; and in 1862 it was 14 cubic yards. Their work was excavating foundations, receiving the granite face-stones for the sea-wall, and setting the granite and concrete blocks of the sea and harbour walls. The stones and blocks were speedily lowered by a single chain 45 feet from the top of the stage by the gantry crabs and a rope-break. This latter was a piece of rope, with rope yarn twisted round, made fast to the frame of the crab and then fastened to the pinion shaft of the single and double gear. The chains employed were of the best charcoal iron $\frac{7}{16}$ inch thick, and they were only used for two seasons.

"The divers' work being ready for the masons, to the level of low water of spring tides, two days before full or new moon, the resident engineer gave orders to commence setting the face-stones in cement, if, in his opinion, the tide fell low enough; otherwise he could stop the works. The masonry of the breakwater on the sea face was of granite and native Alderney stone; on the harbour side it was of native stone, and the space between was filled in with backers and small rubble. The backers were as large as the machinery would lift, and were set in Medina cement and sand, in the proportion of 1 part of cement to 2 parts of sand. Sand suitable for building occurred in abundance in the island. No better cement could have been had for the purpose, for very often the masonry, ten minutes after it was built, was covered with water. No large stock of cement was laid in, as its quick setting qualities were impaired by time. The masons endeavoured to keep the work done at each tide as long as possible out of the water, and to get a course closed up in one operation. When the masonry had been brought up three courses, the tide was considered to be mastered. The helmet divers had two gantries appropriated to their work, and the masons had four. The masons came out at noon and at midnight, and it was always low water at that time; therefore nearly half of the wall above low water was built in the night. The average day's work for each mason was 6⅓ cubic yards, and no piecework was permitted."

The diving work was always ended in August, and the building up to the quay level in September, while the promenade was generally finished before Christmas. The rate of progress gradually increased from 300 linear feet to 600 linear feet of wall per annum.

**Mole or Jetty at Zeebrugge.**[1]—The outer portion of the jetty at Zeebrugge is constructed in solid concrete, and it has a foundation composed

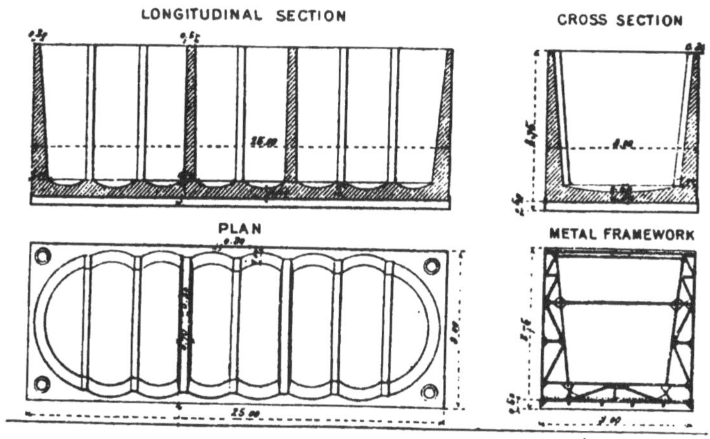


Figs. 146–149.

of monoliths, or blocks, 82 feet long by 24¼ feet wide in the portion flanking the quay, and by 29½ feet in the portion beyond, the heights ranging from 23 feet to 30 feet according to the level of the foundation bed.

These concrete blocks are inclosed in an iron framework caisson, which served as an outer shell. The plating in sides and floor is ⅛ inch thick. Lattice beams 3¼ feet high stiffen the bottom; eighteen frames strengthen the sides; and on the underside is a rim or edge 18 inches deep.

The caissons were manufactured in the workshop adjoining the site, the various parts being conveyed to the block-yard at the inner harbour by means of railway trucks, from which they were unloaded by an overhead land traveller or gantry.

[1] *Vide* "Les Ports et le canal maritime de Bruges," par M. L. Coiseau. *Mémoires de la Société des Ingénieurs civils de France. Bulletin de décembre*, 1904.

This land traveller was a kind of movable bridge running on a double track. Its extreme span was 220 feet, the length comprised between the supports being 66 feet, and the two cantilever arms 66 feet and 88 feet respectively. It served five rows of caissons.

The caissons were put together on wood blocks or packings, but, as soon as the rivetting was finished, these were removed and the caissons lowered to the ground.

At this point, concreting work was put in hand, commencing with the floor. The concrete was composed of 3 parts of broken stone, 3 parts of sand, and 12½ lbs. of Portland cement to the cubic foot. It was mixed by an electric motor mixer, served by a crane which lifted a box containing the dry ingredients and tipped them into the hopper of the mixer. Hence there was a gradual sliding progression into and through a cylinder working with a rotary movement and slightly inclined. The materials were turned over and mixed—dry throughout one-third of the length and wet throughout the remaining two-thirds, the water being administered through a central perforated tube. The output could be regulated at will.

By the time the concrete had reached the extremity of the cylinder, it was thoroughly incorporated and all the stone well bedded in mortar. It was then allowed to fall into compartments on small wagons, and conveyed along double tracks on to a traveller of the same form and range as that already described. The wagon boxes were tipped into shoots attached to the traveller, and set over the caissons.

While the concrete work of the floor of a new block was in hand, a large framework "mould-stripper," consisting of a stage resting on strong cross-beams supported by two upright frames the same distance apart as the frames of the concrete mixer and running over the same track, removed the moulds from the blocks previously finished, with the assistance of two electrically-worked derrick cranes.

The concrete mixer next turned back and concreted the partitions, and, this being done, both it and the mould stripper were free for another block.

The blocks, as was remarked, were constructed in the inner harbour and a branch dock, which was emptied for the purpose of these extension works. As soon as the new lock and its entrance channel had been completed, water was again admitted, and the blocks, being finished, were ready to be towed to their allotted positions.

The first block was set in place on 20th May 1900, and two others succeeded it before the end of the year. The work was then interrupted by a severe storm on 27th January 1901, and was not resumed until 20th October following, in consequence of the damage which accrued to a framework jetty connecting the mole with the shore.

The operation of setting the blocks in position was as follows :—

In the first instance the blocks, as stored in the inner harbour, were allowed to fill with water to keep them stationary. Their sides projected from 15 to

30 inches only above the surface. When about to be moved, they were emptied by a centrifugal pump, which was inclosed in sheet iron casing to protect it from the water, for it had to be sunk below the surface to avoid the necessity of charging it. As the process of evacuation proceeded, two rows of props or stays were inserted in the caisson at right angles to the sides at each counterfort. These stays were destined to resist the external pressure of the water, the side walls of the caisson being insufficiently strong in themselves.

Next was placed on top of the caisson a stout log of timber 60 feet long and 15½ inches square, and it was firmly secured to anchorages in the concrete. The object of this was to assist the alignment of the caisson.

Two longitudinal timber walings were placed along the sides of the caisson and connected by through bolts, forming transverse ties. This was necessary to strengthen the side walls in case of an excess of internal pressure, and to prevent them from yielding outwards in case of invasion by waves in a rough sea.

Finally, a huge steel cable encircled the caisson and formed a means of attachment to the tug.

The block was thus in every way equipped and ready for towage. It floated easily. The hollow spaces in its interior were adequate to counter-balance the weight of the concrete and framing, and to permit it to emerge from 2 feet to 2 feet 6 inches out of the water.

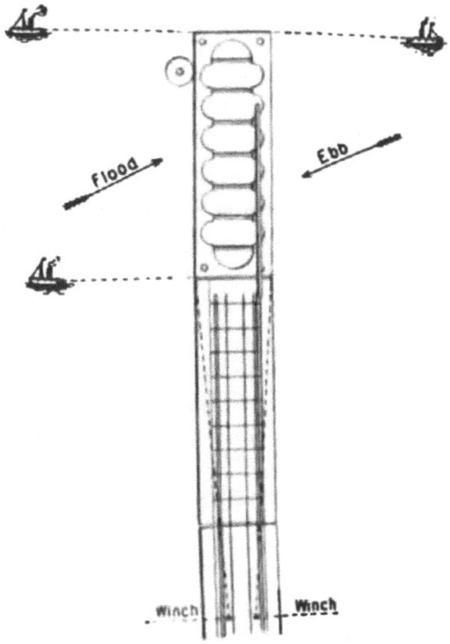


FIG. 150.—Aligning Caisson, Zeebrugge Harbour Works.

It was towed through the lock by a tug of 300 H.-P. At the outer gates a second tug of the same power placed itself in tandem. In the open water a third tug joined the rear of the procession so as to prevent the block from yawing. Departure took place about the time of high water. At this time the flood current is still very strong, and the caisson presenting its broadside thereto tended to be drawn in a direction away from the jetty. About an hour and a half was necessary for each journey.

It was found that on the way the caisson often shipped quantities of water, and so wood coamings were provided to increase the freeboard and prevent the incursion of waves. This was the more necessary during that period of the work when the roadstead was very exposed. As the work progressed, more and more shelter was obtained, and the blocks were finally able to travel in perfect tranquillity.

When a block had reached the extremity of the finished portion of the jetty, one end of it was moored up thereto by cables which led back to winches on the jetty, and the flood-tide commencing to slacken, the block was gradually swung round by the tugs and brought into proper alignment. Its position was accurately determined by two guiding elements: one, on the right hand, the long beam already alluded to, and arranged so as to bear against the jetty structure; the other, on the left, at the outer extremity, an enormous truncated cylinder of concrete 13 feet high, weighing 55 tons and resting on the sea bottom.

As soon as the slack of the tide arrived—it only lasts a little more than a quarter of an hour,—the plugs which closed three orifices in the sides of the block were removed, and water flowed into the interior of the caisson, which gradually foundered. The process of sinking was often hastened by the water pouring over the side walls until the caisson finally disappeared from view amid a swirling sea. The sight was remarkably impressive.

The caisson reappeared above the surface at low water, when 3 feet or so of it became visible. It was freed from its temporary strutting, guide beam, and moorings.

The filling in of the compartments with concrete commenced immediately, and was continued to conclusion without cessation, except during the period of strongest tidal run towards high water.

The concrete was manufactured in a yard situated at the junction of the jetty with the shore. At this point were assembled immense banks of gravel from the Rhine, and quantities of Portland cement from Cronfestu, Niel, Tournai, and Les Laumes. The stores for the cement consisted of three galvanised iron buildings, covering a superficial area of nearly 1100 square yards.

Four electric concrete mixers, each manufacturing 260 cubic yards of concrete per day, were located at each side of the stores; they were fed by an electric crane, and water was laid on to each from a water tower.

Suitable sidings permitted waggons carrying skips to present themselves under the shoots of the concrete mixers. The gravel and cement duly measured were deposited in layers in these skips; the crane lifted the skips, swung them, and emptied their contents into the hopper of the mixer.

The completed concrete was discharged into skips holding 12 cubic yards each. These were carried on special trucks drawn by ropes from electric capstans.

When four skips had been marshalled, a locomotive took them in charge and conveyed them to the extremity of the jetty. Here a Titan crane or a pair of sheer legs lifted them from the trucks, carried them over the caisson, and lowered them to the bottom, where they were automatically discharged without loss or disturbance. After each discharge the skip was withdrawn and replaced on the waggon.

FIG. 151.—Sinking a Caisson at Zeebrugge Harbour Works.

FIG. 152.—Titan at Zeebrugge Breakwater. [*To face p.* 178.

Upon the foundation course thus prepared was raised the body of the jetty, which comprised three courses of concrete blocks, each 16½ feet long, 8¼ feet broad, and 6½ feet high, weighing 55 tons. The concrete of which they were composed was rather richer in cement than that previously described, in that it contained 14½ lbs. of cement per cubic foot instead of 12½ lbs.

**Breakwater at Cette.**—New spurs extending the existing breakwater at this port at each extremity of its length were constructed between the years 1881 and 1895. The foundation is of a very shifty nature, being sand exposed to considerable scour. Such action inevitably tends to settlement and dislocation in any structure built upon it. Still, the problem had to be faced, and the system adopted was as follows. Upon the treacherous base was deposited a mass of small riprap and rubble, the pieces ranging up to a weight of 440 lbs. and forming a core about 80 feet wide by 13 feet thick (fig. 153). Above this there is a layer, one-half that thickness, of larger rubble, the largest lumps of which attain a weight of 4 tons each. This layer extends seaward of the riprap core for a distance of about 60 feet, and thereon is laid in regular horizontal courses artificial blocks having a volume of 700 cubic feet each.

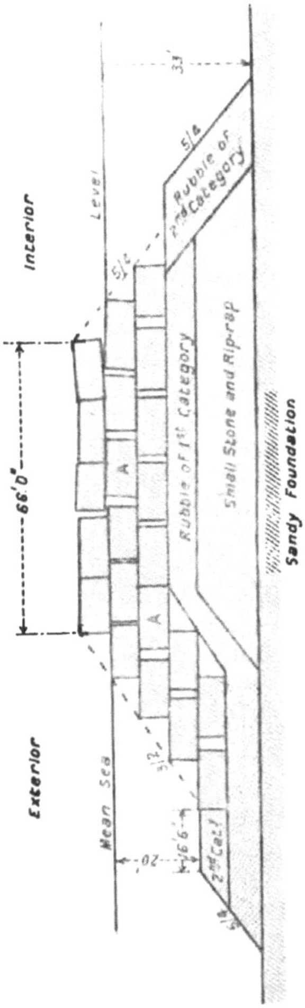

Fig. 153.—Section of Cette Breakwater.

The blocks were laid by floating derricks, and were so placed as to have their longitudinal axes perpendicular to the line of the breakwater. They are not in actual contact with one another, but the spaces of 24 or 30 inches between them have been filled up with masonry and concrete so as to form an unbroken front.

The lowermost two courses of blocks, however, on the sea side were simply tipped into position, the upper surface being roughly levelled by means of rubble filling. Altogether, the blocks were not adjusted with the precision which is characteristic of similar breakwaters elsewhere — at Genoa, for instance.

At first it was intended to surmount the whole structure with a blockwork parapet, but this idea was abandoned, as the addition would probably have reacted detrimentally upon the stability of the outer slope by increasing the recoil of the waves. As it is, the sea overrides the breakwater in rough weather, and this, combined with an unstable foundation, produces movements in the outermost blocks. Voids are created, and these have to be filled with fresh blocks tipped as closely as possible into position. Hence, the seaward face is losing, to a very large extent, its arrangement in regular courses, so that the desirability of creating, or of attempting to maintain, anything of the kind with detached blocks in an exposed position, is open to question.

The cost of depositing the blocks was as follows:—

Tipped overboard, 57s. per block.

Set without regular coursing, 78s. 6d. per block.

Set and coursed regularly, 92s. 4d. per block.

The cost of the breakwater complete ranged from £75 to £90 per foot run. The cost of replenishing the blocks on the outer slope forms a current charge of about eight guineas per lineal yard per annum.

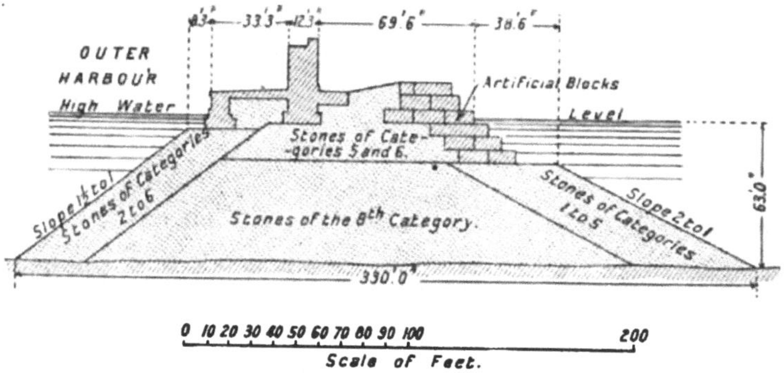


FIG. 154.—Section of West Mole at Genoa.

**Breakwater at Bilbao.**—The protection works in Bilbao Bay, at the mouth of the River Nervion, afford an illustration of the method of construction by caisson monoliths. The example is the more interesting in that the original design for the breakwater was very materially modified under the severe experience gained in the course of its formation.

The original design is shown in fig. 155. It comprised all the features associated with breakwaters of the mixed type, viz., an inner core of small, with an outer layer of large, rubble stone surmounted by large concrete blocks deposited at random, upon which was to be erected a superstructure of mass concrete with blockwork facings, and an upper parapet wall. The foundation consisted of mud and sand except in the parts immediately adjacent to the shore where the rock was exposed. The artificial blocks contained from 40 to 65 cubic yards each, and brought the level of the work up to low water line of equinoctial tides. Begun in 1888, the substructure was allowed to settle for a couple of years before any additional weight was imposed upon the foundation.

In 1891 the superstructure was commenced, and in 1893 indubitable evidence was given of the prejudicial, and even disastrous, influence which it exerted upon the breakwater as a whole. The waves striking against the vertical face of the wall fell back with great force upon the top of the mound, disturbing the blocks and laying bare the rubble core, which was then easily washed away. The experience was renewed and confirmed in the following year by a storm which has already been alluded to.

Accordingly, a change of plan was decided upon. The superstructure was commenced at a level of 16½ feet below its former level. Most probably this in itself would have been insufficient to secure immunity from undermining, resulting from the collapse of waves and their back-draught, had there not been additional shelter afforded by the setting back of the line of the wall

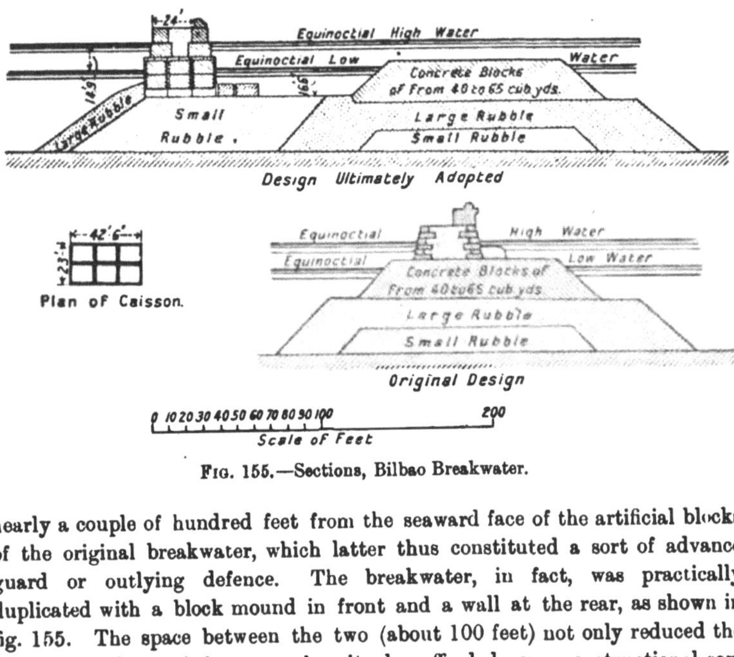


FIG. 155.—Sections, Bilbao Breakwater.

nearly a couple of hundred feet from the seaward face of the artificial blocks of the original breakwater, which latter thus constituted a sort of advance guard or outlying defence. The breakwater, in fact, was practically duplicated with a block mound in front and a wall at the rear, as shown in fig. 155. The space between the two (about 100 feet) not only reduced the force and violence of the waves, but it also afforded some constructional convenience by providing room for a tugboat to work and facilitate the building of the wall.

To render the wall as solid and homogeneous as possible, it was decided to build it with the aid of framed and plated caissons. The dimensions adopted for these were 42 feet 6 inches long by 23 feet wide by 23 feet deep. The weight of the caissons was 30 tons, and they had a light draft of 12½ inches; but before actually towing them into position (they were constructed on the river bank), they were ballasted with a layer of Portland cement concrete 5 feet thick, which increased the draught to a little over 11 feet. When sunk in position, they afforded a margin of 6 feet 6 inches above low-tide level. The material of the caissons was Bessemer steel in plates ¼ inch thick, strengthened internally by longitudinal and transverse bulkheads of lattice-work so as to form six equal compartments. In each of these compartments were subsequently placed two concrete blocks of about 40 cubic yards volume.

The rubble work bed on which the caissons were to rest was levelled ready to receive them through the agency of a diving-bell. The time occupied in this operation varied from one to two days. The caisson, having been ballasted, as already stated, was then towed into position during the last two hours of an ebb-tide, so as to allow as much time as possible for its adjustment and settlement. Sinking was effected by pumping water into the

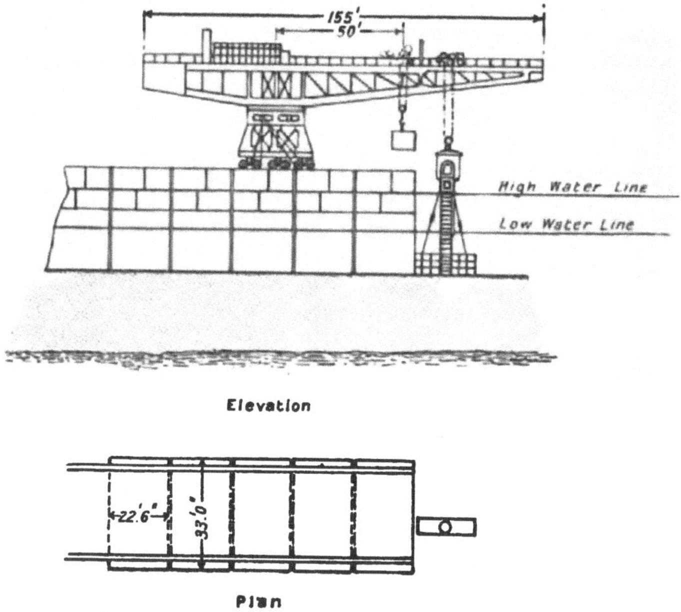


FIG. 156.—Bilbao Breakwater, showing Titan and Caisson.

interior of the caisson through a centrifugal pump suspended from a Titan crane and worked by an electric motor. If irregular settlement took place, the action was reversed and the caisson refloated until its misplacement had been rectified. Once truly placed, eight or ten 40-cubic-feet blocks were deposited in the compartments of the caisson by the same crane. This generally absorbed the whole of the time available on the tide. The following tide, after pumping out the caisson, the remaining blocks were deposited and liquid concrete was run in between them and the others with a solid layer, 20 inches thick, covering the whole.

During the third period of low water the superstructure was taken in hand. It consists of two facings of concrete blocks, approximately 13 feet by

10 feet by 8 feet each, set in two courses as headers and stretchers alternately. The hearting is of mass concrete. Given favourable conditions and fine weather, the superstructure could be finished during the fourth tide with the exception of the parapet wall, which was not undertaken until the full measure of settlement had been obtained. This was considered to have been achieved after the lapse of a couple of winters. From the moment a caisson was first placed to the completion of the superstructure, the settlement averaged about 8 inches. Under the load and vibration of the Titan crane setting blocks further seaward, and under the influence of winter storms, a further settlement of 16 inches took place, making 24 inches in all. When this had been realised, the joints between the caissons, which were 12 inches wide, were made good with cement concrete, and the parapet wall was built.

The total amount of the contract was equivalent to £962,756, and, as the length of the breakwater is 4757 feet, the cost works out to £203 per foot run. The quantities of material in a caisson length (42 feet 8 inches) were as follows:—

| | | Cub. yds. | Cub. yds. |
|---|---|---|---|
| Caisson. | Bottom ballast 42 feet 8 inches × 23 feet × 4 feet 11 inches = | 178·54 | |
| | Twelve blocks each 39·4 cubic yards . . . . | 470·88 | |
| | Filling-in concrete . . . . . . | 180·20 | |
| | Steel bulkheads and timber struts . . . | 3·55 | |
| | | | 833·17 |
| Superstructure. | Eight blocks, each 39·24 cubic yards . . | 313·92 | |
| | Filling-in concrete . . . . . | 166·77 | |
| | | | 480·69 |
| | Total . . . | | 1313·86 |

Say 1314 cubic yards in all.

**Breakwater at Bizerta.**[1]—The type of structure primarily adopted at Bizerta for the converging jetties at the entrance to the port, built between 1889 and 1895, was the rubble mound, consisting of a core of *pierre perdue* of all sizes, surmounted and protected on the sea face by a revetment of natural blocks of large size. The site is not so exposed to violent gales as are other places on the north coast of Africa, and this system of construction was found to answer very satisfactorily. Unfortunately, the local stone (a marly limestone of poor quality, flaking rapidly in salt water) was of such a character as not to commend itself for further use, and a complete departure in design was made in dealing with the new breakwater and the extension of the north jetty, the former of which has a length of 2000 feet, and the latter of 660 feet.

These works, as shown in fig. 157, were carried out by means of metallic caissons with movable upper works, forming ultimately huge artificial blocks faced with marble, which were laid upon a mass of miscellaneous riprap at a level of 26 feet below low water. The blocks measured 102 feet by 26¼ feet by 26¼ feet cubing at 70,000 feet, and having a weight of 5000 tons each. They were set with great precision, and only two blocks out of twenty-three varied perceptibly from the exact line. The settlement, averaging 27 inches, was very regular throughout. The slopes of rubble, however, intended to be

[1] De Joly on Breakwaters, *Min. Proc. Tenth Int. Nav. Cong. Milan*, 1905.

5 to 4 on both sides, had to be reduced on the sea face to 5 to 2, and a further modification of the original design was the extension of the riprap on the port side so as to afford an additional 16 feet of base. The original and modified outlines are shown in the section. The work was completed in 1903. Unfortunately, the subsequent history of the breakwater has not been free from disappointing incidents M. de Joly reports as follows:—

"In spite of their great size the artificial blocks are not absolutely stable. A north-westerly storm which occurred in February 1904, destroyed the regularity and symmetry of the breakwater in a few hours: not only did the blocks settle unevenly and cant slightly towards the interior, but some of them became wedged in after pivotting about their western ends. The displacement of the eastern ends amounted to as much as 6 inches, giving an angular slew of $\frac{1}{200}$.

"An attempt was made to avoid any return by concreting up the vertical

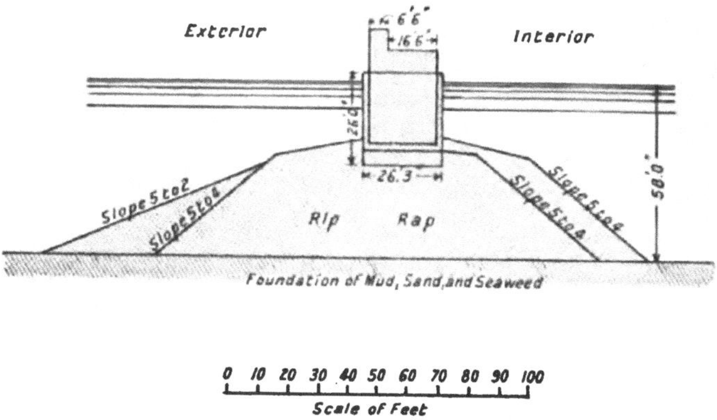


FIG. 157.—Section of Bizerta Breakwater.

joints, which occur every 100 feet or so; simultaneously the interior slope was strengthened as previously described.

"The filling of the joints was just about finished when, at the end of November 1904, a storm arose from the east and occasioned further important dislocations. All the joints were split and some of the blocks seriously damaged. Fears are entertained that any subsequent storm will so increase these injuries as to completely fracture the joints and split the blocks vertically.

"It seems evident from the November experience, which is different from that of the February one, that the width of 26 feet 3 inches is somewhat insufficient for a 5000-ton block under existing circumstances, and that undoubtedly it would have been preferable to increase the width to say 33 feet, leaving the vertical joints between the blocks open so as to allow of settlement freely without tension or shear. Also, it may be questioned whether it would not have been a better plan to make the topmost layers of riprap of harder material, which would more effectively resist crushing and grinding under the combined action of the weight of the blocks and the shock of the waves."

The cost of the earlier type of breakwater (rubble mound, with the same superstructure as that in fig. 157 above the caisson block) was £142 per linear yard, while the cost of the later caisson type was £182 per yard.

**Breakwaters and Piers at Dover.**[1]—The breakwaters and piers are formed of concrete blocks of from 26 to 40 tons each, according to the exigencies of the bond to which the work conforms, and the average weight throughout may be taken at 32½ tons. The blocks were built, in the yard, of 6 to 1 concrete, and those for face-work have granite fronts. All the blocks from foundation level upwards are bonded and dowelled with 4 to 1

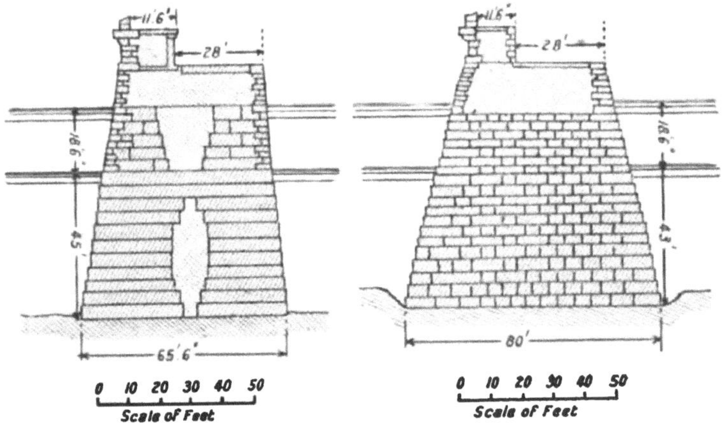


FIGS. 158, 159.—Sections of Breakwater at Dover.

concrete dowels of circular section, while above low water the courses are bedded and grouted in cement.

The blocks were set by means of Goliath overhead travellers running on temporary staging. In the case of the Admiralty works, the roads on the stagings were 27 feet 6 inches above high water of spring tides and 46 feet 3 inches above low water. The piles were arranged in trestles or clusters of six to form spans of 50 feet 3 inches, and the Goliaths, both on the stagings and in the workyards, were all of 100 feet gauge.

The system of operations, carried on simultaneously, was as follows. At the outer end of the work was a stage-erecting machine, followed by a Goliath crane working a grab; then a second Goliath, from which the diving-bell was worked. Following this came a third Goliath, setting blocks under water; and behind that, a fourth, setting blocks above low-water level.

The diving-bells used on the works were 17 feet 6 inches by 10 feet, with 6 feet 6 inches headroom. They weighed about 35 tons out of water, and

[1] *Vide* Matthews on Harbours, *Trans. Am Soc. C.E.*, vol. liv.

5 tons when submerged. They were fitted with the electric light and telephonic communication, but it is recorded that the men preferred mechanical signals.

The foundations generally were carried from 4 feet to 6 feet into the chalk and flint bed, and were protected from the abrading action of sea currents on the outer face by an apron of concrete blocks 25 feet in width.

Excavation for foundation work was carried out by grab-dredging down to within 12 inches of the finished level, and the remaining material removed by the aid of bell-divers. Four men were engaged in the bell, excavating and finishing the foundation ready to receive the lowermost course of blocks. Each shift was of three hours' duration, and two shifts *per diem* were generally worked by the men. When the weather was favourable, work was continuous, night and day.

The greatest depth of the foundations is 53 feet below L.W.O.S.T., the average depth being 47 feet. There was thus an average working head of 66 feet at H.W.O.S.T., corresponding to a pressure of 29 lbs. This head has been found to be a maximum for working under comfortable conditions; on several occasions, when the depth was exceeded for a short period, inconvenience was experienced from the extra pressure.

Very excellent plant was provided by the contractors for the preparation of the concrete blocks. In the workyards six electric portable concrete mixers were used, each capable of turning out about 100 cubic yards of concrete a day. The mixers were of the Messent type, revolved by a motor of 18 H.-P., and driven from the point where the aggregate was received to the block moulds where the finished concrete was deposited, by a 25 H.-P. motor. The gauge of the mixers was 11 feet 7 inches.

# CHAPTER VIII.

## PIERHEADS, QUAYS, AND LANDING-STAGES.

Importance of Pierheads—Forms adopted—Main Features—Lighthouses and Harbour Lights—Examples of Pierheads at Toulon, Sunderland, and Pillau—Quays—Landing Slipways—Stairways and Ladders—Spending Beaches—Entrance Booms—Landing-stages: Fixed and Floating—Pontoons—Conditions of Stability—Centre of Buoyancy and Metacentre—Case of Semi-immersed Pontoon—General Case—Case of the Ballasted Pontoon—Internal Stresses in Pontoons—Liverpool Landing-stage.

**Importance of the Pierhead.**—Whatever the relative exposure of other parts of a mole or breakwater, there can be little doubt that its termination—the **pierhead** as it is called—is subjected to experiences considerably more trying than any which fall to the lot of maritime structures in general. To begin with, it is unfavourably situated; and, in the second place, it is still more unfavourably adapted to its position. Exposed on three sides out of four, it is called upon to withstand and repel the most powerful subversive agencies without that uninterrupted lateral support which constitutes so important and valuable a mainstay of the breakwater proper. The absence of this support renders it a matter of the utmost moment to make a pierhead thoroughly self-sustained and independent—to treat it, in fact, as a perfectly detached and isolated structure, capable of resisting, unaided, all external influences; for in its downfall is involved the destruction of more than is contained within its own limits. The pierhead removed, the section of the breakwater immediately adjoining it has its security materially impaired, and becomes practically defenceless. It enters into that precarious condition which is inseparable from a "scar-end," and the area of damage may be almost indefinitely extended. The pierhead, therefore, should be looked upon as the keystone of a breakwater's stability.

In itself, the pierhead may possibly not exhibit to the eye any specially marked features of height or width—many minor structures do not; but whether these features be in evidence or otherwise, the necessity for greatly increased strength and powers of resistance cannot be gainsaid. As a matter of fact, most moles and breakwaters of any importance are equipped with prominent pierheads of striking outline and substantial construction. There are several reasons why this should be the case. Not only do pierheads, as a rule, project into deep water, demanding a broad and extended base to ensure corresponding stability, but they have also special functions to discharge, which require a certain individuality of treatment. They serve to mark the entrances to ports and harbours, and should therefore be clearly and readily recognisable. It is to be noted, in this connection, that they also run the risk of collision and impact with passing vessels. At night-time, and in misty weather, an entrance, especially if at all narrow, should be efficiently lighted, and, to this end, pierheads are often furnished with a lantern, fixed either to a mast or mounted on a platform, or set in a lighthouse.

Pierheads, therefore, demand consideration from two different aspects: (1) as outlying works in particularly exposed situations, requiring special precautions in regard to design and construction; and (2) as a means of guidance and direction to vessels entering a port, more especially in times of stress and heavy weather.

**Form of Pierheads.**—With respect to the former point of view, there are, in the first place, strong reasons for conferring upon a pierhead a shape which, in plan, is symmetrical about a point or axis. Accordingly, the ends of many moles and breakwaters are expanded as already stated, and given some geometrical form—circular, square, octagonal, rectangular, hammerhead, etc., as the case may be. Of these, the circular may be claimed as the most convenient shape, from the point of view of manœuvring vessels, it being easier, in case of necessity, to warp a ship round a continuous curve rather than to swivel it through an entire quadrant. The widening of the pierhead, however, somewhat interferes with the working of vessels alongside an inner quay, though, on the other hand, it serves to cut off any strong flow of current which might endanger the vessels moored there.

(1) *As regards intrinsic stability*, a pierhead should be, as far as possible, homogeneous throughout, without joint or intersection. This is not always a feasible arrangement, nor in any case easy to achieve. There can be no doubt, however, that an ideal pierhead would be one of the nature of an enormous monolith. For the purpose of constructing such a monolith, a buoyant caisson chamber of the type which has already been described (p. 175) is often constructed, floated into position, sunk, and filled with concrete. Another method is to form an outer ring or boundary, of large blocks securely anchored together, and to deposit mass concrete in the interior. In this instance, it has been found essential to introduce a number of cross ties passing right through the ring, to prevent the blocks from being disturbed under the temporary fluid pressure of the concrete.

Yet another method is the driving of an outer ring of sheeting piles, either of timber or steel, to form the necessary inclosure. But this operation is attended by some difficulties. The satisfactory alignment and driving of piles is not always accomplished with facility, even when the work is straightforward and the conditions favourable. With both these factors acting in an adverse sense, the likelihood of a successful issue is more remote. Curvilinear work in piling is a particularly awkward undertaking.

One word of caution is very necessary on the undesirability of forming anything of the nature of a terrace, or level platform, at the foot of the superstructure. The waves will act upon the upper surface of such a ledge with disastrous effects, both to that and to the recessed pierhead of which it is a part. The latter tends to become undermined in the manner illustrated by the pierhead at Pillau.[1] The best form of pierhead is that which presents an upright front to the sea on all sides, without benching or recesses of any description.

By making a pierhead perfectly self-contained, the problem of connecting it by any system of bond with the breakwater proper does not arise. On several grounds, and especially in reference to breakwaters liable to considerable settlement, it is desirable that pierheads should not be involved in any movement of the adjoining parts. Therefore, while the breakwater and its pierhead may be and commonly are in close contiguity, there should be a vertical joint between them, rendering them independent of each other's action.

(2) *As a means of guidance and direction* for vessels, it is desirable that a pierhead should possess some prominently outstanding feature. This desideratum may be met by a lighthouse (fig. 230, page 264), or mast (fig. 160), or, where these are not required for signalling purposes, by an elevated platform.

A **lighthouse**, if in masonry or concrete, should be stoutly and substantially constructed. Concrete work lends itself admirably to homogeneity, there being, in this case, no difficulties to encounter such as attend subaqueous work. Masonry is more costly, in that bedding and jointing demand the most careful execution; and in many cases there must be introduced elaborate bonding courses containing dovetails and keystones, all entailing considerable expense. Of the forms adopted in connection with either of the systems, the circular is, on the whole, the best, offering least obstruction to, and assisting in, the deflection of breaking seas. A flat or plane face, however, is convenient for window, door, and other openings.

The lantern being situated at the summit, the lower part of the tower should be utilised so as to afford a storeroom and also a shelter chamber for those whose duties necessitate their presence during times of storm. Stress of weather may, in fact, interrupt communication with the shore for several hours, if not whole days, at a time. A spiral staircase in the interior of the tower generally leads to the lantern.

Pierhead lighthouses may also be constructed of steel, either in the form of open framing or with plate sheeting, and also of a cluster of cast-iron columns connected by ties and bracing. Variations in design are, in fact, numerous and almost illimitable.

The **lantern** (fig. 242, page 272) is a glazed chamber, having a pedestal of cast- or wrought-iron, and a framing of gun metal or steel. Glazing is now generally circular rather than polygonal, as was formerly the practice. The glass is polished plate, in most cases $\frac{3}{8}$ inch thick. Spare panes should be kept handy to replace breakages, which are likely to arise from birds flying against the lantern as much as from the effects of storms.

[1] See p. 191

Where the light is of minor importance and lacks the rotative mechanism which demands a protective lantern, a **mast** (fig. 160) may be used for the purpose of elevating it to the required focal plane. A steel column, tapering in form and fitted into a strong cast-iron base, will generally be used in such cases. The base must be securely bolted into the pierhead structure. At the head of the mast there will be a curved bracket, or yard arm, to support the lamp, which may be raised or lowered by a chain or wire rope, leading to a winch at the foot.

Apart from any guiding signals for the night-time, such as the above, or for the daytime, such as semaphores or other indicators, the pierhead needs to be equipped for general purposes with a capstan or two, with snatch-blocks, mushrooms, or fairleads, and also with mooring-posts at frequent intervals. Steps leading down to the water level should also be provided in a sheltered spot. The provision of life-belts and life-lines is almost too obvious a duty to call for mention.

**Pierhead at Toulon.**—The pierhead of the St Mandrier Jetty at Toulon has a diameter of 92 feet, comparable with the 42 feet width

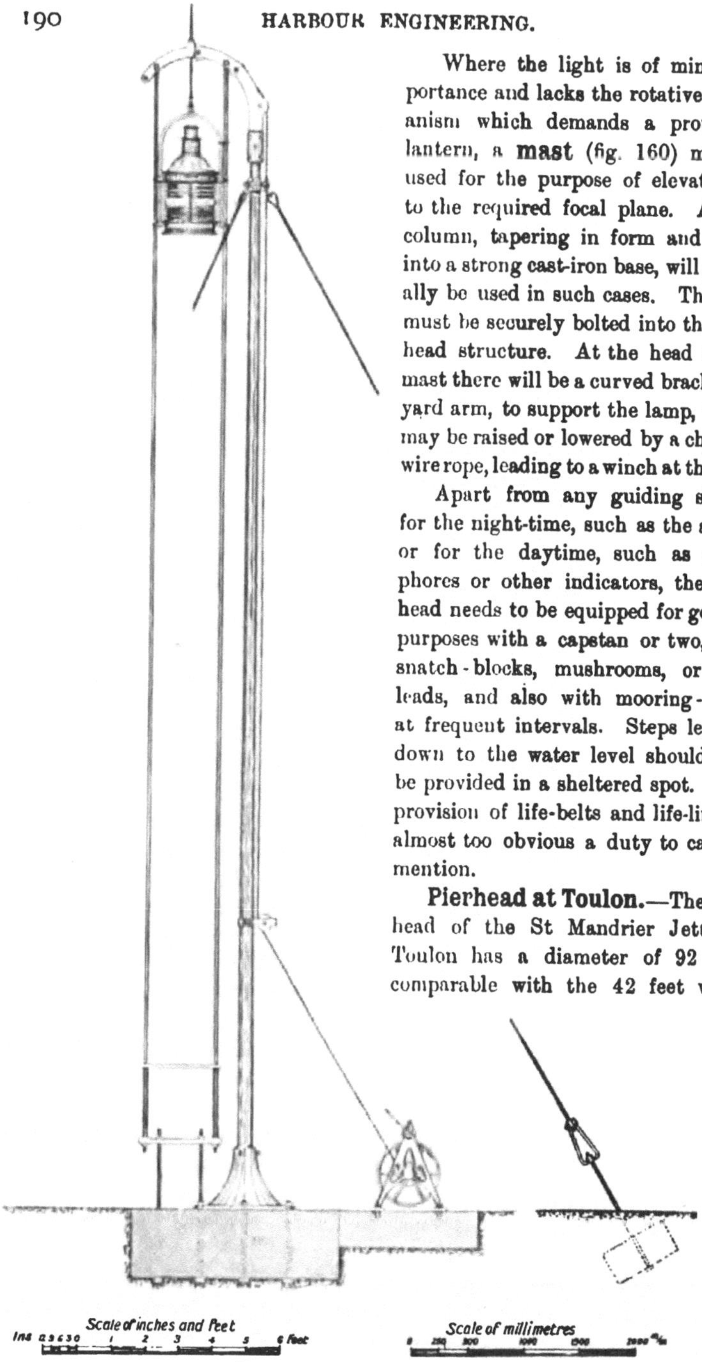


FIG. 160.—Port Light and Mast.

of the mole or jetty itself. Its base consists of three courses of masonry blocks, set in the form of a circle, and inclosing a space which is filled with mass concrete. The base rests upon a rubble-work deposit brought up from the sea bottom to within 14 feet of the surface of the water, and levelled with great care by the aid of divers. The superstructure is of masonry, and consists of a lighthouse with a cellar compartment used as a cistern, absolutely water-tight, so carefully was the concrete work carried out.

The pierhead does not actually join up to the jetty; that is to say, there is no interlocking bond. A narrow vertical joint separates the two structures so as to remove all possibilities of fracture arising from unequal settlement.

**Pierhead at Sunderland.**—The pierhead of the Roker pier was formed by means of an iron-plated caisson 100½ feet long, 69 feet wide, and 26½ feet deep. The caisson, containing 3500 tons of concrete, was floated out into position with a draught of 22 feet on to a carefully levelled foundation bed of concrete bags finished at 23 feet below low water. After being sunk, the caisson was built up with 15-ton and 25-ton blocks, mass concrete, and cement-grouted rubble, until, when completed, its weight amounted to 10,000 tons. On top of this, the pierhead superstructure was constructed in blockwork and surmounted by a lighthouse.

**Pierhead at Pillau.**[1]—The new moles at Pillau were built, during the closing quarter of the nineteenth century, to a type which is common on the Baltic seaboard, viz., that of a rubble mound confined between two lines of sheet piling connected by iron ties and having a brickwork superstructure. The width at mean water level between the sheet piling is 31 feet, the summit width 26 feet, and the height above mean water 10 feet.

Both mole ends have occasionally to withstand very violent attacks by the sea. The pierhead structures, therefore, were given enlarged dimensions, the width being increased to 46 feet. In plan the termination of the north mole exhibits a return in a straight line; at the southern mole, the pierhead front forms three sides of a regular hexagon. The superstructures are set back from the ends of piled work by 18 feet in the case of the north mole, and 30 feet in the case of the south mole. The area of the recessed portion in each case was paved over with brick to a thickness of 3 feet, forming a sort of terrace or platform. The depth of water at the pierheads was 30 feet when building operations were commenced. As the piles were driven down to a depth of 50 feet below water level, no special rubble apron was deemed necessary.

The south molehead was only just completed in the year 1885, when, in the month of September, it had to withstand the onslaught of a severe gale, which blew for two days from the south-west and north-west, gradually increasing in intensity. The whole front portion of the head, the terrace work, and a large portion of the sheet piling, were destroyed. Moreover, the rubble filling inclosed by the lastnamed, in default of restraint, fell away, depriving the superstructure of its support to such an extent that it leaned

[1] Anderson on Prussian Breakwaters, *Proc. Int. Nav. Cong. Milan*, 1905

forward 4 inches out of the vertical, and a deep fissure was formed at its junction with the mole proper.

The cause of this collapse, which has proved most disastrous in its con-

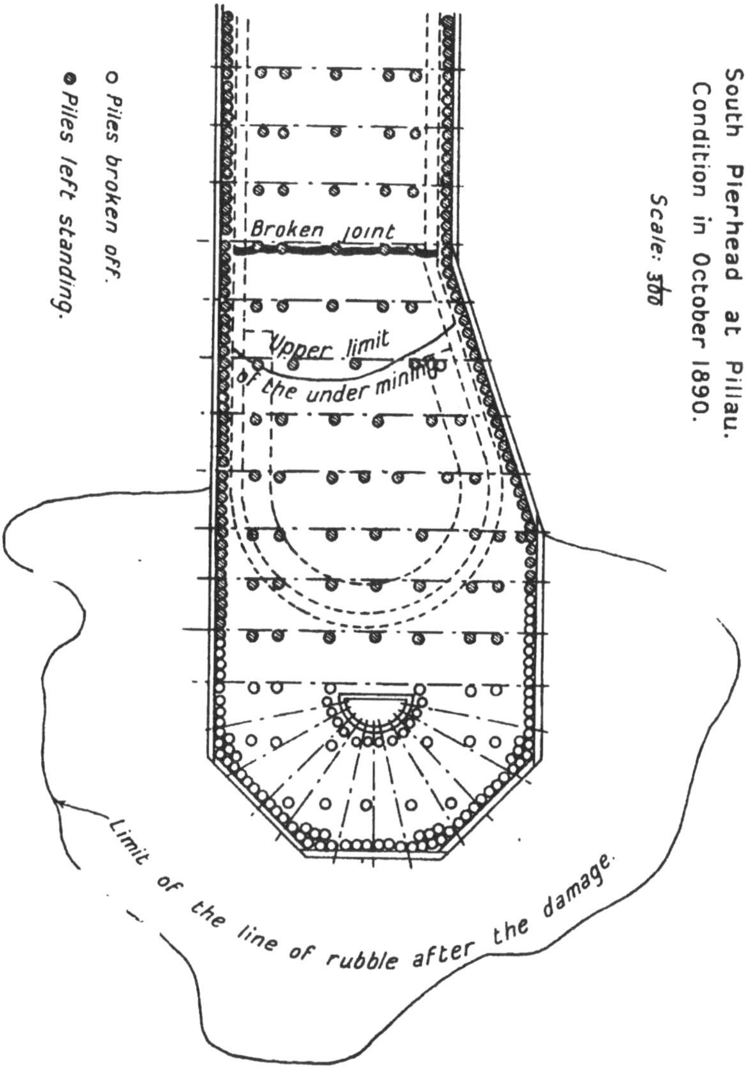


Fig. 161.—Plan of South Pierhead at Pillau.

sequences, is to be found in the great distance separating the superstructural from the substructural ends; in other words, it was due to the terrace. This was alternately attacked from below by the pressure of the water penetrating through the openings between the piles, and through the interstices in the

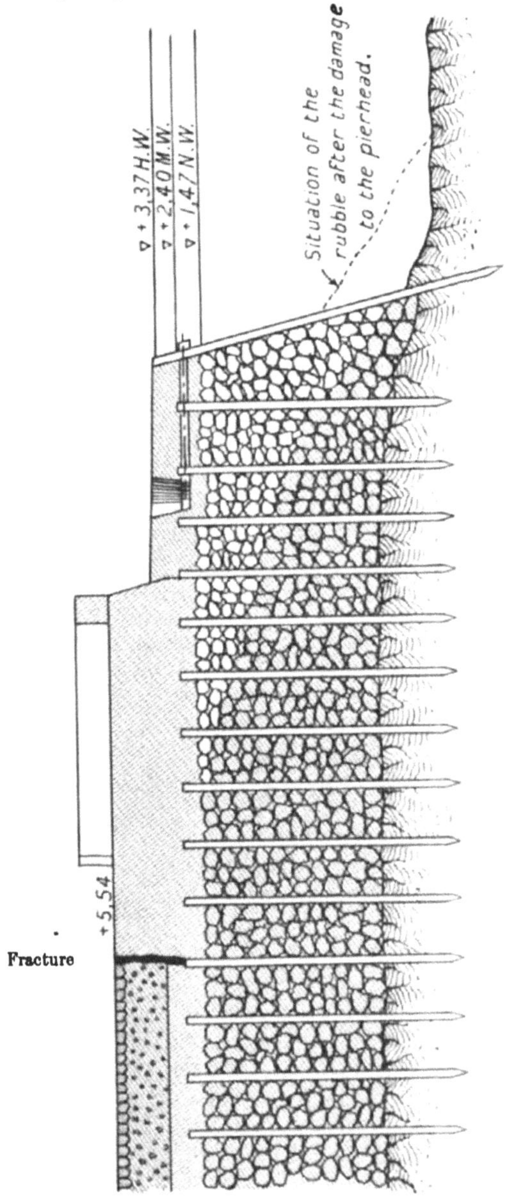


FIG. 162.—Longitudinal Section of South Pierhead at Pillau.

rubble work, and from above by the direct downward stroke of the waves. The floor was not in itself strong, and its great extent did not add to its stability.

There must also be taken into consideration the fact that the continuity of the pavement had been broken by trenches cut to receive the anchorage bars of the head. Moreover, there had probably been some settlement in the rubble work which it covered. All these causes combined to render it an easy prey to the storm, which swept away the floor and battered down the piling, with the aid of the loose masses of stone comprising the hearting.

With the front portion of the pierhead reduced to a mere mass of débris

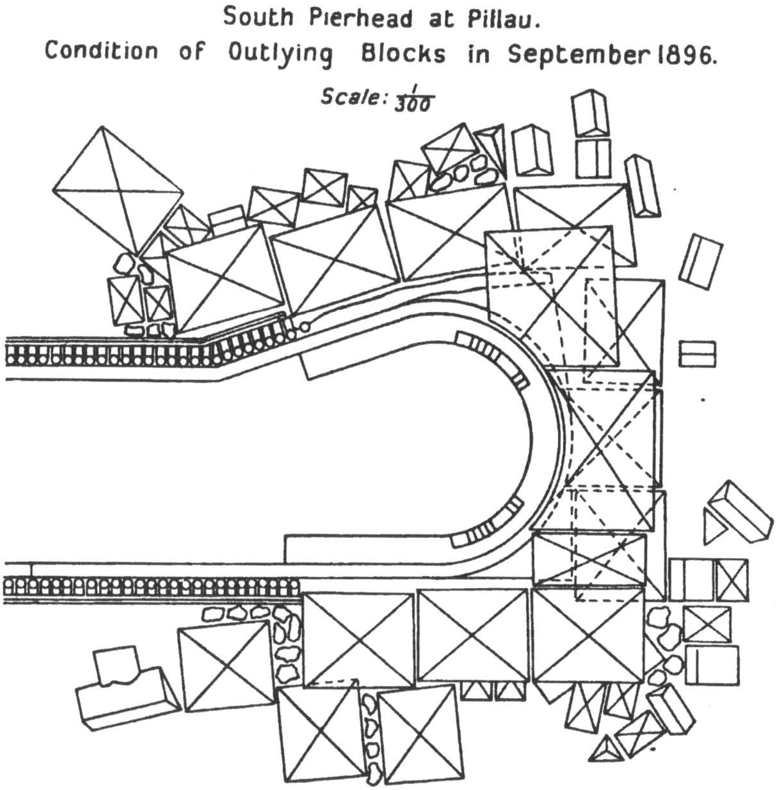


FIG. 163.—Pillau Pierhead—Protection Blocks.

sloping downwards to the sea bottom, it became necessary to devise measures for the protection of the superstructure. Any renewal of the destroyed portion was out of the question, owing to the impossibility of driving piles through the rubble mound. The plan decided upon was to mould and consolidate the mounds into a protective apron.

The first step taken was to deposit a quantity of massive granite blocks on the site of the terrace. The level at this point was 3 feet above mean low water, and it was annually increased. Up to the year 1891, over

140,000 cubic feet of stone, bulking from 14 to 35 cubic feet a-piece, were deposited and formed into a berme 10 feet in width, with a slope of 2 to 1. In order to secure this stone-work in position, large blocks, containing 175 cubic feet each, were placed on top. The blocks were made of granite fragments and cement mortar in the proportion of 1 to 3; they had a length of 6 feet 6 inches, a width of 5 feet, and a height of 5 feet 6 inches, and they weighed 12½ tons.

Work proceeded on these lines from 1892 to 1894. A storm in November 1894, however, once more wrecked the pierhead, destroying the uncompleted berme and its covering. All burrs and blocks lying in less than 7 feet 6 inches of water were dislodged, and some of them swept up the slope.

Subsequent measures for the protection of the pierhead have been directed to the consolidation and maintenance of the mound, but on somewhat different lines, experience having indicated that the largest blocks formed the best covering material. Thus, in addition to a large number of the smaller blocks which served to fill up gaps and to provide a flat surface, in 1895 eleven blocks of 635 cubic feet, and in 1896 ten blocks of from 530 to 1765 cubic feet, were bedded on the rubble work in concrete composed of granite and cement mortar.

Yet these measures were attended by no better success than that of their predecessors. The very largest blocks, weighing 120 tons, were sooner or later dislodged by storms and driven down the slope. They were continually restored, until finally, in November 1899, a gale wrought such serious havoc that even the lower portion of the pierhead foundation was exposed and partially withdrawn. As a consequence, the whole weight of the superstructure came to rest upon the piling and staging which had been erected for pile-driving machines and rail-tracks during the construction of the mole, and which had been left buried in the mound.

The only plan now was to fill up the cavities, and this was done by means of stones and sacks of concrete packed within a circumscribing ring of blocks of 350 cubic feet, each set in a double row. But, before this work could be completed, fresh disasters occurred. The blocks were disturbed, and some of the rubble carried away, in November 1900. No satisfactory repairs could be effected during the winter season, and in the following April a violent westerly gale once more devastated the whole pierhead. The staging-piles broke; the crack which had formed at the junction of the mole and pierhead at the time of the initial destruction of the terrace, widened out on the harbour side to a width of 4 feet 3 inches, and the pierhead, a huge mass of 23,660 cubic feet, and having a weight of 1680 tons, was slewed on its axis through 12 feet towards the sea, and left with its outer edge depressed to the extent of 4 feet 4 inches. In this condition it rested upon a few projecting peaks of the mound.

During these experiences, it was observed that a concrete block of 1553 cubic feet, which had been set on the deep-water side of the head, had not suffered more than a very slight displacement. It was therefore decided t surround the pierhead with a ring of nine concrete blocks of extremely larg size. These were formed on the buoyant caisson principle. Wooden boxe of grooved and caulked planking, 7 inches thick, were formed, at a timber-yar 1¼ miles distant, to a trapezoidal shape with widths of 23 feet and 15 feet, length of 23 feet, and a height of 13 feet, so as to inclose a volume of 565 cubic feet. Stiffened by transverse and longitudinal ties, and filled to a dept of 3 feet with granite rubble concrete, they were floated to the site of th head and there sunk close alongside one another, in about 11 feet of water, level base having been prepared for them by blasting away the protuberanc of the mound, and filling the hollows with stones, débris, and sacks of concret The boxes were then filled up with concrete in a couple of days, in this wa attaining a weight of 400 tons. Of the projected nine boxes, six were delivere in the year 1901, and, in addition, four blocks of 370 cubic feet were set. Th filling of the inclosed space was then put in hand.

But in September of the same year a storm produced a fresh cross fractur in the pierhead, and a further sinking of its front face by 8 inches. The con crete boxes were slewed about 18 inches, but without damage. In th succeeding December a further and more pronounced depression of the hea took place, the settlement amounting in all to 6 feet 6 inches.

By this time, however, the superstructure had found a firmer bearing o the foundation stonework. Such small cavities as remained were closed wit concrete. During the following year three more concrete boxes were deposited and the space between the boxes and the head densely packed with larg stones. Owing to these improvements, the pierhead suffered but little durin the gales of December 1902 and February 1903, which almost destroyed th north breakwater. The boxes, indeed, were considerably shifted, but they di not suffer much damage. The pierhead was again built up to the height o the original summit, and the breaches which had been made between the boxe were closed in 1903 by four additional boxes, and the remaining gaps wer made good in 1904.

At this point the interesting but calamitous narrative ceases. It would b too sanguine a view to conclude that perfect stability had at last been attaine But, as a record of disasters and of remedial expedients, it is most instructiv and the lesson it enforces is complete.

## Quays.

The inner sides of breakwaters and the shore frontages of harbours ar commonly bordered with quays for the reception of merchandise an passengers.

A **quay** is, properly speaking, a paved space or area devoted to th purposes of loading and unloading craft, and it is usually bounded at th water's edge by a wall or wharf founded at a sufficient depth to permit o vessels lying alongside. Vessels do not always remain afloat; in many case with the recession of the tide, they take the ground. A quay and a quay wal are commonly treated as synonymous terms. Yet such is not strictly the case, as indicated above.

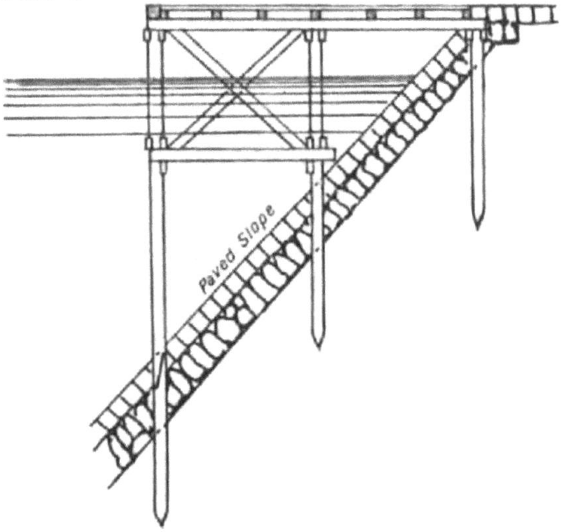


FIG. 167.—Quay and Wharf.

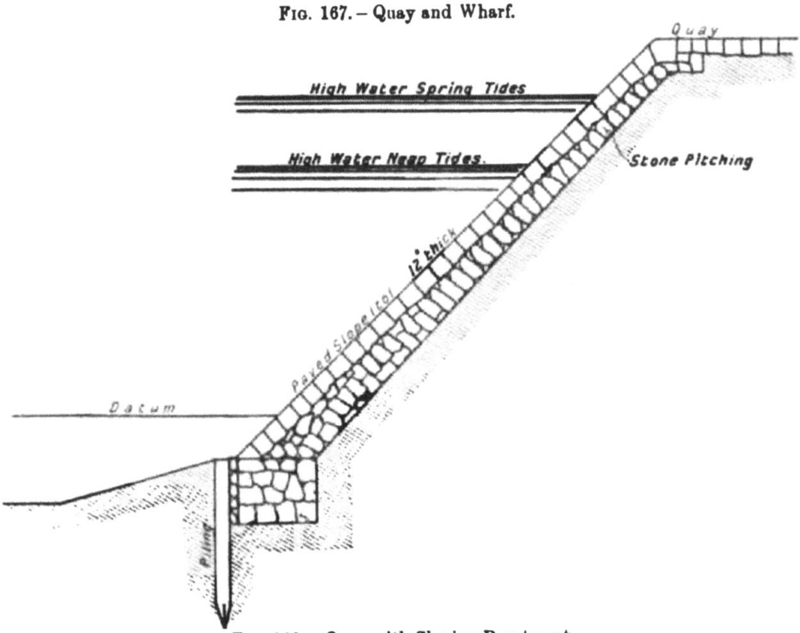


FIG. 168.—Quay with Sloping Revetment.

Quay walls are called upon to discharge very important duties. They act as retaining walls to uphold the solid material which forms the foundation

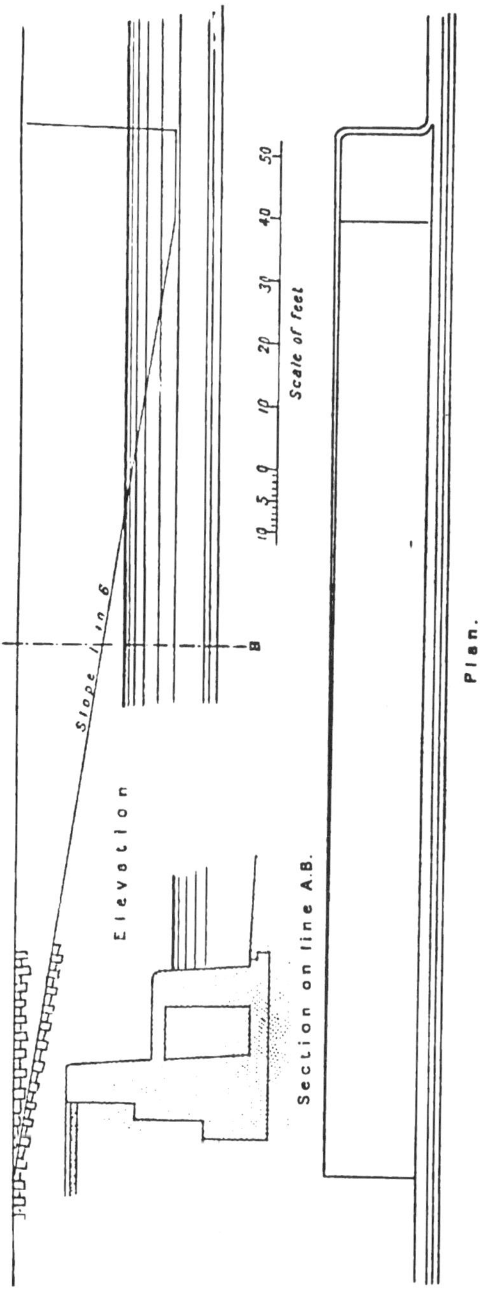


FIGS. 169-171.—Slipway.

upon which the quay surface is prepared and laid. They also serve to present a uniform frontage against which vessels may be moored and warped without damage. As retaining walls, they possess characteristic features and discharge special functions which call for very detailed treatment; and as the author has subjected them to a careful investigation elsewhere,[1] it would be unwise to attempt any further, and necessarily restricted, allusion to them here.

**Landing Slipways.**—In situations where the water does not maintain a fairly uniform level, it becomes necessary to provide sloping ways, or slipways, leading from the surface of the quay down to the lowest water level. These slipways are even desirable under any circumstances, and especially for the purpose of affording access to small craft of shallow draught and rowing-boats, and to facilitate the withdrawal of the latter from the water. Slipways range from about 5 to 15 or 20 feet in width, with an inclination not greater than 1 in 5. They should have a covering of concrete or be paved with large heavy stones, having a flat upper surface and presenting as few joints as possible to the pick-like action of the waves. Such joints as there are will be well flushed and pointed with cement, for it can be readily understood that slipways are subjected in a peculiar manner to the most destructive action of breaking waves, which renders it imperative to present thereto as hard, smooth, and unbroken a surface as possible. Where slipways are paved with cubes, or setts, these latter must be thoroughly grouted and bedded in a perfect manner upon a substantial concrete foundation.

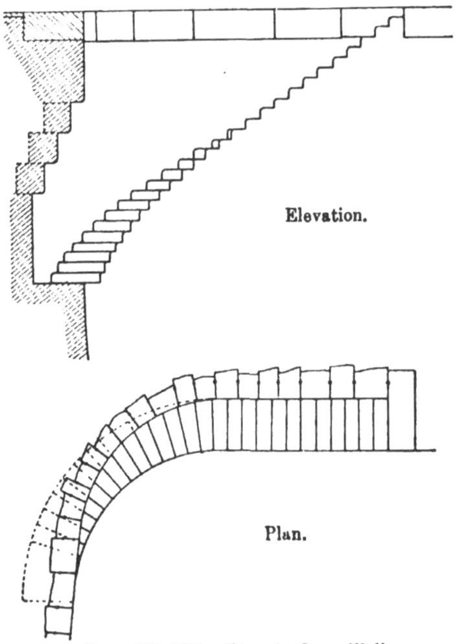

FIGS. 172, 173.—Steps in Quay Wall.

Slipways are provided with a bottom landing, and sometimes with one or two intermediate landings — all level platforms.

**Stairways and Ladders.**—Access to the water-line may also be obtained by stairways and ladders. The former are simply steps set in the wall in the ordinary manner, and not uncommonly at corners, where they are least likely to interfere with the use of the quay by shipping. The latter generally consist of galvanised wrought iron vertical sides and circular rungs, recessed within the face line of the quay so as to

[1] *Dock Engineering*, Chapter V.

avoid forming anything of the nature of a projection likely to produce damage.

Other adjuncts of a quay are **life-chains, mooring-rings,** and **mooring-posts.** The firstnamed are suspended at intervals and festooned, so as to enable persons accidentally immersed to support themselves until succoured. The last two are for the purpose of holding vessels close against a quay wall. Mooring-rings, useful chiefly for very small craft, are now fairly obsolete, as they are awkward of access and difficult to maintain in order. The most conveniently arranged of them are recessed within the quay face, so as to acquire a certain amount of cover. The best type of mooring-post has a lip arranged on the side furthest removed from the quay edge, so as to hold the rope well and keep it from slipping upwards.

A good sloping **beach** is very desirable in the immediate proximity of a harbour. If situated at the entrance, it forms a very useful spending-ground whereon waves may dissipate a very large proportion of their activity. A beach is also desirable in that small craft may ground thereon for repairs. When formed artificially, as is sometimes necessary in rocky localities where the shore descends abruptly, quarry refuse and débris may be used for the purpose. A slope of 1 in 10 or 12 will be found most serviceable.

**Booms.**—Inner basins, or harbours of the smaller class, may be still further protected from the effects of sea swell by means of a temporary closure or boom across the entrance, which would naturally, in such a case, be narrow. Booms take the form of a log partition, set in horizontal layers one above another, with their ends engaging in grooves specially constructed at each side of the entrance. It is necessary to observe that unless the logs extend to the very base or bottom of the passage-way, wave motion will be transmitted beneath them, and they will prove ineffective for the purpose in view.

**Mooring-Buoys and Stages.**—In addition to the facilities afforded by posts, rings, and bollards on the quays for securing vessels, in large basins and in rivers floating and fixed moorings are also provided. The former consist of buoys of various shapes anchored to the bottom of the basin or the bed of the river, and the latter of piled stages suitably braced. These last are sometimes known on the Continent as "Ducs d'Albe."

## Landing-stages.

In all his operations the maritime engineer is more or less in touch with the requirements of the naval architect, and the boundary line between the two professions is by no means easy to define; indeed there is oftentimes a zone within which both practitioners find a common field of action, and where it would be difficult, if not absolutely impossible, to lay down any limitations for one or the other. Thus, in the case of entrance caissons, floating docks, and buoyant structures generally, there are presented to the engineer all the processes and features characteristic of ship design and calculation, and so, too, in connection with the subject of the present chapter, the laws governing

FIG. 174.—Mooring Buoys. [*To face p.* 200.

the behaviour of floating bodies have to be carefully studied and clearly understood.

**Landing-stages,** whether adapted for the use of passengers, vehicles, goods, or cattle, fall naturally into two main types—fixed and floating.

**Fixed Landing-stages.**—Besides serving no apparent purpose, it would be difficult to make an exact differentiation between the functions of a fixed landing-stage and those of a wharf, pier, or quay. Generally speaking, the term stage is limited to framed structures of timber or iron, and fixed landing-stages are most commonly platforms of woodwork on a piled foundation. But even with this limitation, they possess no features which are not common to jetties and wharfs of similar construction (*vide* fig. 167). To pursue this branch of the subject further would be redundant, and therefore unnecessary.

**Floating Landing-stages** are decks or rafts of timber or iron work, either self-supporting or carried upon hollow pontoons which afford the necessary bearing power by reason of the buoyant properties of water. The pontoon is the more usual arrangement, and it is certainly the only system applicable to stages of size and importance.

## Pontoons.

The stability of any statical structure depends, in the first instance, upon the equilibrium of the external forces which act upon it. In the case of a floating body, these external forces are essentially and practically two, and two only, viz., (1) gravity, due to the intrinsic weight of the body, acting vertically downwards, and (2) buoyancy, or the vertical component of hydrostatic pressure, acting vertically upwards. It is true that, in addition, there are the horizontal components of hydrostatic pressure, but, as in still water, whatever the shape and size of the body, these components must always and exactly neutralise one another, there is no necessity to bring them into consideration.

We have therefore only to take into account two resultant forces, opposed to one another. In ordinary statics, equilibrium would be sufficiently assured by the fulfilment of the conditions, that (1) the lines of action must coincide, and (2) the forces must be equal in magnitude and opposite in direction. As regards floating bodies, these requirements are satisfied when the vertical line through the centre of gravity passes through the centre of buoyancy (or centre of gravity of the displaced fluid), and when the weight of the body is just equal to that of the fluid displaced.

But the extreme mobility of water and the absence in a floating body of any perceptible inherent resistance to disturbance, involve another phase of equilibrium in its practical and working sense. It is obvious that there must be not only perfect balance at any instant, but also a disposition on the part of the body to right itself, or to recover its initial position, in the event of a slight or moderate displacement. Both these points have to be taken into consideration in the design of floating vessels.

The calculations necessary for the purpose are much simpler in regard to pontoons than they are in regard to ships and other navigable craft, since the former are generally constructed to some regular geometrical figure which permits of the easy determination of its centre of gravity and also of the centre of buoyancy. The calculation of the weight of an ordinary ship and the point at which it may be assumed to be concentrated, as also of the displacement and its geometrical centre, are matters of great complexity and difficulty, calling for the exercise of no little patience, ingenuity, and skill.

Pontoons, on the other hand, are generally, if not universally, either rectangular, cylindrical, or spherical in form, with centres of gravity and displacement readily determinable by simple geometrical construction.

Thus, in fig. 175 a rectangular pontoon is shown partly immersed in water. The disposition of the principal resultant forces is that shown by the arrows, and the primary condition of equilibrium is manifestly fulfilled.

Now, suppose such a body to have acquired a slight displacement,

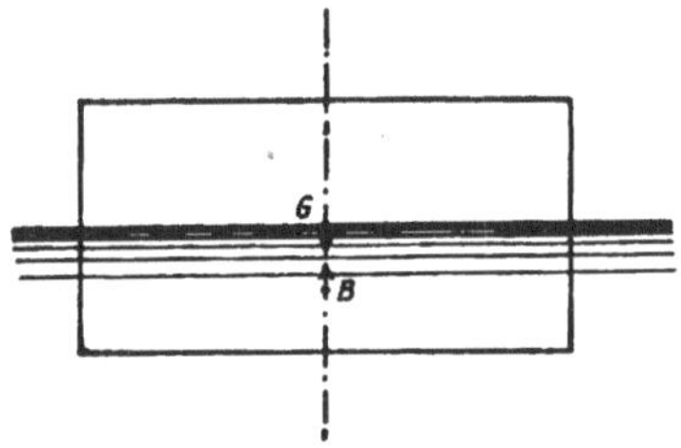


FIG. 175.

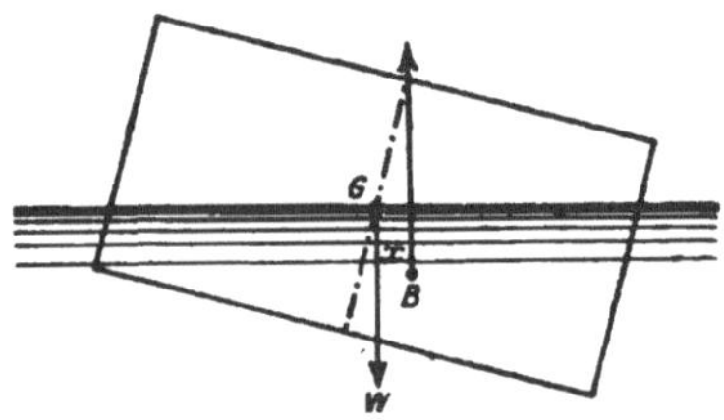


FIG. 176.

with the result that it has taken up the position shown in fig 176. The centre of gravity (G) remains unchanged, but the centre of buoyancy (B) has been removed to a point which does not lie vertically below the centre of gravity. Obviously there now exists a couple, the moment of which, W$x$, is the weight of the body (W) into the horizontal distance ($x$) between the two centres. The moment is a righting moment, and tends to bring the pontoon back to its original position.

Suppose, however, the pontoon to float on its narrower side or end as shown in fig. 177. In the upright condition the primary condition of equilibrium obtains as before. But when a slight displacement takes place (fig. 178), the moment (W$x$) called into existence is an overturning one instead of a righting one, and the pontoon has every tendency to capsize. We see, then, that a different state of things has been produced, and it becomes necessary, therefore, to investigate the relative positions of the centres of gravity and buoyancy a little more closely.

In each of the figures, let the vertical line drawn through the centre of gravity when the pontoon is in its initial position, and also that through the centre of buoyancy in the displaced condition, be continued until they intersect at a point which we will designate M. The technical name for this point of intersection is the **metacentre.**[1]

It will be noticed that there is a very striking difference in the position of the metacentre in the two figures. In one case it lies above the centre of gravity of the pontoon; in the other case, it lies below it. The former arrangement produces a righting moment; the latter, an overturning moment.

The metacentre of a floating body has a variable position dependent upon (*a*) the shape of the body and (*b*) its centre of gravity, and also (*c*) upon the centre of buoyancy; but, from what has been pointed out, it follows, as a general rule, that a pontoon is stable or unstable according as the metacentre lies above or below the centre of gravity.

It would not be strictly correct to say that overturning would absolutely ensue in the latter event, as through the variation in the centre of buoyancy

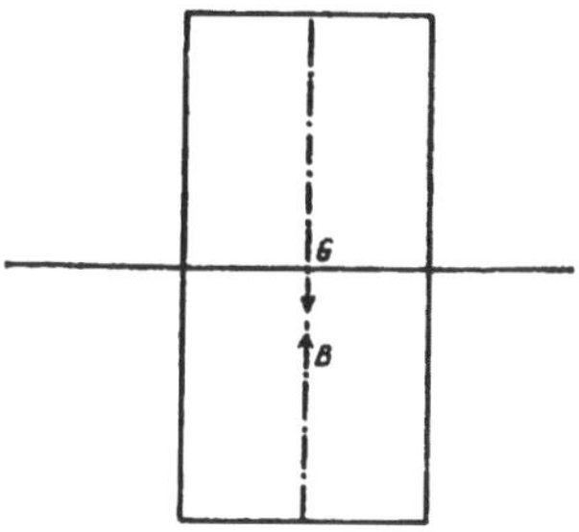


FIG. 177.

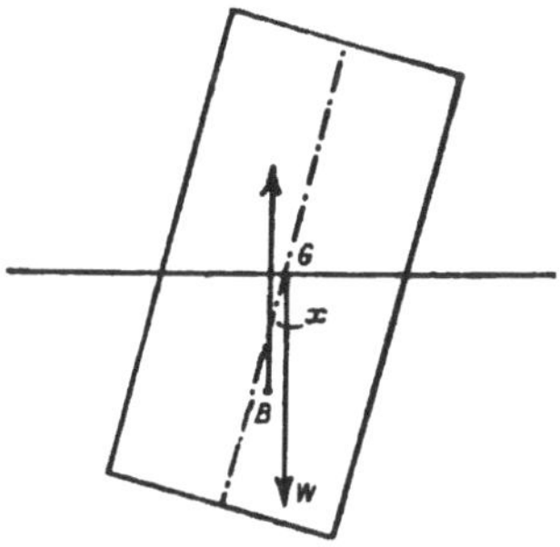


FIG. 178.

an intermediate position might be reached in which the conditions of equilibrium are satisfied.

It will be well, therefore, to go through the process of determining the complete range of position of the metacentre, and to construct a diagram showing all the changes in position of the centre of buoyancy corresponding to various degrees of inclination.

(I.) *Case of the unballasted pontoon immersed to half its depth.*—Let us take the case of a symmetrical pontoon of rectangular cross section floating so as to have a moiety of its volume immersed. Fig. 179 represents such a case: GHNP is the pontoon, and RT the water surface level in the initial position, while FS, HP, and VX are other water-lines corresponding to changes of inclination in the pontoon.

[1] Bouguer, who introduced the term a century and a half ago, employed it to designate a point in a ship's vertical axis above which the centre of gravity of the vessel might not be raised without producing an inclination in the axis. The metacentre must not be confused with any of a series of points on a curve distinguished by Bouguer as the *metacentric*. The metacentric may be defined as the locus of the intersections of successive verticals through adjacent centres of buoyancy as a ship undergoes a series of slight inclinations. In other words, it is the evolute of the curve of buoyancy, or the locus of its centre of curvature.

In the initial position, the centre of buoyancy is at $B_1$ in the vertical line QK passing through O.

Now, suppose the pontoon be acted upon so as to take up an inclination in which the water line is FOS. The immersed section becomes FHNS, and the centre of buoyancy which corresponds to this position lies somewhere to the right of $B_1$. We have to determine its position.

On examination, we see that of the immersed area, a triangular wedge RFO, representing upward pressure both in amount and intensity, has been transferred to the other side of the axis KQ, viz., to TSO. Accordingly, by a principle of mechanics, the centre of gravity of the rectangle RHNT has been moved along a line parallel to the line joining the centres of gravity of the equal triangles RFO and TSO, a distance measured by that between the centres of gravity of the triangles, multiplied by the ratio of the area of one triangle to the area of the original rectangle. To put this in symbols, let $g_1$

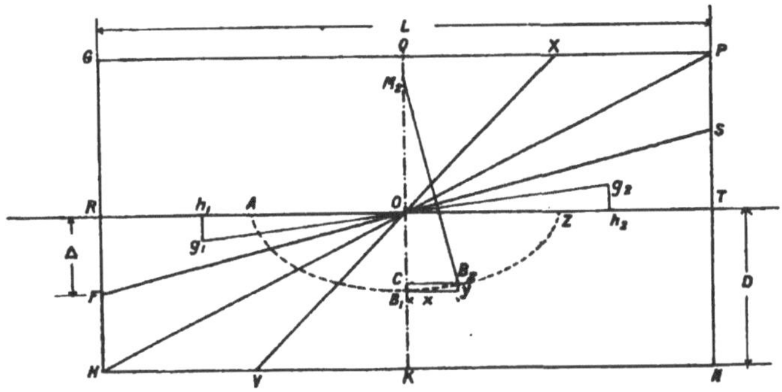

FIG. 179.

be the centre of gravity of the triangle RFO, and $g_2$ the centre of gravity of the triangle STO. Then, if $B_2$ be the position of the new centre of buoyancy, we must have

(1) the line $B_1B_2$ parallel to $g_1g_2$;

(2) the distance $B_1B_2$ equal to $g_1g_2 \times \dfrac{\text{area RFO}}{\text{area RHNT}}$.

Since the alteration in the position of B corresponds to the direction of $g_1g_2$, and therefore is partly an upward movement, we may conveniently find the locus by co-ordinates, with the initial position of B as origin. A line through $B_1$, parallel to RT or HN, will be the axis of $x$; the line KQ, the axis of $y$.

Draw $g_1h_1$ and $g_2h_2$ perpendicular to RT.

The abscissa of $B_2$ is proportionate to $h_1h_2$; its ordinate to $g_1h_1 + g_2h_2$.

Call OR, the semi-width of the pontoon, $\dfrac{L}{2}$; RH, the original depth of immersion, D; and RF, the extent of emergence, $\Delta$.

Then, by the ordinary principles and operations of trigonometry and mechanics, we have

$$h_1h_2 = 2h_1O = \frac{2}{3}L.$$

Similarly,

$$h_1g_1 + h_2g_2 = \frac{2}{3}\Delta,$$

and the ratio

$$\frac{\text{area RFO}}{\text{area RHNT}} = \frac{\frac{1}{2}\text{FR.RO}}{2\text{RO.RH}} = \frac{\text{FR}}{4\text{RH}} = \frac{\Delta}{4\text{D}}.$$

Whence we obtain as values for the co-ordinates,

$$x = \frac{2}{3}L \cdot \frac{\Delta}{4D} = \frac{\Delta L}{6D}$$

and

$$y = \frac{2}{3}\Delta \cdot \frac{\Delta}{4D} = \frac{\Delta^2}{6D}.$$

The equation of the locus accordingly is

$$y = \frac{6D}{L^2}x^2.$$

It is at once apparent that, L and D being fixed quantities, $y$ varies directly as $x^2$; that is, the locus of B is the curve of a parabola whose longitudinal axis is $QB_1$, and vertex, $B_1$.

These equations for $x$ and $y$ hold good up to the point where $\Delta$ becomes equal to D. Giving $\Delta$ the value $\frac{D}{2}$, we obtain

$$x = \frac{L}{12};\ y = \frac{D}{24};$$

and, giving it the value D,

$$x = \frac{L}{6};\ y = \frac{D}{6}.$$

This brings us to the diagonal HOP, at which inclination the side GP commences to be immersed and the side GH is entirely out of water. It will simplify matters now if we regard the pontoon as undergoing disturbance from an initial position in which the vertical axis is RT, and the surface level KQ, for that is the position towards which the pontoon is tending in the continuation of its revolution.

The calculations for the locus of B in reference to the new axis will equally give a parabolic curve, having its vertex at Z, where $OZ = \frac{L}{4}$.

Accordingly, the locus of B resolves itself into a curve consisting of four parabolic arcs touching at the diagonals, HP and GN, of the parallelogram. A moiety of the curve is traced in fig. 179.

Now, to find the metacentre corresponding to any assigned centre of buoyancy, it is only necessary to draw, perpendicular to the water surface, a line from the given point $B_2$ on the buoyancy curve until it intersects the axis QK.

Its position may be located algebraically, thus:—

$$B_1M_2 = M_2C + CB_1$$
$$= M_2C + y.$$

Now, the triangles $CB_2M_2$ and RFO are similar, their sides being respectively at right angles to one another.

Therefore

$$M_2C = \frac{OR}{RF} \cdot CB_2$$

$$= \frac{L}{2\Delta}x,$$

and

$$B_1M_2 = \frac{L}{2\Delta}x + y.$$

But

$$x = \frac{\Delta L}{6D} \text{ and } y = \frac{\Delta^2}{6D},$$

whence substituting

$$B_1M_2 = \frac{L^2}{12D} + \frac{\Delta^2}{6D}.$$

So when $\Delta$ is zero in the initial position, $B_1M_2 = \frac{L^2}{12D}$; and when $\Delta = RH$,

$$B_1M_2 = \frac{L^2}{12D} + \frac{D}{6}.$$

The range of position of the metacentre is accordingly within a length $\frac{D}{6}$ upon the axis QK, and, similarly, within a length $\frac{L}{12}$ upon the axis RT.

A much simpler method, however, may be devised for obtaining the position of the metacentre by geometrical construction.

It will be observed that the equation of the parabola,

$$y = \frac{6D}{L^2}x^2,$$

is satisfied by the values $x = \frac{L}{2}$, $y = \frac{3D}{2}$; that is to say, the curve passes through the points G and P, which are the uppermost corners of the pontoon. Knowing also the vertex, $B_1$, of the curve, it is a very easy matter to construct, by any of the recognised methods, the parabolic arc $GB_1P$,* the portion of which, $\alpha\beta$ lying between the semi-diagonals HO and ON, constitutes a part of the buoyancy curve (fig. 180).

Now, for any assigned water-line FS, take the chord *fs* on this line which

* *Vide* construction, fig. 187, p. 213.

is cut off within the parabola, and bisect it at the point C. Draw $CB_2$ parallel to the axis QK, intersecting the curve at the point $B_2$. Then $B_2$ is the centre of buoyancy for the water-line FS. From $B_2$ draw $B_2M_2$ at right angles to FS so as to cut the axis QK in $M_2$. $M_2$ is the metacentre under the same conditions.

The proof of the foregoing construction lies in the facts that the tangent at the extremity of a diameter of a parabola is parallel to the chords which are bisected by that diameter, and that the tangent to the curve of buoyancy at any point is parallel to the line of flotation corresponding to that point as centre of buoyancy.

Thus far, we have only dealt with centres of buoyancy lying within the section HON; that is for water-lines ranging between GN and HP.

For water-lines between HP and NG, it is necessary to construct the parabola GZH. This being done (fig. 180), the curve of buoyancy is defined from $\beta$ to $\gamma$, and the corresponding positions of the metacentre may be

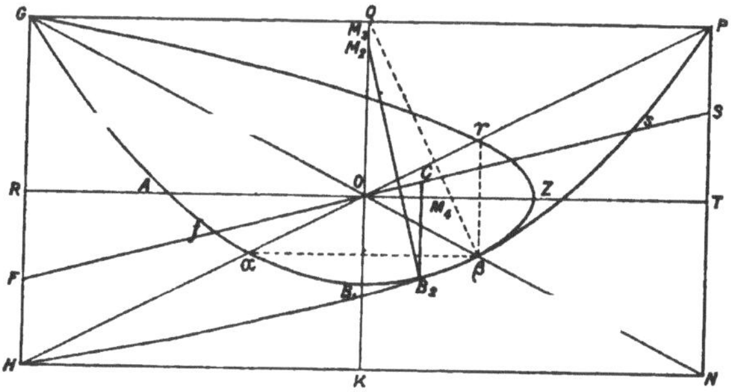


FIG. 180.

determined as already explained. They will lie, however, not on the axis QK, but on the axis RT, parallel to which the lines, from the water surface to the curve, must be drawn.

It is interesting to consider the water-line HP. There are two parabolic chords appertaining to the position, viz., $H\gamma$ and $\alpha P$. The middle point of the former is $\alpha$, and of the latter $\gamma$. Both the line $\alpha\beta$, drawn parallel to the axis RT of the parabola GZH, and the line $\gamma\beta$, drawn parallel to the axis QK of the parabola $GB_1P$, meet at the point $\beta$, which is the limiting position of the centre of buoyancy for both parabolas. The line $\beta M_3$, perpendicular to HP, gives extreme positions for the metacentre on the respective axes: the upper limit $M_3$ on QK, and the lower limit $M_4$ on RT. In fig. 180, owing to the particular ratio of L to D, the point $M_3$ coincides with the point Q.

The curve of buoyancy for the remaining moiety of the pontoon is simply a replica, upon the other side of the diagonal HP, of the curve $\alpha\beta\gamma$.

The parabolas $GB_1P$ and GZH have three point contact; that is, the curves not only touch one another, but cross when continued.

(II.) *Case of the unballasted pontoon immersed to any fraction of its depth.* —The foregoing is a special case in which the pontoon floats with exactly one-half of its bulk immersed.

From the principles enunciated, and by means of the methods which have been described, it is not difficult to determine the locus and draw the curve of buoyancy for the general case in which the pontoon floats with any proportion of its volume under water.

Let fig. 181 represent the conditions in question, the surface level of the water, RT, not being coincident with JX, the horizontal axis of symmetry of the pontoon.

In the initial position, the centre of buoyancy lies at the point B on the axis QK, which is such that $BK = \frac{RH}{2} = \frac{D}{2}$. Taking B as origin and a

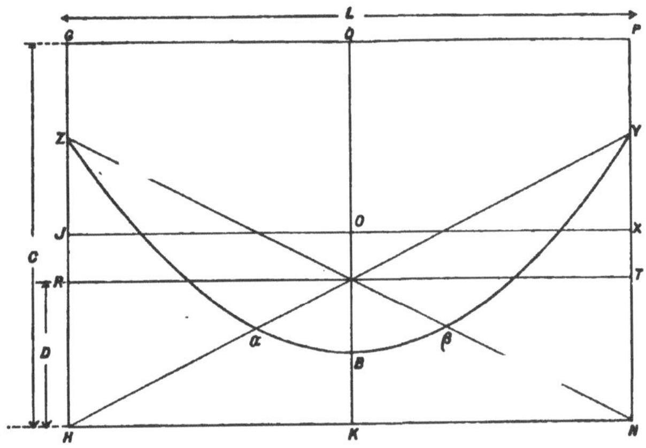


FIG. 181.

horizontal line through B as axis of $x$, we have (as explained in the preceding investigation) for all values of $\Delta$ between zero and D,

$$x = \frac{\Delta L}{6D} \text{ and } y = \frac{\Delta^2}{6D},$$

giving, as the equation of the curve,

$$x^2 = \frac{L^2}{6D} y.$$

The parabola thus defined passes through the points Z and Y, where RZ = TY = D, and constitutes the locus of the centre of buoyancy within the limits $\alpha$ and $\beta$ situated on the diagonals of the parallelogram ZHNY.

So long as the corner, H, of the pontoon remains under water, the immersed section is a quadrilateral. When the point H lies on the surface, the water-line passes through the point Y and the immersed area becomes triangular, remaining in that form until in course of continued revolution the water-line passes through P, from which point onward it resumes the quadrangular shape.

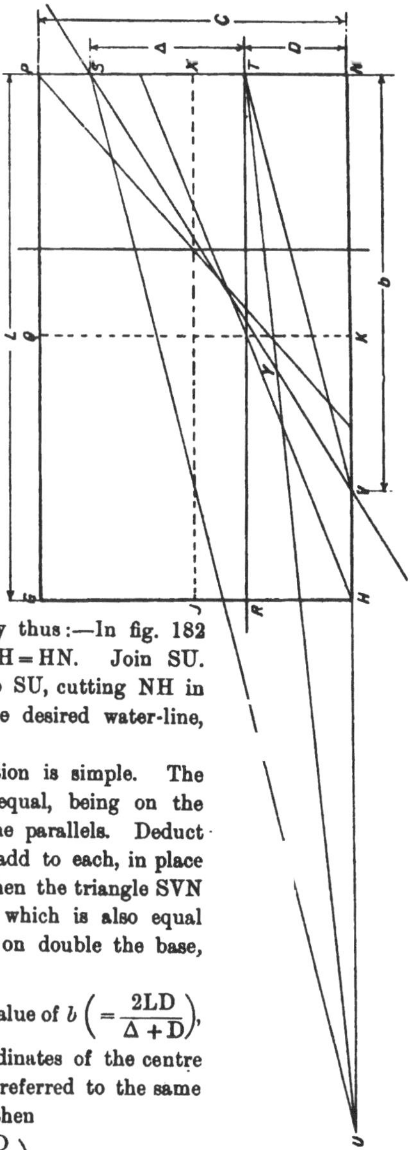

Fig. 182.

The second phase of the problem, therefore, is to deal with values of $\Delta$ between C and C − D (fig. 182), where C is the full depth of the pontoon and D the depth of immersion in the initial position.

First, we must determine the length of base-line on HN corresponding to any assigned value of $\Delta$, say TS.

Let $b$ (fig. 182) be the length of the base required. Then, since the area of the triangle of immersion must equal the area of the rectangle RHNT, we have

$$\frac{b}{2}(\mathrm{D}+\Delta)=\mathrm{LD};$$

$$\therefore\ b=\frac{2\mathrm{LD}}{\Delta+\mathrm{D}};$$

or, we can find $b$ geometrically thus:—In fig. 182 produce NH to U, so that UH = HN. Join SU. Through T draw TV parallel to SU, cutting NH in V. Join SV. Then SV is the desired water-line, and VN = $b$.

The proof of the construction is simple. The triangles SVU and STU are equal, being on the same base and between the same parallels. Deduct the common portion SYU, and add to each, in place of it, the trapezium VYTN. Then the triangle SVN is equal to the triangle UTN, which is also equal to the rectangle RHNT, being on double the base, between the same parallels.

Reverting to the algebraical value of $b\left(=\frac{2\mathrm{LD}}{\Delta+\mathrm{D}}\right)$, and taking $x$ and $y$ as the co-ordinates of the centre of gravity of the triangle SNV, referred to the same axes as before intersecting at B, then

$$x=\frac{\mathrm{L}}{2}-\frac{1}{3}\left(\frac{2\mathrm{LD}}{\Delta+\mathrm{D}}\right)$$

$$=\frac{\mathrm{L}}{6}\cdot\frac{3\Delta-\mathrm{D}}{\Delta+\mathrm{D}}$$

and
$$y = \frac{\Delta + D}{3} - \frac{D}{2} = \frac{2\Delta - D}{6},$$

whence we can obtain the equation for the locus, viz :—
$$36xy + 18Dx - 18Ly - LD = 0,$$

which indicates a hyperbola.

Transfer the origin to $x_1$, $y_1$. Then
$$36(x + x_1)(y + y_1) + 18D(x + x_1) - 18L(y + y_1) - LD = 0.$$

If the coefficients of $x$ and $y$ in this expression be made equal to zero, the values of $x_1 y_1$, corresponding thereto, will give the centre of the curve. Thus
$$36y_1 + 18D = 0,$$

and
$$36x_1 - 18L = 0.$$

Therefore
$$y_1 = -\frac{D}{2},$$

and
$$x_1 = \frac{L}{2}.$$

Accordingly, the point N is the centre.

Referring the equation to N as origin, with NH and NP as axes, it becomes
$$36\left(x + \frac{L}{2}\right)\left(y - \frac{D}{2}\right) + 18D\left(x + \frac{L}{2}\right) - 18\left(y - \frac{D}{2}\right) - LD = 0;$$

whence
$$xy + \frac{2}{9}LD = 0.$$

Therefore NH and NP are the asymptotes of the rectangular hyperbola constituting the curve. Further, since
$$xy = -\frac{2}{9}LD,$$

we see that the locus lies on the conjugate hyperbola.

If the transverse axis of a rectangular hyperbola be $2a$, then
$$4xy = -2a^2$$

is the equation of the conjugate hyperbola.

This may be written
$$xy = -\frac{a^2}{2}.$$

Comparing it with the preceding value of $xy$, we see that
$$\frac{a^2}{2} = \frac{2}{9}LD;$$

that is,
$$a = \frac{2}{3}\sqrt{LD}.$$

Turning again to the general equation,
$$36xy + 18Dx - 18Ly - LD = 0,$$

it will be observed that the curve passes through the point $\frac{L}{6}, \frac{D}{6}$.

Further, taking the equation of the parabola in the first phase, viz.,

$$x^2 = \frac{L^2}{6D}y,$$

where the two curves intersect, we get

$$36 \cdot \frac{6D}{L^2}x^3 + 18Dx - 18L\frac{6D}{L^2}x^2 - LD = 0,$$

or

$$(6x - L)^3 = 0.$$

Therefore the parabola and hyperbola have three point contact at $\frac{L}{6}, \frac{D}{6}$.

The curve may now be traced from the foregoing data.[1] This being done (fig. 183), it remains to find the centre of buoyancy corresponding to any possible water-line within the limits already specified. Take VS as such a

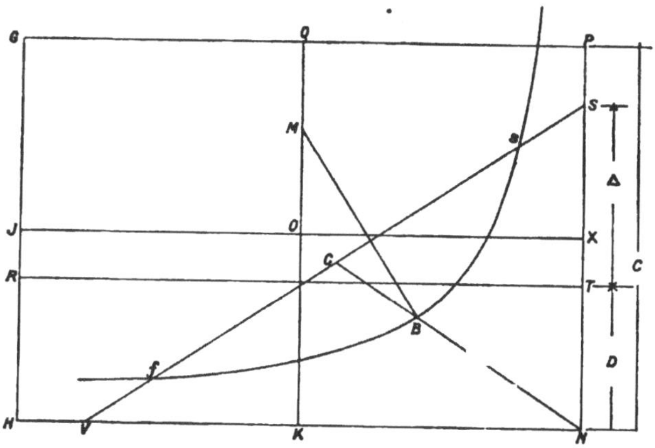


Fig. 183.

line and, as in the case of the parabola, bisect that portion of it, *fs*, which forms a chord of the hyperbola. Join this middle point C to N the centre of the curve. Where the line CN intersects the curve is the centre of buoyancy for the position, and a line drawn therefrom at right angles to the water-line will cut the original vertical axis in the metacentre. Thus B and M are the centre of buoyancy and the metacentre respectively, for the water-line VS.

The last phase of the problem is the same as the first. When the pontoon is immersed as with its axis JOX (fig. 184) vertical in the initial position, we have a similar set of conditions to those in which the axis QOK was initially vertical.

To find the depth ($D_1$) immersed, we have

$$D_1C = LD \text{ or } D_1 = \frac{L}{C}D,$$

[1] The method is shown in fig. 186.

or, geometrically, thus (fig. 184):—Through the point W, in which the diagonal NG intersects the primary water-line RT, draw the line YZ. Then YZ is the new water-line, and $VX = D_1$.

It follows, exactly as before, that the curve of buoyancy for the latest

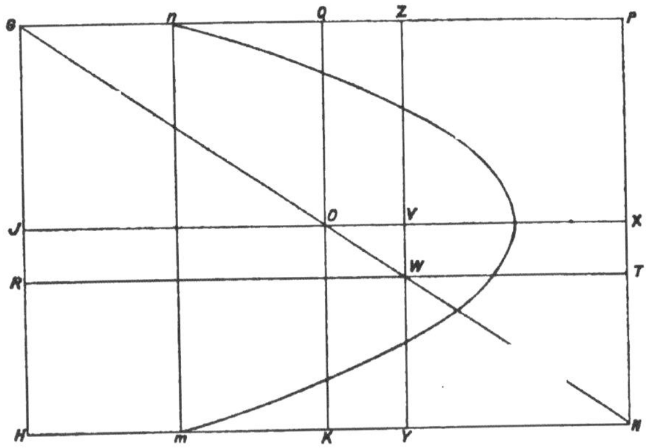


FIG. 184.

condition of things is a parabola, the vertex of which is in JX at a distance $\frac{D_1}{2}$ from the point X, and which passes through the points $m$ and $n$ in the lines HN and GP, where $Nm = Pn = 2D_1$.

We have now traced the curve of buoyancy through rather more than half of its path, and the remaining portion lies symmetrically about the same

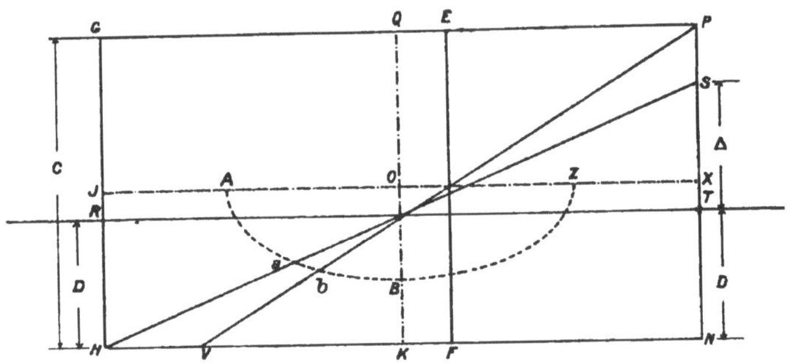


FIG. 185.

axes, so that it is quite easy to draw the entire curve. A moiety is shown in fig. 185. In the left hand quadrant the parabolic arc extends from A to $a$, to be succeeded by a hyperbolic curve from $a$ to $b$. From $b$ to B the curve is parabolic once more.

Hence, the complete locus of the centre of buoyancy is a curve made up of four parabolas and four hyperbolas.

When $D = \frac{C}{2}$, the hyperbolic portion vanishes and we get the special case of a pontoon immersed to one-half its depth, the investigation of which occupied our attention at the outset.

**Method of tracing parabolic and hyperbolic curves.**—*Parabola.*

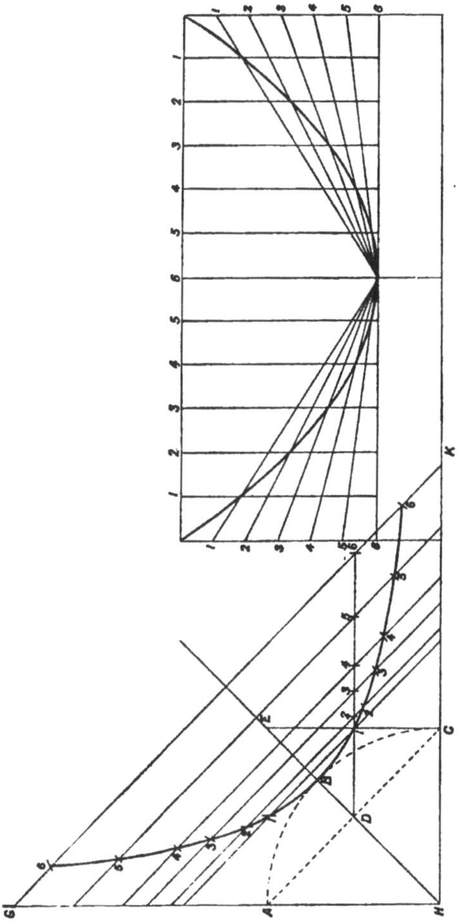

FIG. 187.—Parabola.

FIG. 186.—Hyperbola.

Divide the semi-base and the height into an equal number of parts, as shown in fig. 187. Draw lines parallel to the axis through each of the points on the base, and join the points on the sides to the vertex. The intersections give points on the curve.

*Hyperbola* (see fig. 186).—The curve required is the rectangular hyperbola, and the distance *a* from the centre H to the vertex B is given. Draw the axis HX bisecting the angle GHK. With centre H and radius *a*, describe the quadrant ABC. Join AC. From C draw CE at right angles to HK, cutting the axis in the point E, which is the focus. Through D draw a horizontal line of indefinite length.

Take any number of parallel lines intersecting the axis at right angles. Measure along the horizontal line through D the distance from D to each point of intersection. With these distances as radii respectively, and the focus as centre, mark corresponding intersections on the parallel lines, on both sides of the axis. These intersections are points on the curve.

*Note.*—In the investigation on p. 210, the value of $a$ was found to be $\frac{2}{3}\sqrt{LD}$. To obtain the value graphically, take a line (fig. 188) whose length is L + D, and divide it into two parts equal to L and D respectively. Upon the line describe the arc of a semi-circle, and from the point of section draw a vertical line to meet the semi-circle. This line has a length $\sqrt{LD}$, and two-thirds of it will give the value $a$.

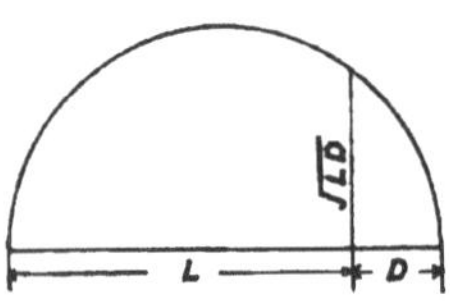

FIG. 188.

(III.) *Case of the ballasted pontoon.*—We have so far only considered the pontoon as an empty box—independent and self-contained. We have now to regard it in its working aspect. It is intended to carry a load, and for purposes of insertion and withdrawal from position it is oftentimes ballasted with water. The object of admitting water to the interior of the pontoon is to enable it to be lowered temporarily to a deeper draught, from which it can be raised again by pumping out the water.

We will deal first with the matter of ballasting.

When the pontoon is a single box, without compartments, the introduction

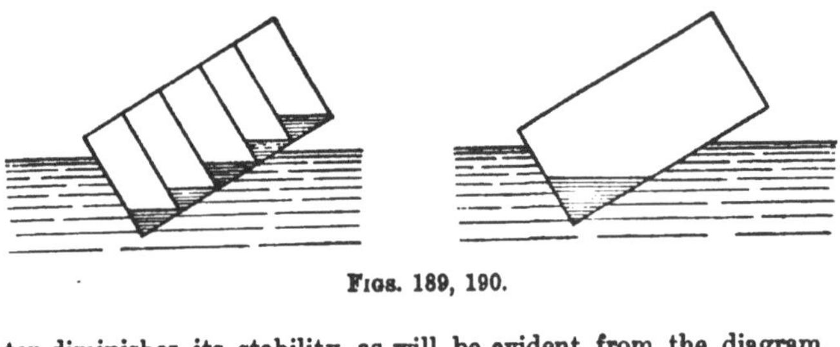

FIGS. 189, 190.

of water diminishes its stability, as will be evident from the diagram. The fluid, instead of remaining equally distributed under disturbance, immediately flows to the deeper side and there assists the overturning moment by its impetus, or, at least, impedes the righting effort.

The buoyancy, moreover, of the pontoon is reduced by the occupation of internal space.

The drawback of shifting ballast may be to some extent mitigated by subdividing the pontoon into compartments. When disturbed, the distribution of the water becomes less markedly unequal, as is evident from figs. 189 and 190. The greater the number of compartments the more uniform the ballasting will become; but, of course, there are limits of economy and accessibility to be considered.

Before investigating the case of the ballasted pontoon, it is desirable to reconsider the meaning to be attached to the term "centre of buoyancy" which we have hitherto found an important and essential feature in the statics of the unballasted pontoon.

After displacement through an angle, the forces tending to restore a ballasted pontoon to its original position, or to move it further from that position, are:—

(*a*) The weight of the displaced water, acting upwards through its centre of gravity.

(*b*) The weight of the ballast water acting downwards through its centre of gravity.

(*c*) The weight of the pontoon acting downwards through its centre of

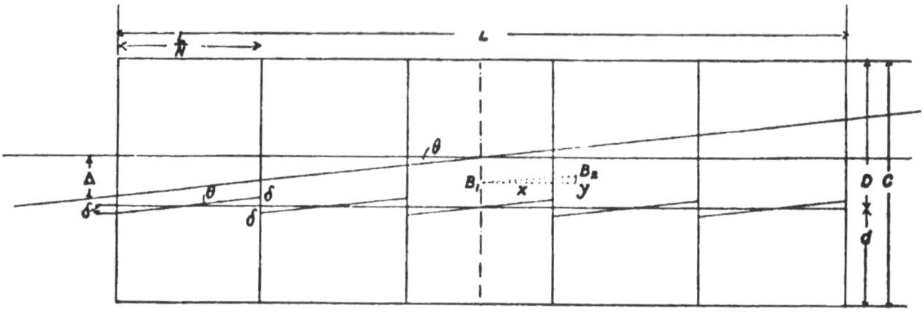


FIG. 191.

gravity, which, in the case of a rectangular pontoon, is the centre of that figure.

The resultant of the first two forces is a force equal in magnitude to the third force, but opposite in direction, and acting through a point which may be described as the centroid of the buoyancy area.

This resultant and the force (*c*) form the righting couple, and henceforth the term centre of buoyancy must be understood in the sense of **centroid of the buoyancy area,** which it really is. To avoid misconception on the point, it will perhaps be as well to adopt the expression "centroid of buoyancy."

Resuming the investigation of the stability of the pontoon, it is, in the first place, desirable to consider the alterations in position of the centroid of buoyancy in a pontoon which is subdivided into compartments. Let fig. 191 represent such an arrangement, the number of compartments in this case being five.

Under a slight displacement, the pontoon takes the position shown with reference to the inclined line in fig. 191. The buoyancy area has been changed from a rectangle into a series of parallelograms. A little inspection will, moreover, show that the portion of buoyancy area deducted from the extreme left-hand compartment has been added to the compartment on the extreme right, and that the adjoining compartments of each have likewise experienced a similar transfer of area.

The most obvious way, therefore, of determining the new position of the centroid of buoyancy is to sum up the products of the individual areas transferred, into the distances of their respective transferences, and divide by the whole buoyancy area. This will give us the proportionate transference of the centroid of buoyancy of the original rectangle.

Also, it will be well to proceed by means of the horizontal and vertical components, as before, which yield the co-ordinates of the locus.

Let D (fig. 191) be the depth of buoyancy as originally immersed, and $\Delta$ the extent of emergence or submergence of either side under an angular displacement $\theta$. The emergence or submergence in any compartment of N compartments we will call $\delta$. If $d$ be the original depth of water inside the pontoon, then $D+d$ equal the depth of the pontoon.

We have

$$\Delta = \frac{L}{2} \tan \theta .$$

Also from similar triangles,

$$\delta = \frac{L}{2N} \tan \theta .$$

Therefore
$$\delta = \frac{\Delta}{N} .$$

The area of buoyant section transferred from the extreme left-hand to the extreme right-hand compartment (1st to 5th) is

$$\left(\Delta - \frac{\Delta}{N}\right)\frac{L}{N},$$

and the horizontal distance between the centres of gravity of the two compartments is

$$\frac{N-1}{N} L .$$

The product of these two is

$$\left(\Delta - \frac{\Delta}{N}\right) \cdot \frac{L}{N} \cdot \left(\frac{N-1}{N}\right) L,$$

which, for the purpose of forming a series for summation, may be written

$$\Delta\left(\frac{N-1}{N}\right) \cdot \frac{L}{N} \cdot \left(\frac{N-1}{N}\right) L .$$

The similar product in the case of the 2nd and 4th compartments is

$$\Delta\left(\frac{N-3}{N}\right) \cdot \frac{L}{N} \cdot \left(\frac{N-3}{N}\right) L,$$

and, in the event of there being additional compartments, we could write as the next term,

$$\Delta\left(\frac{N-5}{N}\right) \cdot \frac{L}{N} \cdot \left(\frac{N-5}{N}\right) L .$$

If N be odd, the displacement for the middle compartment is zero. Accordingly, we have the following series to summate :—

$$\frac{\Delta L^2}{N^3}\left\{(N-1)^2+(N-3)^2+(N-5)^2+\ldots\ldots\right\};$$

which comes to

$$\frac{\Delta L^2}{N^3}\cdot\frac{N(N^2-1)}{6}=\frac{\Delta L^2}{6}\left(1-\frac{1}{N^2}\right).$$

The area of the original rectangle is LD, and, dividing the preceding expression by it, we obtain the following value for the horizontal component of transference :—

$$x=\frac{\Delta L}{6D}\left(1-\frac{1}{N^2}\right).$$

Similarly, it can be shown that

$$y=\frac{\Delta^2}{6D}\left(1-\frac{1}{N^2}\right).$$

Comparing these equations with the values on p. 205 previously given for the ordinates in the case of the undivided and unballasted pontoon, we see that they differ only by the factor $\left(1-\frac{1}{N^2}\right)$, which is a constant for any assigned case. Therefore, so long as the displacement is confined within limits such that the bottom of the pontoon is not exposed either inside or out, nor the upper corners immersed, the curve of buoyancy is parabolic as before.

The metacentric height, measured above the primary centroid of buoyancy, is

$$\left(\frac{L^2}{12D}+\frac{\Delta^2}{6D}\right)\left(1-\frac{1}{N^2}\right);$$

or, measuring from the centre of depth O of the pontoon,

$$\left(\frac{L^2}{12D}+\frac{\Delta^2}{6D}\right)\left(1-\frac{1}{N^2}\right)-\left[\frac{C}{2}-\left(d+\frac{D}{2}\right)\right].$$

When in the upright position, $\Delta=0$, and the latter expression reduces to

$$\frac{L^2}{12D}\left(1-\frac{1}{N^2}\right)-\left[\frac{C}{2}-\left(d+\frac{D}{2}\right)\right]$$

from which it is clear that with an increase in the number of compartments, the metacentric height increases, and therefore the greater the number of the compartments the greater the stability.

When $N=1$, the metacentric height is

$$-\left[\frac{C}{2}-\left(d+\frac{D}{2}\right)\right];$$

and as the metacentric height (measured, as assumed above, from the centre of depth) is proportional to the moment of the righting couple, the pontoon is in stable, neutral, or unstable equilibrium, according as

$$d + \frac{D}{2} \gtreqless \frac{C}{2} .$$

We turn now to the question of load, which only concerns the pontoon in its empty condition. When in position a pontoon carries kelsons, or main

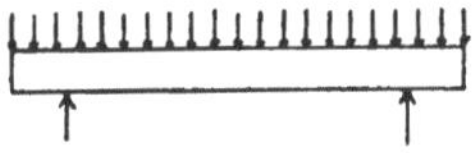

FIG. 192.

FIG. 193.

girders, which, in turn, receive the load of deck-beams and stringers. These imposed loads, beyond raising the centre of gravity, do not affect the external conditions of equilibrium; their immediate interest is in regard to the conditions of internal equilibrium.

The problem of the internal stresses to which pontoons are subjected is not one which need cause any difficulty in the way of solution, the same methods of procedure being applicable as in dealing with ordinary beams. There is

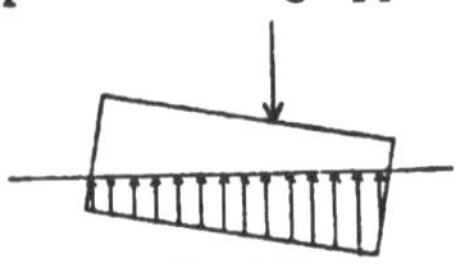

FIG. 194.

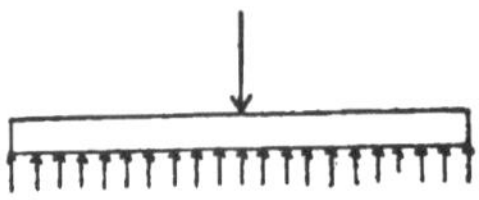

FIG. 195.

but one difference, albeit a striking one, between the two cases; but the effect of this is not nearly so embarrassing as might at first sight appear. In a beam the upward reaction is concentrated at isolated points of support; in a pontoon the reaction is distributed over the whole of the immersed area. A very simple expedient serves, however, to put the two cases on an equal footing.

Take fig. 192, representing a beam uniformly loaded and supported beneath at any two points. Now, invert the diagram, as in fig. 193, and we

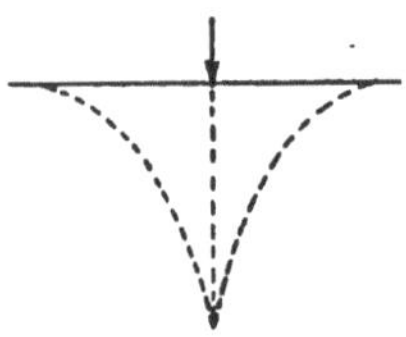

FIG. 196.

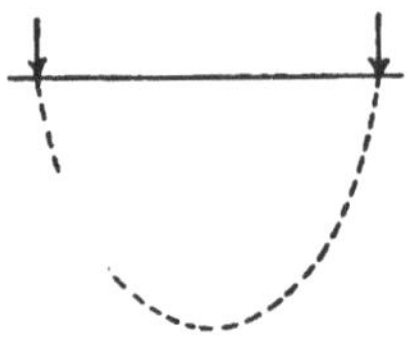

FIG. 197.

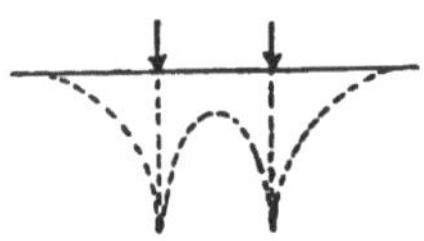

FIG. 198.

have the case of a floating pontoon carrying two concentrated loads. Obviously, the same diagrams of shearing stress and bending moment will serve in both cases, and there will be no difficulty in proceeding by this method in most cases. Even when the pontoon does not float upon an even keel, a measure of the exact distribution of the upward force is given by the area of the buoyant section (fig. 194). Figs. 196–198 are consequently typical bending moment diagrams.

When a pontoon supports three or more concentrated loads, the conditions become identical with what is known as the case of the continuous beam, but, in this instance, one of the chief obstacles to a ready determination of the

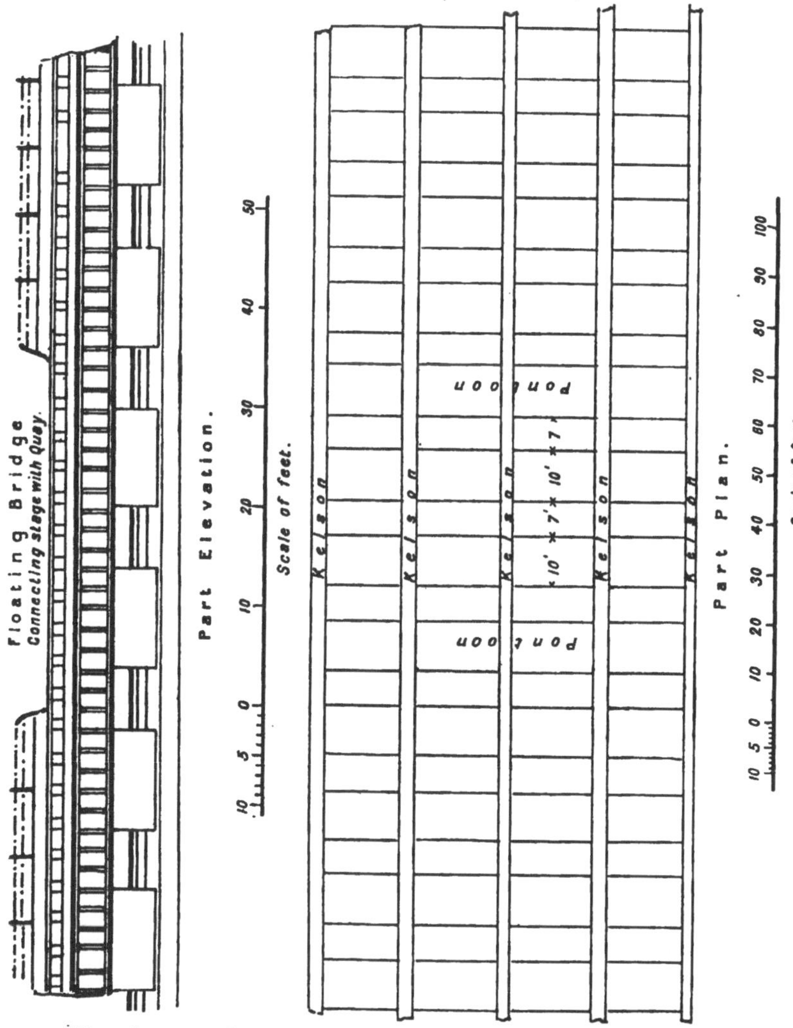


FIGS. 199, 200.—Liverpool Floating Landing-stage.

problem is removed, in that the exact amount and dispositions of all the external forces is known.

Such being the conditions, it is unnecessary to pursue the matter further, as it simply resolves itself into an application of the ordinary principles of mechanics to structural stresses in general, and lies, therefore, outside the special province of this volume.

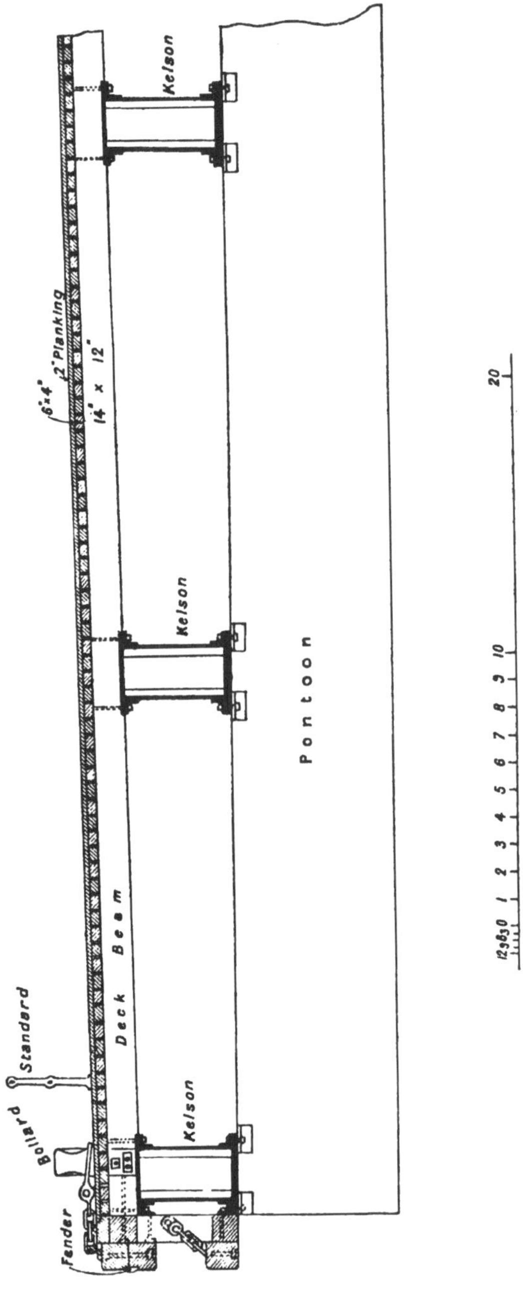


FIG. 201.—Liverpool Floating Landing-stage—Part transverse section.

**Liverpool Floating Landing-stage.**—The Liverpool stage, of which illustrations are given in figs. 199–201, showing the arrangement of the pontoons and decking, is 2478 feet long by 80 feet wide. It has eight bridges connecting it with the shore, and, in addition, a floating bridge 550 feet long by 35 feet wide, by means of which an easy incline for goods traffic is maintained at all states of the tide, which, during springs, has a maximum range of about 30 feet. The constructive arrangement of the stage is evident from the figures.

# CHAPTER IX.

## ENTRANCE CHANNELS.

Variation in Conditions—Features of a Tidal Régime—Blind Channels—Variable Channels—Fixed Channels—Accretion and Reclamation—Navigable Routes—Bars and their Origin—Training Works for Channels—Groynes—Walls—Fascines—Wave Traps—Height and Extent of Training-walls—Dredging Appliances—Mechanical Eroders—Rock-cutting—Suction Dredgers—Sluicing—Instances of Channel Regulation Works at the mouth of the River Weser, Germany; at Tampico Harbour, Mexico; at Westport Harbour, New Zealand; at the mouth of the Richmond River, New South Wales; and at the Ports of Ostend, Belgium, and of New York, U.S.A.

**Variable Conditions of Entrance Channels.**—The regulation, and, where necessary, the rectification of entrance channels, are matters of extreme moment to all ports which are situated otherwise than upon the open seaboard, and especially do they call for attention in connection with harbours located within an estuary or upon the banks of a river within tidal range. In the case of ports lying either upon a river flowing into a tideless sea, or upon a tidal river above the limits of tidal access, the agencies determining the form and direction of the river bed are, comparatively speaking, fixed and constant. The stream follows certain well-defined laws which, if not thoroughly understood, are at least clearly enunciated and expressed. It is a matter of general knowledge that in pursuing its sinuous course to the sea, the current of a river, as is indicated in fig. 202, impinges alternately against each bank, scouring the concave side of a bend and being thence diverted to a similar concavity on the opposite side at the next deflection of the river-bed: all this being in accordance with the principles of centrifugal force and action. The tendency, therefore, is for the navigable channel to form and maintain itself along the line of current, and there are few or no conflicting agencies to interfere with or modify this tendency.

**Features of a Tidal Régime.**—Within the limits of tidal influence, however, the conditions are of a totally different character, and the dispositions arising therefrom become much more intricate and complex. The current does not always flow in a seaward direction. For a very considerable portion of the time it is completely reversed, and a very large bulk of sea-water is directed inland with a velocity which is sufficient to overcome the fresh-water discharge, and to raise the level of the surface for a very considerable distance. The main body of the incoming current, moreover, does not by any means necessarily pursue the same course as that of the downward stream; neither does it essentially confine itself, more especially near the river mouth, to one definite bed or channel. The influences at work upon littoral currents at the entrances of large coastal inlets are manifold and powerful. Gales and storms arise at irregular intervals from varying points of the compass, and the pressure exerted thereby is inevitably felt by the tidal flow which is accelerated or retarded, augmented or reduced, to an appreciable extent. The action of the wind, moreover, produces certain changes of direction. Hence, the course of the incoming tide, though generally established, is subject to some mutation, and the volume of its flow fluctuates very considerably, not only in consequence of irregular meteorological phenomena, but also in conformity with the natural cycle of springs and neaps.

In the tidal region, therefore, there are two conflicting agencies: first, the downward stream, with its relatively uniform flow and its tendency to establish a definite bed, and secondly, the tidal current with exactly opposite characteristics. The resulting feature of the tidal estuary is accordingly unstable channels amid shallows and sandbanks. Through the latter, the river, taking the line of least resistance as it presents itself at the moment, ploughs its course to the sea in routes which the succeeding tides break through, destroying them in succession as they are formed, before they have time to become confirmed. In the natural order of things, the fluvial current

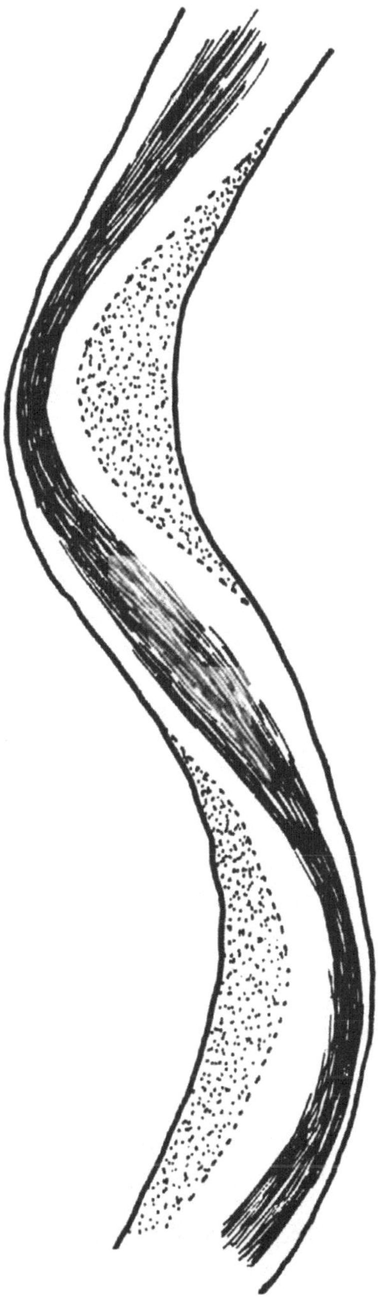

FIG. 202.—Type of Normal River Flow. Main bed of stream shown by dark lines; shoaling shown by tinting.

has acquired a sinuous or spiral motion, deflecting it from side to side, while the flood-tide sets inward in a straight line, curbed by none of the influences which control the river. The tendency of each is to obliterate the traces of the other where they diverge, and to accentuate the common bed where they coincide. At certain points, the river follows a course along one bank, while the main tidal stream favours that opposite, with the result that there are intermediate zones of slack water conducing naturally to the formation of shoals. A typical example of this, if it be necessary to select one, is furnished by the redoubtable "James and Mary" shoal, which constitutes the

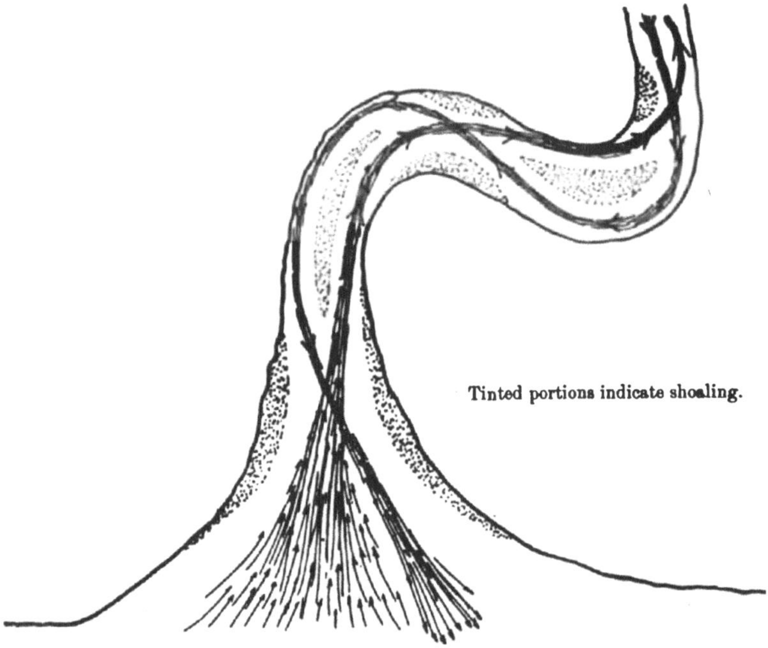


FIG. 203.—Diverse Courses of Inward and Outward Currents in Rivers.

most dangerous obstacle to navigation in the River Hooghly, and which has the evil reputation of being one of the most fruitful sources of shipwreck and disaster of any river in the world. The late Professor Vernon-Harcourt, who made a special and exhaustive study of the River Hooghly in 1901, thus describes the circumstances of the formation of the shoal[1]:—

"The descending current of the freshets in the rainy season scours out a deep channel alongside the concave left bank, and, diverging only slightly from this bank on passing Nurpur Point (figs. 204–206), it goes straight across the river into the very deep channel along the concave right bank below the bend a little beyond Gewankhali; whilst the deep flood-tide

[1] Vernon-Harcourt on the River Hooghly, *Min. Proc. Inst. C.E.*, vol. clx.

ULABARIA

channel along the right bank, from Mornington Point towards Shipgunj Point, becomes more or less silted up during the prevalence of the freshets. On the other hand, during the dry season the flood-tide channel, or Western Gut, along the right bank, in the lower part of the reach, opens out again, till the scouring energy of the flood-tide current is dissipated to some extent on approaching Shipgunj Point, where it spreads out and passes across to the left bank between Ninan and Fulta Point. At the same time, the Eastern Gut near the left bank, which depends on the ebb-tide almost entirely for its maintenance throughout the dry season, is reduced in depth; and a bar is formed between the ebb-tide channel near Hooghly Point and the deep channel at right angles to it in front of Gewankhali, by the conflicting action of the flood-tide running up this latter channel, thereby joining the James and Mary Shoal to the Hooghly Sand below."

**Blind Channels.**—This conflict of routes, resulting in the temporary predominance of one or other, is also responsible for the formation of what may be termed "blind channels" or cul-de-sacs, such as will be noticed in greater or less prominence on the charts of all important estuaries. These are deep depressions in the river-bed extending for some distance without any apparent outlet, terminating simply in a ridge or stopped end. The Sloyne in the River Mersey is a notable example, as also the Bog Hole off Southport in the estuary of the Ribble, Mostyn Deep at the entrance of the Dee, and the Great Nore Channel at the mouth of the Thames. These and many other instances are undoubtedly due to the antagonistic tendencies of the upward and downward streams.

**Variable Channels.**—The roving disposition of channels in sandy estuaries is manifestly the cause of much waste of physical power. The energy possessed by the stream in virtue of its momentum, which might profitably be expended in maintaining a deep clear channel and in removing or preventing any obstruction of the nature of a bar, is dissipated in the effort of eroding and displacing huge volumes of sand. It is observable in the river Mersey, for instance, that the estuarine channel rarely occupies the same position for a week consecutively. Between Hale Head and Garston, where the estuary is three miles wide, the channel has been diverted within a period of twelve months across the entire width from the Cheshire side to the Lancashire side, and *vice versâ*. Generally the changes are found to be coincident with upland floods, which bring a considerable accession of water; but so trifling are the initiatory causes sometimes, that barges grounding against a side of the existing channel have been known to produce a most marked deflection. The agitation arising from the process of erosion must inevitably cause a considerable quantity of sand to remain in suspension and to be transported to the mouth of the river, where its deposition in more tranquil waters is only a matter of time.

**Fixed Channels.**—A constant channel, on the other hand, where such can be assured, has all the advantages attaching to fixity and stability. It entails no frequent surveys with alterations of buoys and lights; it does its own maintenance work, and it acts generally on the lines of an ideal stream.

One forcibly impressive claim which has been put forward on behalf of a roving channel is, that by its constant change of course it deters the estuary from silting up in any part. This contention is one which has no little weight, because, with the reduction in capacity of a tidal basin or compartment, there is a corresponding reduction in the quantity of flood-water admitted, and a loss of scouring effect on the subsequent ebb. The confinement of a channel within restricted boundaries inevitably leads, in the case of water heavily charged with silt, to accretion in the adjacent submerged area. In other words, channel-training is a preliminary to land reclamation, and land reclamation is the general outcome of channel-training. Land reclamation is not an unmixed benefit; it may be attended by serious consequences to ports situated between the locality of reclamation and the sea, and it may entail other physical disabilities not altogether easy to foresee. Considerable discretion is therefore required both in planning and in carrying out undertakings embodying any such scheme.

**Fixed v. Variable Channels.**—Taking the question of river-training, however, as a whole, on its intrinsic merits, it seems to turn on the point of relative advantages—whether, in fact, it is preferable to have a deep, narrow, well-defined, constant channel, with adequate energy for its own maintenance, but with none utilisable for counteracting any silting tendencies elsewhere, or, on the other hand, to have a shifting channel with a more sluggish flow, sluicing a large expanse of sand so as to keep it from consolidating in any part, and so affording a broad waterway of greater sectional area, but of inferior depth, and subject to all the inconveniences of a shallow bar. The first undoubtedly represents the ideal condition, but, as indicated above, there are practical and circumstantial grounds in some cases constituting a preponderating argument in favour of the latter.

**Accretion.**—Although it is oftentimes assumed that accretion is the inevitable consequence of confining a channel within narrow limits, yet such an assumption is not legitimate on all occasions. Accretion can only arise from the deposition of suspended sediment, and this sediment can only be forthcoming from a supply in excess of that which the outgoing stream can carry. Now, there is nothing to show that any additional detritus is forthcoming from the upper reaches of a regulated river. But even supposing that there be an augmentation, the increased velocity of the stream renders it capable of transporting a larger percentage of solid matter than before. Evidently, therefore, any deposition which takes place is hardly attributable to detritus brought down by the upland waters.

The more likely and, as a matter of fact, the only possible source of accretion, is a tidal flow laden with the harvest of coast erosion. The flood-tide, entering estuaries on a sandy coast, is almost universally heavily charged with mud and fine particles which have every tendency to deposit themselves at the period of slack water, unless the down stream be so directed as to bear

TRAINING WALL
AS RECONSTRUCTED.

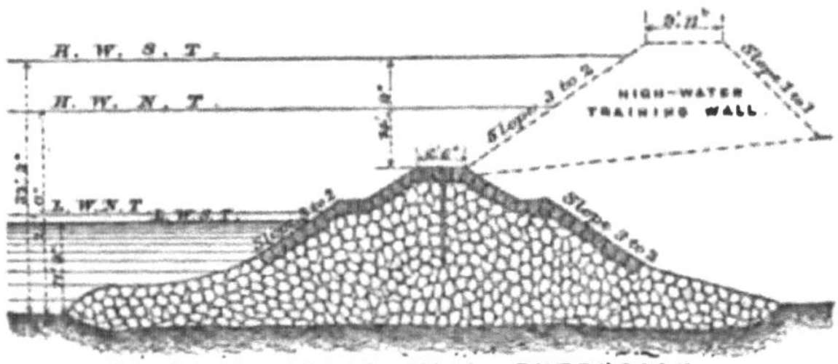


SOUTH TRAINING WALL EXTENSION.

e dotted lines.

upon the area of settlement, and this cannot be the case with a channel limited to one part of it.

The River Seine constitutes a typical illustration of the effects of training a channel through an estuary. Fifty years ago the outlet exhibited all the usual vagaries of estuarine channels in regard to alteration in position and irregularity of depth. From that time regulation works have been in hand, and the channel is now clearly defined from Rouen to some distance beyond Berville (fig. 207). At the outset, the probable results were entirely miscalculated; little or no consideration seems to have been taken of the question of silting, or rather, its potentialities were so under-estimated as to be deemed negligible. It was not long, however, before the consequences began to make themselves felt. Huge volumes of alluvium settled in the external vicinity of the training-walls, and the quantity increased rapidly as the capacity of the estuary to receive tidal water was diminished. Land reclamation followed as a natural sequel. But these processes, though beneficial in some respects, and by no means disadvantageous to the port of Rouen situated 74 miles up the river, became seriously prejudicial to the port of Havre at its mouth. The entrance channels of this latter port began to shoal, sandbanks formed in the approaches, and Havre, as a port, was threatened with extinction. The training-works were arrested for a time. The gain to Rouen had been undoubtedly great; a serviceable channel was promoted and assured, so that, whereas formerly vessels of between 100 and 200 tons navigated the distance from the sea with difficulty, vessels of ten times that tonnage now effected the journey with ease. Moreover, the gain of land had appreciable advantages from a national point of view. Still, it was manifestly mistaken policy to consider that these benefits outweighed a depreciation in the prosperity of the port of Havre.

The difficulty was met by providing Havre with a sheltered deep-water approach direct from the open sea, entirely beyond the influence of accretion in the estuary of the Seine. With this step, involving the construction of two breakwaters of considerable extent, inclosing a new harbour and the formation of an entrance facing south-west, and outside the estuary altogether, freedom has been gained for prosecuting the training-works of the Seine, and these seem destined to be continued to the river's mouth.

**Navigable Routes.**—It must be pointed out, from a navigational point of view, that the vagaries of a shifting channel do not always entail an entire change of route for shipping. Deep gullies and guts may be excavated on the site of former shoals, and adjacent gullies may be silted up; but vessels entering and leaving a port do not necessarily follow the line of greatest depth. Such a line may, in fact, be associated with the blind channels already alluded to. A navigable channel, as a rule, consists of a series of deeps separated by intervening ridges or shoals, and the serviceability of the channel is governed by the depths of the latter. When any one of the ridges becomes unduly high for the draught of passing vessels, then, in the absence of remedial measures, it becomes necessary to lay down another route; but so long as the depth of water is adequate, this step need not be taken.

## Bars.

The rectification and improvement of harbour approaches involves not only the training of channels, but in many cases also the removal of a bar—in part, at least.

A **bar** is a ridge or narrow plateau, or even a series of several ridges or plateaus, lying across the entrance to a river or coastal inlet, and rising up above the general level of the sea or river floor in its immediate neighbourhood, on both sides of it. When the altitude of the bar is sufficiently great to reduce the depth of water over its summit to an extent exceeding the limits imposed by the requirements of vessels using the entrance, it becomes an obstruction to navigation, and, in any case, it acts as an impediment to the development of the port or ports to which it is the threshold, and detracts from the navigable possibilities of the inlet.

Bars are to be found mainly in connection with tidal rivers; less often

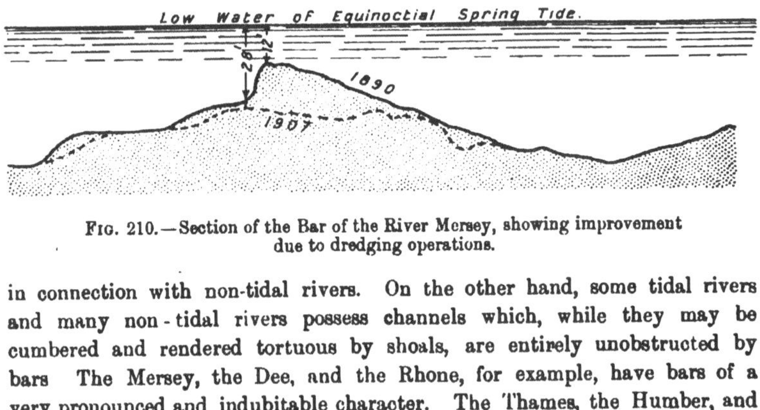


FIG. 210.—Section of the Bar of the River Mersey, showing improvement due to dredging operations.

in connection with non-tidal rivers. On the other hand, some tidal rivers and many non-tidal rivers possess channels which, while they may be cumbered and rendered tortuous by shoals, are entirely unobstructed by bars The Mersey, the Dee, and the Rhone, for example, have bars of a very pronounced and indubitable character. The Thames, the Humber, and the Severn have channels which enter the sea without any marked obstruction at all.

The **origin of bars** has been the subject of some controversy. It was formerly pretty generally held that a bar was due to the detritus brought down by inland waters, and deposited at a spot where the effluent, by reason of its reduced velocity, was no longer able to retain the material in suspension. This argument may indeed hold good in the case of non-tidal rivers, where it has also been advanced to account for the formation of deltas; but in tidal waters the fluctuation of ebb and flow at the river's mouth should obviously result in a dispersal of any such deposit as soon as it had formed, or even before the material had time to settle.

Another view was, that the source of the material being the same, its deposition is brought about by the meeting of conflicting currents, which created zones of slack water. Since, however, the meeting-places of such currents must necessarily vary to a considerable extent from time to time in accordance with the mutable conditions of tide, wind, and weather, the contention does not seem powerful enough on its own hypothesis to account for a fixed bar; and most bars are fairly stationary.

A third theory is that the widening mouth of a river, combined with a constant cross sectional area, naturally entails a reduction in depth. Against this it is to be urged that bars are almost universally abrupt mounds standing at slopes far steeper than would be the case above water, having regard, that is, to the angle of repose for the material, and that, therefore, they bear no apparent relationship to the much more gradual widening of an estuary.

The opinion now most generally held is that bars are the outcome of littoral drift, and that the chief causes of their formation are tidal currents and storms. Of these, the former agencies are more constant in action, and therefore perhaps more influential. The flood-tide, travelling along a shore which is being subjected to secular denudation, carries or rolls along with it a quantity of gravel, sand, and shingle, the motion of which is arrested when it comes in contact with a counter-current issuing from the mouth of a river. This theory does not altogether account for the existence of prominent bars in localities where littoral erosion is not an evident process. In this case, it is contended that the natural tendency of wave motion is to produce irregularities in the bed of the sea, and that these irregularities in certain places have culminated in definite ridges and depressions. But here again the explanation seems to be inadequate, since a bar is a special ridge peculiarly associated with river mouths, and not by any means ubiquitous; though, at the same time, it must be admitted that there are bars in existence off the coast-line, where no river finds its outlet, as, for instance, at Portland Bill.

Finally, it is to be noted that there are bars of indurated material, which are evidently of a permanent character and primeval origin, being due to the denudation of the sea floor and the attrition of its softer portions. Such ridges consist either of rock, tough boulder clay, or conglomerate, and they manifestly constitute features attributable to no transporting agency whatever.

The problem is one attended by some difficulties; and it apparently does not admit of a single solution only. In many cases there are indirect causes, some of which are obscure; while in general, the predominant tendencies are recognisable. On the whole, it seems fairly well established that the formation of the majority of bars, especially those in shallow, sandy estuaries, is attributable to the conflict between the external and internal physical agencies, and constitutes the régime under which the forces are in a state of equilibrium, more or less stable. Hence, if bars of this character be removed, there is every likelihood of their recurrence unless special preventive measures be taken. Bars of indurated material, on the other hand, are such as to give no ground for any apprehension of this kind.

## Training-works.

For the purpose of training navigable channels, any or all of the following measures may be adopted.

(1) **Training by means of Groynes.**—Groynes are narrow jetties generally of timber, occasionally only of stone or concrete, projecting from the bank into the bed of the river at right angles to the direction of its flow. In some cases, the groyne is formed by sheet-piling driven continuously and bound together by horizontal runners; in other cases, detached piles are driven in a straight line so as to form, with longitudinal walings, a series of bays or panels, ranging in extent from 5 feet upwards to 20 feet or more. These bays are filled in with planking, laid horizontally on edge, and spiked

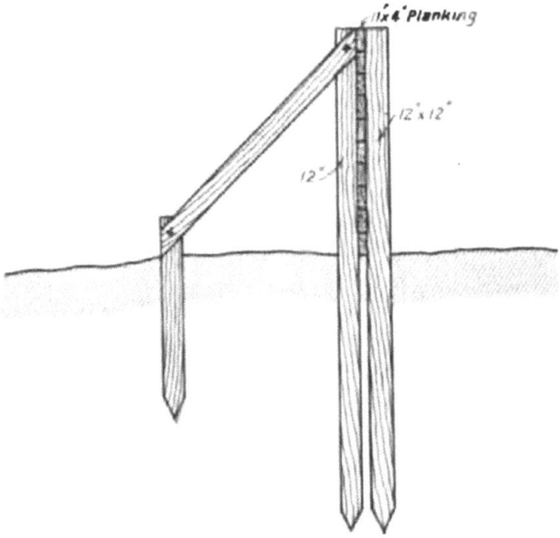


FIG. 211.—Timber Groyne.

to the piles, or by means of bundles of brushwood bound with wire, the interstices between the bundles being packed with clay and shingle. Bags of sand may be used for the same purpose.

The piles for groynes need not be of any great length; a depth of from 10 to 20 feet into the ground will generally suffice. As regards height, they will advisedly be brought at least to the level of the river bank, and as much above it as will serve to indicate the position of the groyne in times of flood. The body-work of the structure need not be carried higher than ordinary high-water level, if indeed so much as that.

Groynes are spaced at varying distances apart: sometimes, at intervals equivalent to their own length, sometimes more or less than that standard, according to the special requirements of each case. They have the effect of warding the current off the bank to which they are connected, and of urging it towards the centre of its bed, and so producing a contraction in width. This contraction or concentration of the flow results in increased scouring action, and consequently in a deepening of the river bed.

As regards the river sides, the spaces between the groynes become gradually warped and accreted, and the accumulation of material intercepted by the groynes leads to the formation of continuous embankments marked by a series of crescent-shaped embayments. These embayments are due to the eddying action of the current, which also has a tendency to denude and undermine the outer ends of the groynes. The extremities, therefore, should be specially protected, or, at least, constructed in a very secure and substantial manner.

Groynes have been extensively used both in this country and abroad: notably on the Clyde, the Tyne, the Tees, and the Danube. They constitute a useful initiatory measure, in that they do not enforce too rigid repression upon a stream. Constructed primarily in short lengths, capable of extension by easy stages, they deflect the current gradually and with an absence of those violent changes of environment which are so liable to produce untoward results.

When accretion has been proceeding for some time, and the current has been induced to occupy its intended channel, the outer ends of the groynes will advisedly be connected by means of a continuous wall so as to form an unbroken front. This leads us on to the second class of work.

(2) **Training by means of Walls.**—The term "wall," though in common use in this connection, is not strictly applicable to the whole class of structures included within its category. In the majority of cases the so-called walls are merely mounds of rubble stone; sometimes the rubble does no more than form a rough surface paving or pitching to a slope, from 2 to 3 feet thick, or even less; at other times it stands up to some height to a wedge-shaped section with a broad base. Moreover, the wall, whether a pitched slope, revetment, or upright mound, is far from being universally constructed of stone. Fascine mattresses, either singly or in layers, have been most successfully adapted to all the functions of a training-wall. Slag and clay are also used, and, in minor cases, bags of sand.

The formation of a **stone training-wall**, though apparently a simple process, is attended by certain difficulties. Rubble, when deposited in a heaped mass, has every disposition to subside in a foundation of soft, saturated sand and mud, particularly when the action is fostered and assisted by the scour of a current along the base. The loss incurred in this way has to be made good, and further material deposited until a firm bearing is obtained; and this result is not achieved, in many cases, without considerable outlay in supplies of stone.

In general practice, the rubble is thrown or tipped overboard by hand from punts and barges; but the process is slow, and, if the undertaking be at all extensive, it will prove a more expeditious and economical course to discharge from hopper barges. When dealt with in this way, the stone takes an initial slope of 1½ or 2 to 1, which subsequently may become modified to 2 or 3 to 1.

**Fascine work** has been largely practised as a substitute for stone in cases where the bed of a river consists principally of quicksand incapable of supporting any great intensity of pressure. And as most estuaries are of a sandy nature, more or less uncertain and treacherous, it is a system which naturally suggests itself, in those cases, for adoption. Circumstances are particularly favourable to fascine work, for instance, in the sodden, low-lying land on the shores of the Netherlands, and at the outlets of the fenland on the east coast of this country.

The nature of fascine work has already been alluded to in connection with its employment for jetty construction (p. 171). For that purpose it is chiefly built up in the form of *mattresses* which are equally suitable for covering a large area of sloping bank, and for being raised in tiers. Where mattresses are not essential, faggots, or "kids," as they are locally called in Lincolnshire, consisting of 6-feet lengths of thorn branches, cut from hedgerows, and made

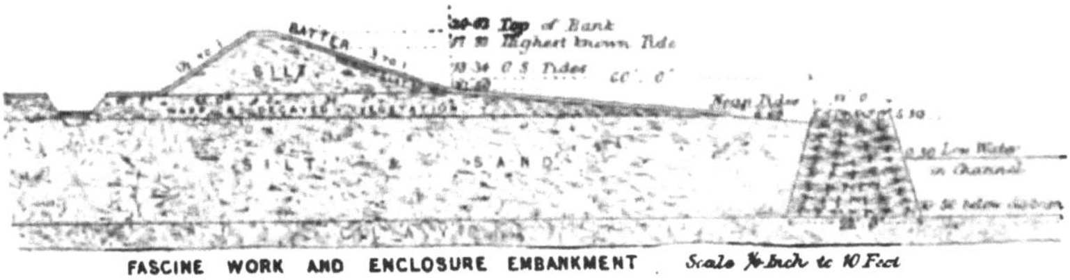


Fig. 212.—Fascine Work in the Wash.[1]

up into bundles 3 feet in girth, may be utilised. These are lighter to lift and easier of manipulation. They are placed overboard, and weighted with sods and clay until they sink, the wall being built up in this way, with the kids overlapping each other in transverse layers.

The interstices of fascines in a waterway rapidly fill with a deposit of earth and detritis, which soon solidifies, and the whole becomes a tough, composite bank, closely cohesive, and, at the same time, fairly flexible; so that if any undermining should happen to take place, no sudden, abrupt fractures would be produced, but the mass would settle uniformly, and no part of it would have any tendency to slip out of position into the fairway of the channel, as sometimes happens in the case of rubble walls. Moreover, the tenacity of brushwood offers effective protection, not only from the ordinary scour of streams, but from the wash of passing vessels and the discharge of heavy rainfalls during periods of low water.

**Arrangement of Walls.**—Training-walls are either single or double. Single walls only are necessary when the nature of the flow is such that erosion is confined to one side of a river, as is the case at bends. In intermediate positions and straight reaches, and also in places where it is desirable to direct a stream across from one bank to that opposite, two parallel walls

[1] *Min. Proc. Inst. C.E.*, vol. xlvi., Plate 8.

are requisite; otherwise the stream will exhibit a tendency to spread, and the channel to shoal.

At the mouths of rivers, double retaining walls may be either parallel or splayed, and the splay may be inwards or outwards, so that the walls either converge or diverge as they approach the sea. Parallel retaining walls serve to maintain the downstream current unimpaired in strength and velocity; but if they are carried up to any height in tidal estuaries, they lead to an accretion which obstructs the flood-stream and excludes a considerable portion of the water which would otherwise enter the estuary. Another danger attaching to such walls is the likelihood of shoaling in the neighbourhood of the entrance, due to the arrest of littoral drift by the walls. This drawback has manifested itself in a number of cases, and at Dunkirk, for instance, the jetties have been extended outwards from time to time, in order to reach deep water and to scour away the intermediate deposit which threatened to destroy the accessibility of the port. Moreover, parallel walls do little or nothing towards the dissipation of storm-waves passing in from the sea. It is from this point of view that converging walls have been designed, the inclosed area being of the nature of a basin containing a relatively larger mass of water, upon which external agitation has less effect. These walls, in fact, are sometimes adapted so as to form compartments called **wave-traps** (fig. 213). The drawback of the system is the same as that mentioned in connection with parallel walls, viz., the reduction in volume, and consequently in scouring efficacy, of the influent waters. This objection, of course, only applies to tidal seas.

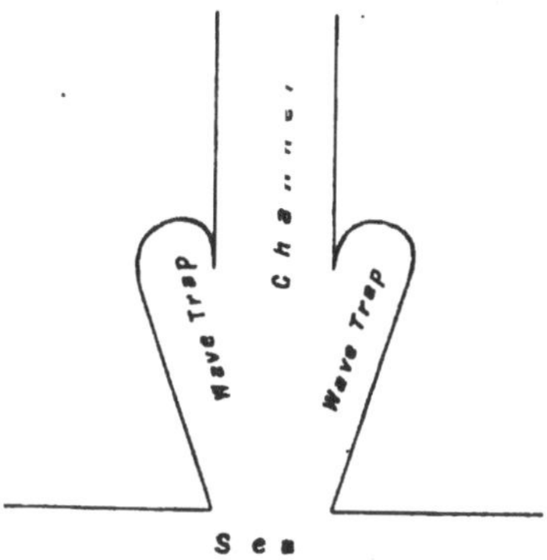

FIG. 213.

From this last standpoint, divergent walls are preferable, for with their splayed arms they admit the flood-tide freely and the outward flow of the ebb maintains the channel in that gradually widening form which is the ideal régime of an estuary. The contraction of the sides must not be too rapid, or there will be a tendency to throttle the inward flow, and pile up the tidal wave until it forms something of the nature of a "bore"—the term applied, in certain rivers, to an influx of water possessing a steep face and moving with considerable rapidity. This is dangerous alike to navigation and to the stability of the banks. It must be admitted that no great uniformity is exhibited in the expansive ratios of natural estuaries. They fluctuate exceedingly, and range, in parts, from something like 2550 feet to the mile in the Humber to little more than 100 feet to the mile in the Severn. On the whole, however, it may be said that a ratio of 2000 feet to the mile constitutes a suitable standard for adoption.

**Height of Training-walls.**—The height to which training-walls should be raised is a moot point. If nothing more than the mere rectification of a channel be in view, the wall will only be of the nature of a low revetment, confirming and protecting the edge of a newly-formed bank, and need not be raised above low water level. It has been urged against this, that in a sandy estuary, a channel so formed would soon be silted up with sediment washed in from adjacent banks over the top of the walls. There is, however, no more reason why silting should take place under the new conditions than there is under the old, and it may be safely assumed that the stream is powerful enough to maintain its own bed.

If it be desired to form an entirely fresh channel, or to radically divert an existing one, something more definite than a mere revetment becomes necessary; scarcely anything less than a half-tide wall will suffice to confine a stream within arbitrary limits and guide it through a novel environment. The tendency to resume a long-established course must always remain a powerful influence, if ever the compelling forces be modified or removed.

When land reclamation is definitely aimed at, training-walls will be first laid up to mean tide level, and then gradually raised until the level of highest high water is reached.

(3) **Training by means of Dredging.**[1]—Of all the agencies at work for the regularisation of channels and for the removal of natural impediments, there is none so effective and so powerful as dredging, exemplified, as it is, at the present day by machines of enormous size and tremendous capabilities. Natural scour is serviceable enough in its way, but it is only effective in soft, friable material. It is quite powerless to remove indurated ground within any reasonable time, and it has no influence whatever on huge boulders and rock. To all training-works, of whatever description, dredging is a most useful auxiliary, and there are few ports the entrance channels of which can be maintained without the aid of continuous and systematic dredging. It is, in fact, the recognised medium for the removal of bars and shoals. Nor are its operations confined to any one class of material. Dredging in rock is as feasible as dredging in alluvium, and boulders are removed as easily as sand.

At its first introduction, dredging was carried on by small and insignificant agencies, but the scope of present-day operations has become so vast and extensive as to necessitate the employment of extremely powerful plant and appliances. Such primitive expedients as "the bag and spoon," "the aquamotrice," and "the rake," except in very insignificant localities, have given way to large and imposing vessels, self-propelled, navigable, and specially equipped with machinery for dealing with something like from one to five thousand tons of material per hour.

The *conditions of dredging* work are exacting. Formidable obstacles are frequently encountered. Not only is the material dealt with of a very uncertain and varied character, ranging from impalpable mud to adamantine

[1] Dredging appliances are fully dealt with and illustrated in Chapter III. of *Dock Engineering*.

rock, and from the most friable sand to the stickiest clay, but vicissitudes of climate, weather, tide, current, and wind, have also to be reckoned with, and operations must generally be conducted in such a way as to cause the least possible disturbance to the existing conditions of navigation. All these matters cause frequent and expensive stoppages and delays. In some cases, the actual useful working time only amounts to one-fourth of the whole year, and it is never safe, under any circumstances, to reckon upon more than 200 working days per annum. A very large proportion of time is taken up with repairs; breakdowns are a common occurrence, and the expense arising from this cause is no inconsiderable sum.[1] Yet, in spite of all these drawbacks, dredging is an institution of untold value. By its means ports are brought into commercial prominence and saved from extinction. No other system can vie with it.

The *principle of dredging*, originally that of digging and dragging, has been extended to include pumping, so that modern dredgers are divisible into two types: first, those in which the action is mechanical erosion, and secondly, those in which it is hydraulic suction. In the most recent machines, both actions are combined.

**Mechanical eroders** comprise scrapers, cutters, picks, buckets, and grabs, singly or in combination.

Scraping implements, apart from suctional adjuncts, have only a very restricted application. They are intended to disturb and comminute material to such an extent as to render it readily removable by the force of the current. But the power of a current to maintain material in suspension is strictly limited, and it soon becomes laden to its fullest capacity. When this point has been reached, it can absorb no increment without an increase in velocity, and at the first diminution in its speed it deposits a portion of its load. Hence, mechanical scouring rarely produces more than a slight displacement, and it certainly is not capable of sustaining operations on a scale of any magnitude.

Combined with a suction tube and pump, however, it is a most useful agency. Experiments have demonstrated that, with the aid of suitable cutters and scrapers, marl, stiff clay, and adhesive material generally, may be separated and dissected to a degree compatible with its removal by pumping. The cutters employed are, generally speaking, cylindrical in shape, with straight or spiral blades mounted concentrically round the extremity of the suction tube. The efficiency of a cutter depends very largely on its design, on the size, number, and shape of the blades and their positions relatively to one

[1] Twelve months' record of U.S. dredger "Gedney," working at entrance of New York Harbour:—

| | |
|---|---|
| Actual working time, parts of 112 days, equivalent to . . . . | 92¼ days |
| Work prevented by weather (fog, storm, etc.) . . . . . . | 29¾ |
| Occupied in general repairs during winter . . . . . . . | 154 |
| Occupied in minor repairs . . . . . . . . . . | 21 |
| Lost from other causes . . . . . . . . . . | 10 |
| Sundays and holidays . . . . . . . . . . | 59 |
| | 366 |

another, and to the suction nozzle. Many of the earlier experimental forms were far from successful in their attempts to remove plastic material. The blades become clogged, and a very small proportion of solid matter found its way into the discharge pipe. Substantial improvements have, however, been effected of late years, and a modern suction cutter dredger is quite capable of dealing with the most adhesive and tenacious materials.

**Rock-cutting** involves dredging appliances of a different type—those allied to the pick or hand-drill. A long, heavy cylinder of steel, fitted with a hard cutting-point, is raised, and allowed to fall by its own weight upon the surface of the rock, which it splinters and pulverises. The hardest rock yields to this treatment, and the blows are repeated until the fragments are reduced to the size of ordinary ballast ready for removal by a bucket or grab.

The **Bucket Dredger** is to be found either in the form of a continuous band of buckets, called the *ladder dredger*, or of a single bucket, worked at the end of a long arm or lever, and called the *dipper dredger*.

The first of these stands foremost in importance. The principle on which it is constructed is that of an endless chain connecting a series of buckets, which revolve continuously around two pivots, or tumblers, at different levels. The buckets excavate material at the lower tumbler, and discharge it into a shoot while passing over the upper tumbler. Dredgers of the ladder type present two varieties: those in which the ladders are centrally situated, and those in which the ladders are set at each side of the dredger.

The bucket dredger can remove sand, clay, shingle, and marl, with equal facility, and it can even deal with the softer kinds of rock. In harder varieties of rock it follows in the wake of blasting operations, or of a rock-cutter. It will lift boulders of a moderate size. A dredger at Bristol, on one occasion, raised a boulder weighing 2½ tons without the least damage to the bucket. Most dredgers working in glacial clay have had some experience of boulder-lifting.

The **Dipper Dredger**, with a single fixed bucket at the end of a long lever arm, is almost exclusively an American type. It is used mainly on river beds and channels where the working depth is not very great; for sea work in deeper and more exposed water, the ladder dredger shows to better advantage. Mounted on a barge, and working either from one end or through a well-hole in the centre, the lever makes a curved upward cut, and the contents of the bucket, after slewing, are dropped into a scow or hopper ranged alongside.

The **Grab** consists of two or more curved plates, or jaws, capable of opening and closing in response to suitable mechanism. It is worked, to a very large extent, with the aid of gravity. Suspended by a chain or chains from the head of a crane jib, the bucket is allowed to fall freely by its own weight, with open jaws, until it buries itself in the ground. The jaws are then brought together, and the inclosed mass of earth is lifted. The economical scope of grab dredgers is limited to confined situations where other forms of dredger are unworkable.

The pumping principle is represented by one type only—the suction dredger.

The **Suction Dredger** has proved itself to be unquestionably one of the most remarkable contrivances ever devised for the removal of subaqueous material, both in regard to the enormous extent of its output and the low cost of its operations. It is to some extent, of course, a special machine. There are, naturally, conditions and circumstances to which it is not applicable; but they are few. It would be useless to expect it to dredge hard rock or to lift massive boulders. In all other cases, the efficacy of the suction dredger has been demonstrated beyond question.

The suction dredger consists essentially of a continuous pipe or tube, through which, by means of suitable pumping machinery, material is sucked up and discharged, either into a hopper forming part of the vessel itself, or into a scow ranged alongside, or through a shoot or tube leading to an adjacent bank or shore, which last arrangement lends itself very conveniently to land reclamation purposes. In the case of sand and light material, no preliminary treatment is necessary, but clay and marl have to be disintegrated by the cutters already alluded to, before they are in a condition to be drawn up the tube.

In exposed situations, such as prevail along the seacoast, the suction dredger possesses a marked advantage over apparatus of other types, the working of which is often materially interfered with by the motion of the waves. Equipped with telescopic pipes and flexible joints, the suction dredger readily adjusts itself to the rise and fall of the sea, and is quite independent of variations in level, either momentary or prolonged.

There is a great deal to be said, in extension of the foregoing remarks, on the relative advantages of the various types of dredgers, and there are many interesting features in connection with their working which might usefully claim our attention; but space will not permit us to pursue the matter further here.

We turn now to the last item in our series.

(4) **Training by means of Sluices.**—The principle of sluicing is based on that of the ebb-tide current, which, flowing out of a coastal indentation, scours its passage as it goes. The application of sluicing, however, is restricted to channel deepening and maintenance. It is rarely, if ever, employed in channel-making.

In practice, a large basin or receptacle is provided, within which the tidal water, entering up to the time of high water, is impounded and subsequently discharged through sluices or outlets at or about low water. The most effective period for sluicing is during spring-tides, when the flood waters are large and the ebb level is low.

The method has been largely used in ports bordering on the English Channel and the North Sea, such as Dunkirk, Calais, Boulogne, Dieppe, Ostend, etc., where the discharge of a large volume of water in this way has been found highly serviceable in keeping the harbour entrance channels free from silt. The system has its drawbacks. The retaining basins tend to silt up themselves during the quiescent period of retention. To obviate this, the basin is, in some cases, as at Honfleur, only filled about the time of high water, when the influent is comparatively clear. In other cases, as at Ramsgate and Dover, the basin has been divided into two compartments, one of which is used periodically to cleanse the other.

Some harbours are equipped with a natural sluicing basin. Such is the case at Santa Ana, Curaçao, which is probably one of the finest natural harbours in the world. The Schottegat lagoon, behind it, forms a tidal basin 2½ miles in length, with a depth of 50 to 60 feet. At Yarmouth there is a magnificent backwater, receiving various tributaries and forming an immense reservoir of fresh and salt water, which serves to keep the harbour fully open, and even deepens the approaches.

In cases where the sluicing basin is fed with fresh water, it is desirable to note that the specific gravity of fresh water being less than that of salt water, there is a marked tendency for the lighter liquid to flow over the denser; and this phenomenon, which is a matter of ordinary observation, detracts somewhat from the scouring effect of fresh water.

A coastal inlet or estuary may be transformed into an automatic sluicing basin by the construction, as at the mouth of the Liffey, of a low retaining wall, which becomes submerged above half-tide level. When the tide falls again below this level, the ebbing water converges to a contracted outlet, which sluices the harbour entrance.

Compared with dredging, sluicing is an agency not nearly so powerful or so effective. The head or pressure under which it acts is rapidly dissipated by the resistance which it encounters, and at some little distance from the source its scouring effect is greatly reduced, and rendered but slightly appreciable. Indeed, it may be said that sluicing, as a means of channel maintenance, has practically been entirely superseded by dredging.

So far as it is possible to deal in a single chapter with a subject which is capable of being expanded to an entire volume, the foregoing represents an attempt at a fair review of the methods by which the entrance channels of ports are improved and maintained. It only remains to append a few examples of work actually carried out in various parts of the world, in order to afford some illustration of the manner in which general principles are applied to particular cases, and the modifications which have to be introduced to meet special local conditions.

## Instances of Channel Regulation Works.

**Regulation Works at the Mouth of the River Weser.**[1]—The estuary of the Weser has been undergoing a course of improvement since the year 1891, when Herr Franzius designed works consisting chiefly of two training-walls for the removal of a bar, caused by a division of the current, which had existed for about thirty years, and had finally attained a length of

[1] Franzius and Thierry on River Regulation Works in Germany, *Min. Proc. Inst. C.E.*, vol. cxxxv.

WESER ESTUARY
REDSAND LIGHTHOUSE
ROBBEN PLATE
EAST EVER SAND
NORTH EVER SAND
WURSTER WATT

PANUCO

nearly 7 miles. One training-wall was situated on the left bank about 4½ miles long, opposite Bremerhaven, and lying between that port and Imsum, and the second wall was on the right bank, 1 mile long, between Imsum and Wremen. These walls had the effect of reducing the excessive width of the river, and the increased velocity of the water produced a scour which removed completely 7 million cubic yards of sand in two years, and changed the position of another 4¾ million cubic yards—all in a length of 12 miles. The movement of material, however, simply led to the formation of banks elsewhere, and a suction dredger of 500 I.H.-P. was ultimately obtained to keep open the channel to Bremerhaven for deep-draughted vessels, which it has done in a perfectly satisfactory manner. A second dredger of greater capacity was added, in 1898, as a substitute in case of damage to the earlier boat, possibly entailing lengthy repairs, and also in order to be in a position to deal with simultaneous shoaling of distant parts of the estuary.

The Robben Plate divides the outlet of the Weser into two streams, and there is a further subdivision at the West Ever Sand. In consequence of these obstructions, an outer bar had formed at the north end of the right-hand channel. To remedy this defect, in 1896 the branch stream between the East and West Ever Sands, which has a breadth of about two-thirds of a mile, was dammed by depositing a layer of weighted fascines, 3¼ feet in height, and in the following year a second layer was added. Simultaneously, protection works were carried out along the frontage of the Wurster Watt. The object in view during these operations was to attain a low-water depth throughout of 26 feet, so that there might be no restriction to the passage of deep-draughted vessels. This object was in due course attained.

It was also at one time proposed to block the channel lying to the left of the Robben Plate by means of a training-wall and dam, as indicated in fig. 214. The works, however, have not been carried out, as the channel has manifested a natural tendency to silt up, and it is considered that any artificial assistance in this direction is superfluous.

Writing in 1906 to the author, Herr de Thierry remarks:—

"The works begun in 1896 and 1897 between the East and West Ever Sands, and for the protection of the Wurster Watt, have been carried out and maintained. The two dredgers, "Columbus" and "Franzius," have been nearly constantly at work deepening and regulating the channel. The dredging of a deep channel caused difficulties between Imsum and Wremen, owing to a bank of boulders bedded in very stiff clay. The dredgers have worked chiefly at this place, and lower down, where the velocity of flow is sensibly decreased, on account of the branching off of the stream between the East and West Ever Sands. A low-water depth of 28 feet throughout was, however, notwithstanding these difficulties, achieved two years ago, and has since been easily maintained."

**Training-jetties at Tampico Harbour, Mexico.**[1]—The improvement works at the mouth of the river Panuco, for the port of Tampico in Mexico,

[1] Corthell on Tampico Harbour Works, *Min. Proc. Inst. C.E.*, vol. cxxv.

consist of two parallel jetties which have been built out from the shore-line into the Gulf of Mexico. They are about 6700 feet long, and extend into 24 feet of water, their direction being E.N.E., and they are 1000 feet apart between centre lines. Following the precedent of the work at the mouth of the Mississippi, Dr Corthell, their designer, constructed them of brushwood mattresses consolidated with rubble stone and detritus. The brushwood was obtained locally, either from the adjacent banks of the river, in which case it was conveyed by barges, or from near the railway, when it was transported by waggons provided with side-posts to retain the material, which, of course, though light, was bulky. The railway was specially extended from the town of Tampico to the mouth of the river for the purpose of conveying materials not only to the site, but also to their place of deposit. To this end, a trestle pier was constructed, which carried a double line of rails with several cross-overs. The mattresses were slung from the pier, between the underside of the pile caps and the surface of the water.

"For building the mattresses, supports of pine scantling, about 3 inches by 8 inches and of a length equal to the width of the mattress, were suspended athwart the jetty line from the caps and stringers of the pier, by means of ropes so arranged that they could be easily and simultaneously released. On the skids were laid other lines of scantling 3 inches by 6 inches, for about 60 feet, the length of the mattress lying longitudinally with the jetty. In these scantlings, forming the bottom framework of the jetties, there were inserted, before being laid on the skids, iron rods ⅝ inch in diameter and of the length required for the thickness for the mattress, which ranged between 4 feet to 7 feet. These longitudinal strips were placed 5 feet apart on the suspended skids, with the rods upright; the brush was then brought to the work, either in a barge alongside when the sea was smooth, or by cars overhead if the sea was rough. It was packed as closely as possible, first in a layer athwart the jetty, and then in a layer lengthwise with the jetty, and so on, until the required thickness was obtained. Mattress strips, or scantlings, of the width of the mattress, were then placed over these rods; and by means of heavy mallets and powerful "grip" levers, with an iron jaw to take hold of the rod, a pull of 3000 lbs. was brought to bear, and the mattress was compressed about 20 per cent. The rods were then bent down over the strips to hold them securely."

The character of the brush was not altogether satisfactory; it was generally crooked and very stiff, and did not yield to compression during construction, or give way to form solid work until it had been heavily loaded a long time with stone under water. For this reason, the final compression after loading was nearly 50 per cent. of the bulk of the mattress on completion.

"Between six and twelve waggon loads of riprap stone, each car carrying about 12 cubic yards, were then usually hauled by the locomotive to the point over the mattress. The ropes suspending the mattress were released, and the stone from the waggons thrown on to it, causing it to sink out of sight in a few moments. Mattress work was thus carried on when it would have been impossible to do so with a floating equipment. By this method of construction, which, with some change in details, was followed throughout the work, the mattresses were built round the piles, of which there were between four and eight in each bent, the bents being 15 feet apart. The only modification was in the varying thickness and width of the mattresses, and in often substituting ½-inch rods for the smaller rods in the corners or outer sides. Only one or two of the mattresses were injured by the waves."

As the south jetty was on the opposite side of the river from the railway terminus, it was necessary to ferry the waggons of rock and brushwood across. This was done by a "model" barge, with two tracks holding six waggons, aprons being arranged, adjustable to the tide, at the end of a short pier on each side of the river. A locomotive for hauling the waggons on the south side was ferried over, and used between the barges and the work.

The total amount of brushwood used on the jetties until the close of their construction in 1892, was 390,532 cubic yards; of rock, 373,048 cubic yards; and of pine piling, 253,347 lineal feet.

Dr Corthell furnishes the following details of actual cost:—

| | *s.* | *d.* | |
|---|---|---|---|
| Uncreosoted piles from the United States, | 1 | 4½ | per linear foot. |
| Uncreosoted Palma or other approved native piles, | | 11 | ,, |
| Creosoted piles from the United States, | 2 | 5½ | ,, |
| Mattress work, | 6 | 2 | per cubic yard. |
| Brush work, | 4 | 7½ | ,, |
| Large stone (not exceeding 3 cubic yards), | 9 | 3 | |
| Small stone (not exceeding ½ cubic yard), | 7 | 5½ | ,, |
| Concrete blocks, | 41 | 8 | ,, |

The prices include not only the materials named, but also all iron, straps, fastenings, ties, scantling, framework, etc., required for their use.

**Wave Basin at Westport Harbour, New Zealand.**[1]—"Westport Harbour is situated at the mouth of the Buller River, on the west coast of the middle island of New Zealand, and is the most important coal port of that colony. The river discharges into the Tasman Sea, nearly at right angles to the coast-line. About 6 miles to the westward of the entrance a natural shelter from the prevailing south-westerly winds is formed by Cape Foulwind and the Steeples."

The external works, designed by Sir John Coode and completed in 1893, consists of two converging breakwaters of granite rubble affording an entrance width of 700 feet in the clear. The breakwaters were splayed inwardly in plan, in order to provide wave-basins on each side of the river for the dissipation of heavy seas during gales. It has been found, however, by Mr Rawson, the engineer to the Westport Harbour Board, that the basin on the west side was unnecessary, and a training-wall, the expense of which might have

[1] Rawson on Westport Harbour, *Min. Proc. Inst. C.E.*, vol. cxxxvi.

16

been saved if the western breakwater had been run straight to a point, 1000 feet higher up the river bank, has since been constructed. "The entrance is open to gales from the north-north-west to the north-north-east, but the

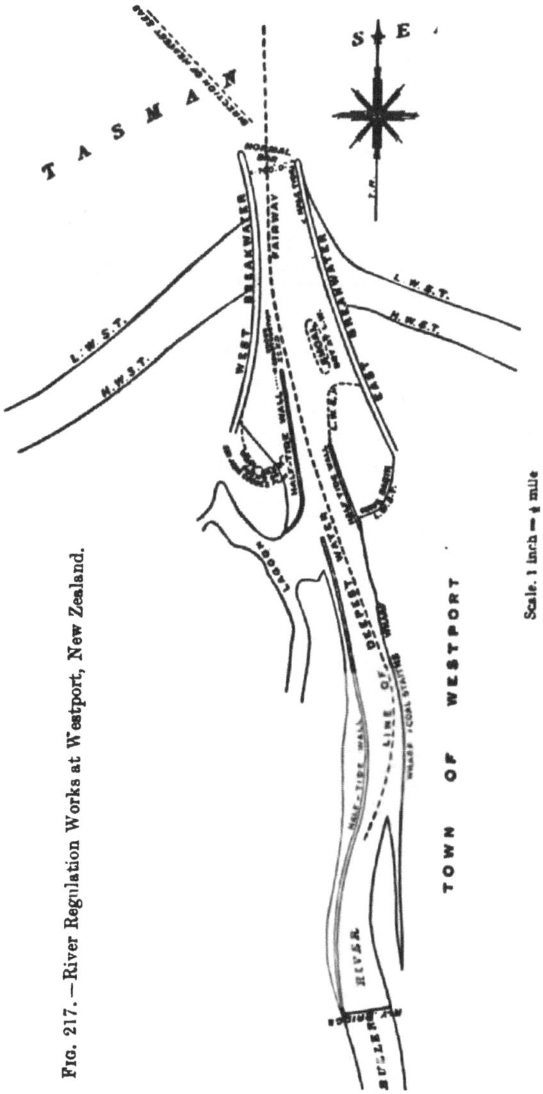


FIG. 217.—River Regulation Works at Westport, New Zealand.

heaviest seas are experienced from the north-west; in exceptional cases, occurring perhaps not once in a year, these seas break about one mile out in about 8 fathoms of water, and from thence to the bar is a mass of broken water. On ordinary occasions, however, the break is close to and on the bar; and in what is considered a very heavy sea, the waves range between about 10 feet and 12 feet in height." From whatever quarter they come, the seas are broken on the shoal within the harbour and completely dissipated in the eastern wave basin.

**Entrance to Richmond River, N.S.W.**[1]—The Richmond River is one of the most important rivers in Australia, serving a large agricultural and farming district. Owing, however, to the cultivation carried on along its bank, the flood waters are heavily charged with silt, and this naturally results in considerable deposits in the vicinity of the entrance. This deleterious action is somewhat counteracted by occasional heavy floods, which scour the channel, although only a small percentage of the rainfall over the basin finds its way into the river, owing to the permeability of the soil and the natural reservoirs formed by swamps. The great drawback to the Richmond River has been the shifting character of its entrance, combined with a shallow bar and adjacent shoals. These evils were intensified by the conflict of the waters of North Creek with those of the main river at their point of meeting, between East and West Ballina.

The position of the entrance has shifted through a distance of more than $1\frac{1}{4}$ mile, and it has also been noted that floods have caused the channel to break through the Southern Spit on four occasions in thirty-five years. Thus, the navigation of the port has been dangerous at all times, and on many occasions impossible.

In 1888, Sir John Coode was invited to report upon such means as were available for fixing the channel and regulating its width, so that the scour might be confined to a definite track of proper proportions; to neutralise the obstruction offered by certain rocks near the mouth of North Creek; and to prevent the conflict of the waters from the North Creek with those from the main river. The remedial works executed, and in course of execution, are shown in fig. 218. Some of these works have been carried out as designed; others have been somewhat modified in accordance with experience gained during the course of operations. The main features are sufficiently intelligible, and the only point calling for particular notice is the somewhat unusual addition of a middle training-wall.

"The construction of the new middle training-wall," says Mr Burrows, "which reaches ordinary high-tide level for the greater part of its length, was determined upon by Mr Darley, then engineer-in-chief, at a time when the unfinished condition of other works in progress allowed a large sand-spit to form across the area between the mouth of North Creek and the south wall, and it was found necessary to train the tidal currents of the river at this place, so that the discharge at ebb-tide would tend to prevent the spit increasing the obstruction to navigation at the river entrance. Gaps or openings were left in the wall for the preservation of the old navigable channel

[1] Burrows on Improvements at Entrance to Richmond River, *Min. Proc. Inst. C.E.*, vol. clx.

along the south wall, until a new channel should be dredged along the north

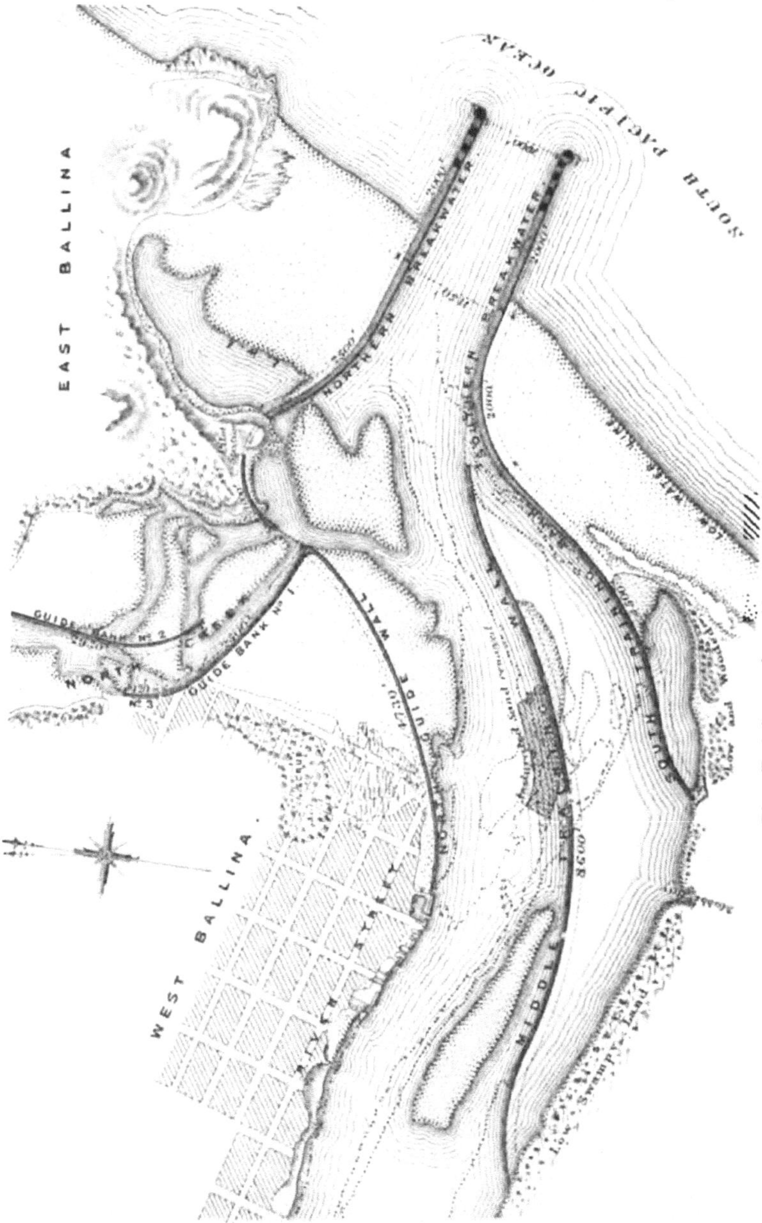


Fig. 218.—River Training-works at Mouth of Richmond River, N.S.W.

side of the middle wall." The results obtained have proved satisfactory.

The dredging of a channel through indurated sand affords an interesting comparison of the relative efficiency and economy of a suction dredger and a ladder dredger respectively, both working in the same material. The suction dredger "Dictys" was fitted with revolving cutting gear, without which she would have been useless, and the spoil was deposited over an adjacent wall, by means of pipes laid on pontoons, into a blind channel which had to be filled up. The bucket dredger "Alcides" followed in the wake of blasting operations, which were carried out cheaply and expeditiously, as the material was easily bored by a water-jet working at a pressure of about 80 lbs. per square inch, and a hole could be put down in this soft rock at the rate of 1 foot per minute. Nobel's Glasgow dynamite was used, and four holes were exploded simultaneously by an electric battery, the ladder dredging following and lifting the débris into hopper barges for conveyance elsewhere.

The following is a tabular statement of the result of a period of working extending over a year:—

| Name of Dredger. | Type of Dredger. | Period of Test. | Effective Work done. | Rate per Month. | Cost per Month. | Cost per Cubic Yard. | Minimum Rate per Cubic Yard per Month. |
|---|---|---|---|---|---|---|---|
| | | Months. | Cubic yds. | Cubic yds. | £ | s. d. | s. d. |
| Alcides | Single ladder, with boring-punt, diver, explosives, and tug-boat | 12 | 17,580 | 1,465 | 400 | 5 5½ | 3 9 |
| Dictys | Pump with cutting gear | 15 | 15,163 | 1,011 | 250 | 4 11¼ | 3 0 |

The comparison as regards output lies in favour of the bucket dredger; but as no towing of dredged material was necessary in the case of the suction pump, the cost of working by the latter system was lower. Possibly it might have been lower still, as at the commencement of the period some experimental work was being carried out with various kinds of blades in the cutting gear, to ascertain the most effective form, with the result that the original cutters were retained with but slight modification. The cost given in the statement includes wages, stores, repairs, etc., but excludes any interest on capital cost or charge for depreciation.

**Entrance Channel to the Port of Ostend.**—The fairway leading to the quays of the port of Ostend is maintained by a dual system of dredging and sluicing. Up to the year 1898 sluicing alone was in vogue, and its effects were deemed satisfactory and adequate. This was perhaps more particularly the case in the interior of the channel, which was subject to silting of a very light nature, the material being chiefly mud. At the entrance and in the external fairway, the results were not quite so pronounced, owing to the more compact and sandy nature of the deposit.

When, in 1898, new works were undertaken for the development of the accommodation of the port, two out of the three existing sluicing basins were withdrawn from use, and have since been demolished, their sites being used for other purposes. A new sluicing basin of much larger area has been designed to take their place.

At the same time, it was recognised that with the increased depth required for modern shipping it would be impossible to realise an effective maintenance service by means of sluicing operations alone. Dredging, therefore, was introduced as an auxiliary. The peculiar conditions appertaining to the port of Ostend are thus set forth by Mr Van der Schueren in his communication to the International Navigation Congress.[1]

"We have pointed out that the method adopted at Ostend for preserving the navigable depth of its channel, consists of a combination of sluicing and dredging.

"It may be objected that what can be obtained by sluicing can be equally well obtained from dredging, and that it is not necessary to have recourse to the combined system. At most, it would be a question of cost. It would be necessary to ascertain whether the mixed system is more economical than that of dredging alone. Yet, it is not certain that from this particular point the advantage would lie with the combined system

"But, in our opinion, the preceding considerations are of secondary importance, and ought not to furnish a basis for the solution of the problem.

"In point of fact, under the conditions in which the sluices were installed at Ostend, these latter not only serve to maintain the inner channel, but also, and specially, they maintain the deep berth in front of the new tidal quay, where navigation requires 26 feet of water at low tide.

"Owing to the prevalence of mud in the port, a rapid diminution in depth may be expected to take place, unless very powerful counteracting agencies are brought into play, combatting the silting tendency without relaxation or discontinuance.

"In default of sluices, dredging would be essential at the foot of the tidal quay; this would entail the occupation of the quay berth by cumbersome vessels, as inconvenient from the point of view of navigation as from that of trade.

"There is therefore every reason for limiting dredging operations at the tidal quay, and, from this point of view, sluicing has its advantages. It reduces the inconveniences to a minimum by considerably diminishing the quantity of material requiring to be dredged."

The new sluicing basin has an area of nearly 200 acres, and its contents are discharged through six openings each 16 feet 6 inches in width.

**Regulation Works on the Rhine.**[2]—"The Rhine, between Mainz and the Dutch frontier, has been systematically regulated in wide, shallow reaches, chiefly by projecting spurs or dykes, generally extending into the river from one bank, but occasionally from both banks where the conditions are unfavourable, as, for example, in a wide reach between two bends; and sometimes

[1] Van der Schueren on Curage des Ports Maritimes, *Proc. Int. Nav. Cong. Dusseldorf*, 1902.

[2] Vernon-Harcourt on Dusseldorf Congress, 1902, *Min. Proc. Inst. C.E.*, vol. clii.

longitudinal dykes have been resorted to, connected usually with the river bank by cross dykes. These dykes consist chiefly of earthwork mounds, protected on the face by rubble or pitching, with a rubble mound on the exposed toe; and occasionally fascines are employed in conjunction with

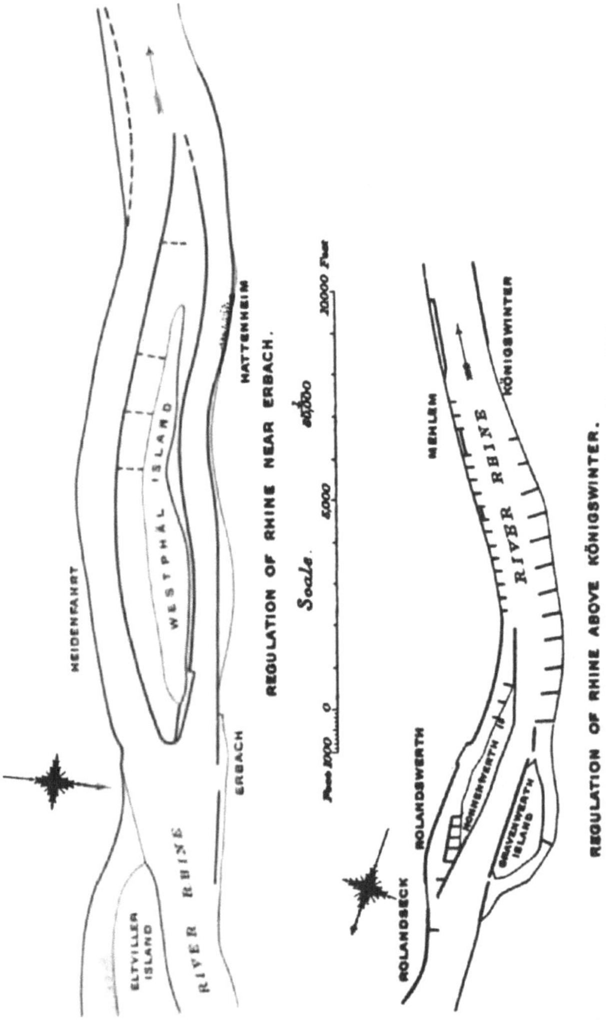


Figs. 219, 220.—Regulation Works on River Rhine.

stone, or a rubble mound alone. These regulation works have, for the greater part, been gradually carried out during the latter half of the nineteenth century; and two examples of somewhat recent and extensive regulation works, constructed about half-way between Biebrich and Bingen and just above Königswinter respectively, are illustrated in figs. 219, 220. These works, in conjunction with dredging, where necessary, have provided a navigable depth, at the ordinary low stage of the river, of 10 feet from the Dutch frontier up to Cologne, and 8½ feet between Cologne and Caub at the base of the steep slope below Bingen; above which point the available depth at average low water is reduced to 6½ feet, which is maintained up to Philippsburg about 22½ miles above Mannheim.

**The Port of New York.**—There are two navigable entrances (fig. 8) to the port of New York. One, which is frequented by transatlantic liners and ocean-going vessels generally, is flanked by Sandy Hook on the south and by Coney Island on the north, and opens directly on to the Atlantic Ocean; the other is an arm of the sea hemmed in between Long Island and the mainland, known as Long Island Sound as far as its junction with East River at Hell Gate, and forming the principal route for coasting vessels trading to and from the northern states and the Canadian provinces.

These entrances present features of direct and striking contrast, both in regard to their nature and the means adopted for their amelioration. The main entrance is broad and spacious, and, fronting the Atlantic, is exposed to all its storms. Moreover, it is beset with shoals and sandbanks. Sandy Hook itself is but a low-lying bank, more or less submerged, and varying from time to time in form and extent.

Obviously, for such a régime, the process of suction dredging forms the proper system of treatment, and this has been carried on for a number of years past with eminently satisfactory results. There is at present a minimum navigable depth of 30 feet at low water along the main ship channel, and operations are well advanced towards the attainment of a depth of 40 feet along a new and shorter route known as the Ambrose Channel.[1]

The tidal current is moderate. At the crest of the bar it rarely exceeds 1½ knots, and within the limits of the inner channels its maximum rate is from 2 to 2¼ knots.

The Long Island entrance is characterised by a sinuous course, undergoing frequent and abrupt changes of direction. It is comparatively sheltered, but has to wind its way amid the intricacies of an archipelago of inlets and rocky reefs, some of the latter rising above the water level, but many of them totally submerged and fraught with danger to navigation. The currents, moreover, are rapid, reaching at certain points a speed of 10 knots, and eddies are numerous. The impetus thus generated, combined with the irregularities of the course, have been, in times past, the cause of numerous disasters to shipping, particularly in the neighbourhood of Hell Gate, where the stream is deflected at right angles past Hallet's Point, to be split up into a multitude of rivulets amid the hidden reefs which abounded at that point. By means of blasting operations, however, on an extensive, not to say gigantic scale, the worst of these obstructions have been removed, and the channel is now navigable in comparative ease, and, at anyrate, with safety.

[1] A ruling depth in the Ambrose Channel of 35 feet at lowest ebb tide, throughout a width of 1000 feet, was reported to have been realised in August 1907.

EXAMPLES OF TRAINING-WALLS.

| Locality. | Material. | Method of Formation. | Length. | Cost. | Remarks. |
|---|---|---|---|---|---|
| Tampico. Mexico. | Fascine mattresses about 60 feet × 15 feet × 4 feet to 7 feet thick. | Slung during construction from overhead staging and deposited. Weighted with stone. | 1¼ miles. | Mattress work, 6s. 2d. per cubic yard; brush work, 4s. 7½d. Large stone, 9s. 3d.; small stone, 7s. 5½d. | |
| River Weser from Bremen to sea. | Fascine mattresses, 66½ feet × 33⅓ feet × 3⅓ feet, bound by galvanised wire. | Made and towed 6 miles to destination by tugs. Sunk in position and weighted with stones. | 43 miles. | 3s. 9d. per cubic yard for materials; 2s. 8d. for labour = 6s. 5d. total. Stone, 6s. 8d. per cubic yard. | Walls left 6 inches above low water to allow for extension of tidal water, and to ensure against damage by waves and ice. |
| Seine Estuary. | Chalk rubble from adjacent cliffs. | Tipped from barges and levelled to an even face above low-water line. | | Stone, 1s. 1½d. per cubic yard, *in situ*. | Walls carried up to high-water level, 6½ feet wide at top. Slopes 1 to 1 on land side and 1½ to 8 to 1 on river side, according to current. |
| Ribble Estuary. | (*a*) Stone from North Wales.<br>(*b*) Local stone.<br>(*c*) Slag. | Deposited by hoppers and trimmed.<br>Discharged from barges by hand.<br>Discharged from barges by hand. | 3½ miles. | 4s. 9d. per cubic yard.<br>1s. 4½d. per cubic yard.<br>2s. 9d. per cubic yard. | Sectional area of mound, 12½ square yards. Side slopes, 2 to 1. Base width about 30 feet. |
| Tees Estuary. | Slag from local iron-works. | Discharged by tipping from hopper barges and also by hand from keels and punts. | 24 miles. | 10½d. per cubic yard. Ironmasters paid 4d. per ton for removing their slag. | Sectional area of mound, 26 square yards. Side slopes, 1 to 1. Finished height about 5 feet above original surface. |

# CHAPTER X.

## CHANNEL DEMARCATION.

Value of Systematic Demarcation—Regulation and Supervision of Channel Marks—Independence of Authorities—Fundamental Characteristics of Signals—Beacons—Buoys—National Systems—Trinity House Regulations—Design of Buoys—Channel Lighting—Luminous Buoys—Wigham Burner—Pintsch System—Lightships—Suspension of Floating Lights—Luminous Beacons and Lighthouses—Incandescent Burners—Light Concentration—Reflectors and Lenses—Catoptric, Dioptric, and Catadioptric Systems—Range of Light—Identification of Stations—Sound Signals—Audible Buoys.

**Importance of Channel Demarcation.**—One of the most essential features of a modern port is a clear and systematic demarcation of the channels by which it is approached from the open sea. Be the channels long or short, winding or comparatively straight, the necessity is universally and incontrovertibly evident, since, in the absence of such guidance, ships run the risk of grounding on the shoals and banks which fringe the coast-line of nearly every maritime country. Few ports are endowed by nature with an illimitable expanse of open fairway, and, in the majority of cases, restrictions and precautions of no inconsiderable perplexity have to be observed. This is more particularly the case with those ports which lie in deep coastal and estuarine indentations, or inland upon the banks of some navigable river within the range of tidal influence. Fluctuations of depth, combined, in many instances, with the eccentricities and vagaries of currents, are a source of continual apprehension to the mariner, who has most generally to fall back on special local assistance in order to reach his destination. Yet there are circumstances under which such assistance may not be forthcoming; and, apart from this, there is always the desirability of according harbours and ports the fullest possible measure of safe and convenient access. Too much importance, therefore, can hardly be attached to the proper and effective delimitation of navigable channels.

In a maritime country one would naturally expect to find a matter of such vital interest to the community dealt with on broad and systematic lines, and the methods adopted carried to a very high state of perfection. Uniformity of practice and treatment would appear to be the most obvious of desiderata. Yet it must be confessed that, until comparatively recently, the demarcation of approach channels was regarded, to a very great extent, as a matter of almost purely local importance, and it was largely left in the hands of district authorities with little, if any, attempt at national supervision. The inevitable consequence was a diversity of practice, which served to puzzle and confuse the navigator rather than to assist him. Each port adopted a system of its own, without reference to the broader interests of the country at large, and different rules and regulations were laid down in various quarters, which oftentimes proved as conflicting as they were involved.

This lack of proper and effective centralisation is, of course, no uncommon feature of British administrative methods, being due, in a great measure, to the spontaneous origin and independent growth of the national institutions. The fact is none the less regrettable, in that while the attendant evils do not always manifest themselves so prominently as to attract public notice, and bring about much needed reform, they invariably result in extravagance and confusion. Fortunately, in such cases, the natural trend of events is towards the establishment of a hegemony of some kind or another, even though it be imperfect and ill-defined. This tendency, which is manifestly one to be fostered and encouraged, has shown itself in the present instance.

There are still in existence, within the limits of the United Kingdom, three separate bodies endowed with the control of the lighting and buoying of the British coast-line.[1] These are Trinity House, London, *primus inter pares*, the jurisdiction of which extends from Berwick-on-Tweed round to the Solway Firth; The Commissioners of Northern Lights, who administer the coast-line of Scotland and the Isle of Man; and the Commissioners of Irish Lights, formerly the Dublin Ballast Board, who discharge similar duties in respect to Ireland. Apart from these corporations, however, though under their respective suzerainties, there are numerous local authorities exercising control within the limits of their several boundaries. Thus, the demarcation of the approaches to the river Mersey is in the hands of the Mersey Docks and Harbour Board; while the higher reaches of the same river are administered by the Upper Mersey Navigation Commissioners. Hull Trinity House supervises the Humber and its precincts; and the Corporation of Lynn, the channels of the Wash. Trinity House, London, looks after the Thames.

Some little time ago there was held a conference which was attended by representatives of Trinity House, the Admiralty, and other interested parties. At this conference a series of regulations were formulated, and recommended for general adoption by all port authorities in this country. These regulations will be referred to in detail later. It is interesting, however, here to note that this step towards the general standardisation of channel marks has met with approval and success. In fact, a similar congress, but representing far wider interests, and international in character, assembled in Washington in 1899, and drew up certain principles for the regulation of navigable waterways in general, and these principles have become recognised on the continent

[1] In September 1906 a commission was appointed by the Government to report on the respective functions of these bodies, with a view to some method of co-ordination or amalgamation.

of Europe and in America as a definite basis for the establishment of a systematic code of channel signals.

**Fundamental Characteristics of Channel Signals.**—Dealing with the question, *ab initio*, it will be evident that the essential features of any satisfactory system of demarcation are—

(1) Conspicuousness, by which the marks or signals may be seen from a considerable distance.

(2) Individuality, by which they may be definitely recognised and distinguished, combined with

(3) Simplicity in regard to their signification, and

(4) Invariableness or unalterability of character.

In the daytime, these conditions are generally fulfilled by beacons and buoys of definite shape and hue; and at night, by lights of a certain range, intensity, and colour.

To avoid confusion, it will be desirable, as far as possible, to restrict the use of the term "beacon" to fixed, and of the term "buoy" to floating structures. The surmounting signal of a buoy, however, is commonly also designated a beacon. Both buoys and beacons may be used as a means of illumination, but it is solely in regard to their construction and outline that they will be considered in the first instance.

**Beacons.**—Beacons, then, are prominent objects or structures on the

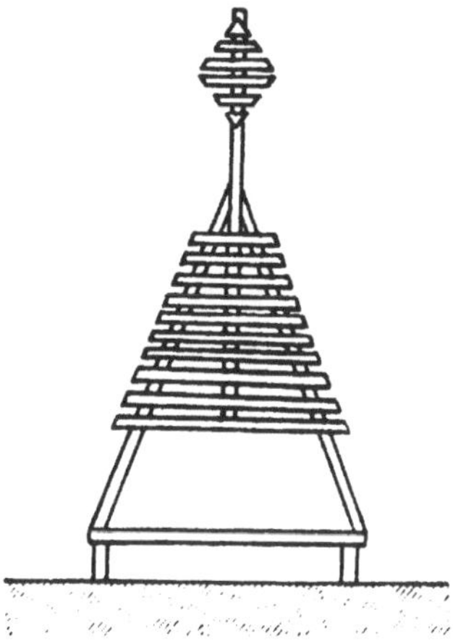

FIG. 221.—Beacon.

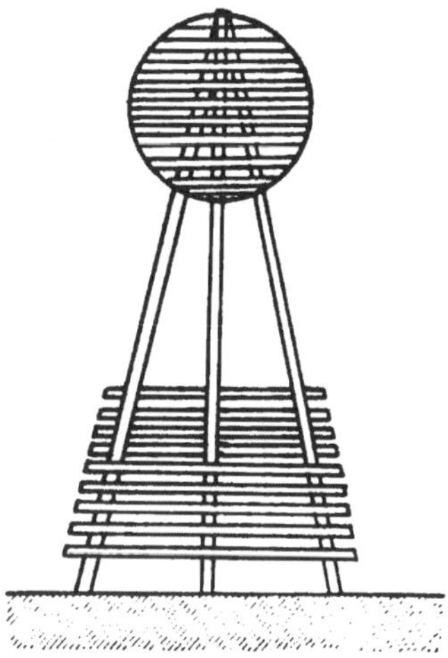

FIG. 222.—Beacon.

coast-line or on a river bank, capable of acting as a means of alignment, or as an indication of change of direction. Natural objects, such as lofty isolated trees; topographical features, such as the edge of a cliff, or the summit of a hill; and prominent structures of any kind, such as windmills, factory chimneys, and church steeples, may all be used as beacons. When special erections have to be made, they generally take the form of a wooden framework tapering from a wide base to a narrow top, or forming some distinctive geometrical figure, such as a triangle or lozenge. It is essential, of course, that the beacon should stand out clearly against its background, and the steps necessary to secure this end will vary according to circumstances. One method is to paint the front surface chequerwise in different colours; another to paint it all one colour, and so on.

**Buoys.**—The difficulties attending the design of a beacon, evident as they are in many instances, are not so great as those involved in the case of a buoy. Steadiness and erectitude are qualities not easily conferred upon floating structures, while the same precision in regard to locality is impossible of attainment. Buoys have to be moored to sinkers, and the length of cable varies from two to three times the maximum depth, which in itself, in tidal situations, is susceptible of considerable fluctuation, so that a buoy is capable of mobility within a circle of not inconsiderable diameter. This renders buoys unsuitable for imparting accurate guidance in regard to alignment. As a general rule, their utility is limited to indicating the proximity of shallows in their immediate neighbourhood.

The limiting width of channels is indicated, in fact, by two lines of buoys, one along each boundary. These are termed starboard and port hand buoys, according as they lie to the right or left of the mariner who is approaching a port from seaward. Generally speaking, the maximum distance between two consecutive buoys, on either hand, is a mile or a mile and a half in wide estuaries, and the minimum, perhaps 300 yards in narrow channels, exclusive, that is, of turning-points.

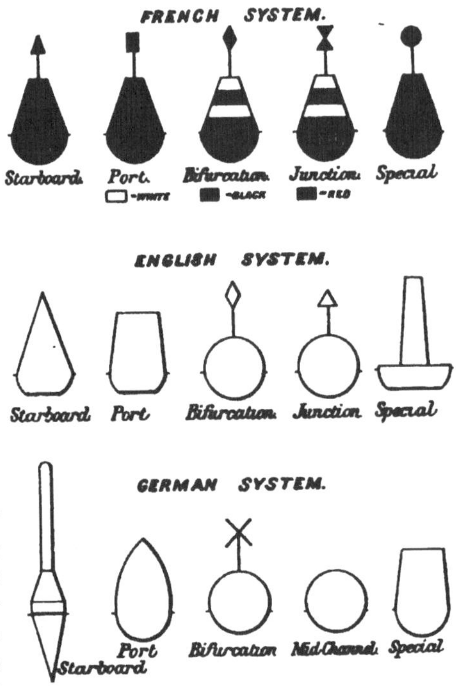


FIGS. 223–225.—National Systems of Buoyage.

**National Systems of Buoyage.**—It is interesting in this connection to compare the practice of this country with that of France and Germany. In English practice special stress is laid upon the shape of the buoy structure; more so than in French practice, where colour is every whit as essential as form. Climatic conditions have something to do with this, for on the English coast the state of the atmosphere is often unfavourable to the ready perception of colour at a distance. It is true that colours may be, and are, used in this country as an additional indication, but their use is entirely subsidiary and may vary locally; while in France colour takes precedence of shape. Shape is not entirely disregarded, but the distinction is confined to a surmounting signal, and does not affect the buoy structure as in England. German signals differ from both French and English signals. The series of diagrams in figs. 223 to 225, have been arranged in juxtaposition so as to illustrate the divergencies in type of all three nationalities.

**Trinity House Regulations.**—The following is a transcript of the regulations adopted in this country in accordance with the uniform system of buoyage approved by the General Lighthouse Authorities of the United Kingdom:—

"1. The mariner, when approaching the coast, must determine his position on the chart, and must note the direction of the main stream of flood-tide.

"2. The term *Starboard Hand* shall denote that side which would be on the right hand of the mariner, either going with the main stream of flood or entering a harbour, river, or estuary from seaward; the term *Port Hand* shall denote the left hand of the mariner, under the same circumstances.

"3. Buoys showing the pointed top of a cone above water shall be called **Conical**, and shall always be Starboard Hand buoys, as above defined.

"4. Buoys showing a flat top above water shall be called **Can**, and shall always be Port Hand buoys, as above defined.

"5. Buoys showing a domed top above water shall be called **Spherical**, and shall mark the ends of middle grounds.

"6. Buoys having a tall central structure on a broad base shall be called **Pillar** buoys, and, like other special buoys, such as Bell buoys, Gas buoys, Automatic sounding buoys, etc., etc., shall be placed to mark special positions, either on the coast or in the approaches to harbours, etc.

"7. Buoys showing only a mast above water shall be called **Spar** buoys.

"8. Starboard Hand buoys shall always be painted in one colour only.

"9. Port Hand buoys shall be painted of another characteristic colour, either single or parti-colour.

"10. Spherical buoys at the ends of middle grounds shall always be distinguished by horizontal stripes of white colour.

"11. **Surmounting beacons**, such as Staff and Globe, etc., shall always be painted of one dark colour.

"12. Staff and Globe shall only be used on Starboard Hand buoys; Staff and Cage on Port Hand; Diamonds at the outer ends of middle grounds and Triangles at the inner ends.

"13. Buoys on the same side of a channel, estuary, or tide way, may be distinguished from each other by names, numbers, or letters, and, where necessary, by a staff surmounted with the appropriate beacon.

"14. Buoys intended for **Moorings**, etc., may be of shape or colour according to the discretion of the authority within whose jurisdiction they are laid; but for marking submarine telegraph cables, the colour shall be green, with the word 'Telegraph' painted thereon in white letters.

"15. **Wreck** buoys in the open sea, or in the approaches to a harbour or estuary, shall be coloured green, with the word 'Wreck' painted in white letters on them.

"16. When possible, the buoy shall be laid near to the side of the wreck next to mid-channel.

"17. When a wreck-marking vessel is used, it shall, if possible, have its top sides coloured green, with the word 'Wreck' in white letters thereon, and shall exhibit:—

"*By day:* Three balls on a yard 20 feet above the sea, two placed vertically at one end and one at the other, the single ball being on the side nearest to the wreck.

"*By night:* Three white fixed lights similarly arranged, but not the ordinary riding light.

"18. In narrow waters or in rivers, harbours, etc., under the jurisdiction of local authorities, the same rules may be adopted, or, at discretion, varied as follows:—

"When a wreck-marking vessel is used, she shall carry a crossyard on a mast with two balls by day placed horizontally, not less than 6 nor more than 12 feet apart, and two lights by night similarly placed. When a barge or open boat only is used, a flag or ball may be shown in the daytime.

"19. The position in which the marking vessel is placed with reference to the wreck shall be at the discretion of the local authority having jurisdiction."

**Design of Buoys.**—The design of a buoy should obviously be such that it will always float upright and be subject to the least possible disturbance of equilibrium in boisterous weather and from drifting ice. Long, narrow buoys, constructed on the principle of the angler's float, are best adapted to withstand currents and rough seas, provided they be not moored from the nether apex, in which case, unless heavily weighted, they tend to heel over considerably. The mooring is preferably attached by a saddle or bridle arrangement at a considerably higher level. Elongated buoys are specially characteristic of German practice, on account of the great quantities of floating ice which obstruct the Baltic Sea and its influents during the winter season. Broad-based buoys are suitable for smooth, shallow waters: they take the ground satisfactorily in the event of exceptionally receding tides. In sheltered positions, flat bottoms with rounded bilges make a good arrangement; in a heavy seaway the rounded bottom is to be preferred, or the hollow cone, as in Admiral Herbert's design.

**Size of Buoys.**—Buoys are classified as first- or second-class, according to their size. The following table shows the generally accepted dimensions of buoys. There are some few examples which exceed these figures—for instance, on the river Mersey,—but they are of local interest only.

BUOY DIMENSIONS.

| Type. | First Class. | | Second Class. | |
|---|---|---|---|---|
| | Diameter. | Height. | Diameter. | Height. |
| | ft. | ft. | ft. | ft. |
| Conical . . . . | 8 | 10 | 6 | 7½ |
| Can . . . . | 8 | 8 | 6 | 8 |
| Spherical . . . | 8 | 7 | 6 | 5¼ |
| Pillar . . . . | 10 | 15 | 8 | 12 |

The material used is steel plating about ¼ inch thick. There should be two water-tight compartments in each buoy, so that in the event of collision with a passing vessel, the risk of foundering may be diminished. Mooring-chains for first-class buoys are about 1¼ inch diameter, and they are attached to sinkers weighing about 25 cwts. For second-class buoys, the chains are usually 1 inch diameter, and the sinkers weigh 15 cwts.

**Mooring Buoys** constitute a special class of buoys with functions quite outside the sphere of the present chapter. Such brief reference to them as is necessary is to be found in Chapter VIII.

**Channel Lighting.**—Having dealt with those features of channel demarcation which are available for use in the daytime, we now turn our attention to means adopted for guidance when such signals are no longer naturally visible. Recourse has then to be had to some artificial source of light.

Of the value of luminous signals to the mariner there can be no question. He approaches his destination without reference to day or night, and during the hours of darkness, while in close proximity to land, he is often without any other reliable indication of his position, or trustworthy guidance in his course. At the same time, it must be avowed that in many respects artificial lights, as they exist at present, are far from constituting an ideal system of localisation. The range of visibility is extremely variable under different atmospheric conditions, and in times of dense fog, and even in squally weather, may become of no appreciable value whatever. Then, again, a very powerful light, while serving admirably as a beacon to shipping at a great distance, is a source of some perplexity and confusion at close quarters, dazzling the sight, projecting deep and dark shadows, and obscuring the position of objects which lie outside the illuminated zone, and especially those immediately beneath the source of light. Thus, lighthouses which act as *landfall* or advance lights, giving the mariner timely warning of his approach to the coast-line, form a different class from those which are used to indicate navigable channels. In the former case, striking brilliance and extensive range are matters of fundamental importance. In leading and harbour lights, a much lower degree of illuminating power is all that is necessary or desirable. Few channels present any lengthy stretch of straight fairway; most are characterised by sharp bends and intricate windings, necessitating the employment of numerous lights, which need not have more than a very moderate range, say 5 or 6 miles at the outside.

Only in times of fog and squalls is there any real necessity for a penetrating light of high calibré, and under these adventitious circumstances, it is possible to bring into action some special reinforcement, and so temporarily intensify the normal power of the light.

Channel lighting is effected by buoys, beacons, lightships, and lighthouses.

**Luminous Buoys.**—Buoys are lighted by means of oil or gas. The former is most usually petroleum; the latter almost invariably vaporised paraffin. Coal gas is unsuitable. One of the first necessities of buoyage illumination is compact storage of the illuminant; and coal gas, when subjected to the pressure which is necessary for this purpose, loses a very considerable proportion of its illuminating power, and burns with the bluish flame of the Bunsen burner. Moreover, it distils a tarry liquid, which obviously causes obstruction and inconvenience in pipes and tubes.

Oil gas is not only free from this defect, but it is also relatively a very powerful illuminant at high pressures. Its candle-power, when compressed to 150 lbs. per square inch, is from 40 to 45, comparable with 55 to 60 at ordinary pressure.

The difficulty of using petroleum oil as a direct illuminant lies in the fact that, in the course of combustion, wicks become rapidly charred or coated with carbon, to such an extent as to obstruct and ultimately arrest the capillary attraction necessary for raising the oil from the reservoir to the burning point. This means that the light will be extinguished unless there be someone at hand to dress the wick. Now constant, to say nothing of skilled attention, is impracticable in the case of buoys. They have necessarily to be left to take care of themselves between certain times of inspection, which can only be frequent at the expense of economy; and whether the period be long or short, there is the same risk of failure of the light.

In France, carbonised wicks, *i.e.* wicks specially prepared by a uniform deposit of carbon upon them, have been introduced. But these, while satisfactory in achieving their special purpose, entail corresponding difficulties of another kind. The wicks not only require careful preparation, but they call for adjustment of the utmost nicety, and they are used with burners of a very complex character. Constant watching is therefore still an essential feature of the system, and this fact discounts their use in connection with buoys.

An English burner, however, known as the **Wigham burner**, has been contrived to meet the conditions of the case, and the following is a brief description of its mode of action.

In an ordinary petroleum lamp, the wick is set perpendicularly to the oil reservoir from which it draws its supplies, and there would be considerable difficulty in making such a wick automatically raise itself as combustion proceeded. Mr Wigham, therefore, designed his wick to burn horizontally, passing it slowly over a small roller, the light being obtained from the flat side instead of from the end or edge. One end of the wick passes up through an oil-tight brass tube, receiving its supply of oil from the main reservoir[1] by means of feed-holes, and the other end of the wick is brought down through another tube soldered or otherwise secured at its lower end, and standing above the level of the oil in the lamp. A circular float, to which this end of the wick is attached, rests upon the surface of the oil in a copper cylinder at the foot of the lamp. The oil in the cylinder is slowly withdrawn, drop by drop, through a valve of special construction, and the float, in descending with the falling level of the oil, draws the wick in its train, and so causes a constant change in the part of the latter exposed to the action of the flame. The light may thus be arranged to burn without attention for periods of one, two, or even three months. The consumption of oil for both illumination and automatic working, together, is at the rate of about half a gallon per day of twenty-four hours.

Turning to the alternate system of vapour lighting, we find that oil gas is manufacturable from shale oil, petroleum, or other oils. Heavy oils generally produce a smaller quantity of gas, but of richer quality than light oils. One gallon of oil yields from 70 to 90 cubic feet of gas, and the cost of production per 1000 cubic feet varies (subject to fluctuations in the price of materials) from 6s. 6d. on a large scale to 10s. on a small one.

As manufactured on the **Pintsch system**, the gas is produced in two D-shaped cast-iron retorts, arranged one above the other, connected by a double mouthpiece and set in a suitable furnace. The furnaces are heated by coal, coke, or other fuel, until the retorts have become cherry red. The oil, previously stored in a wrought-iron tank, is pumped into a small vessel, or cistern, near the furnaces, from which it flows by gravitation in a thin stream regulated by a micrometer cock, through a syphon into the upper retort. In order to protect this retort and somewhat retain the oil, a sheet-iron tray is inserted, into which the oil drops and is immediately converted into a brownish vapour. Passing through the connecting mouthpiece and along the heated sides of the lower retort, this vapour is further decomposed and made into permanent gas, full of impurities. The only outlet of the lower retort is a short pipe, called the descension pipe, through which the gas passes into the hydraulic main, depositing here a certain amount of tar, and thence into a circular condenser. Issuing from a small pipe into the large space of the condenser, the gas cools down and frees itself from the lighter tarry matters. It then passes into the washer, where it is forced through about an inch of water, and, afterwards, through two or three layers of lime and sawdust in the purifier. In small installations, the washer and purifier are generally combined in the same apparatus, one above the other. In the

[1] Divisions are made in the reservoir to prevent the oil from flooding the wick during momentary disturbance, such as is inevitable in the case of buoys and other floating vessels.

washer and purifier the gas is freed from carbonic acid and sulphur, impurities which are not only detrimental to a good gas, affecting the illuminating power, but which, moreover, form an injurious deposit in the fittings, regulators, and burners. The gas, when pure, passes through a meter to be registered, and thence to the gasholder.

The compressors are double-acting pumps worked by steam or hand.

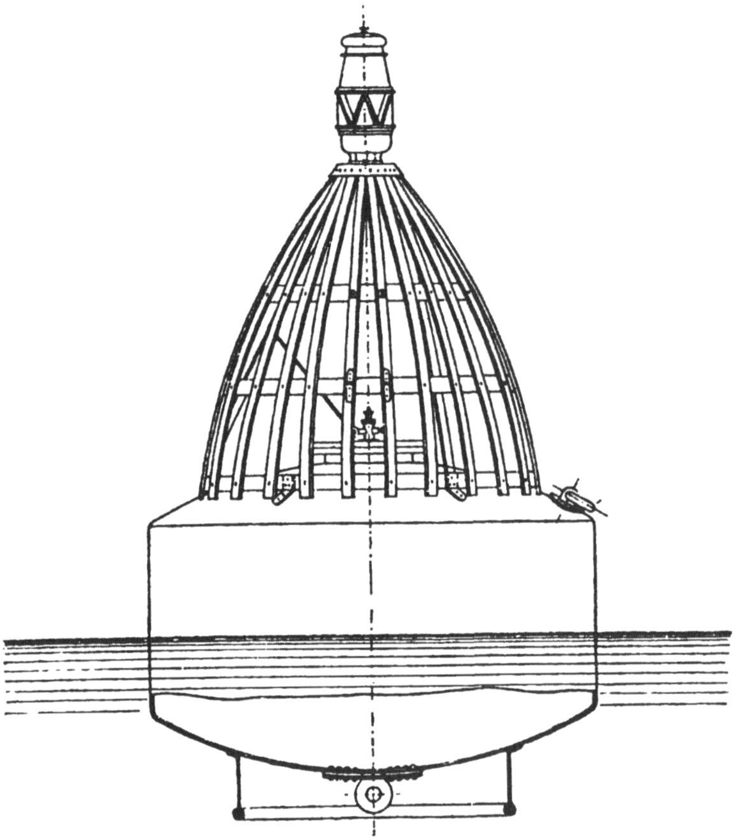

FIG. 227.—Gas Buoy.

Before the gas is passed into the welded, high-pressure storeholder, it passes through a small vessel and there deposits some hydrocarbon, which is drawn off by a small valve.

The storeholder is placed within the buoy to be lighted, where it is connected with the burner, and supplies it with gas for periods ranging from two to six months, without further attention.

In the case of gas buoys a special form of construction is rendered desirable. Rivetted joints, however well caulked, tend to leak, especially if the gas be compressed at pressures over 100 lbs. per square inch. Mild steel

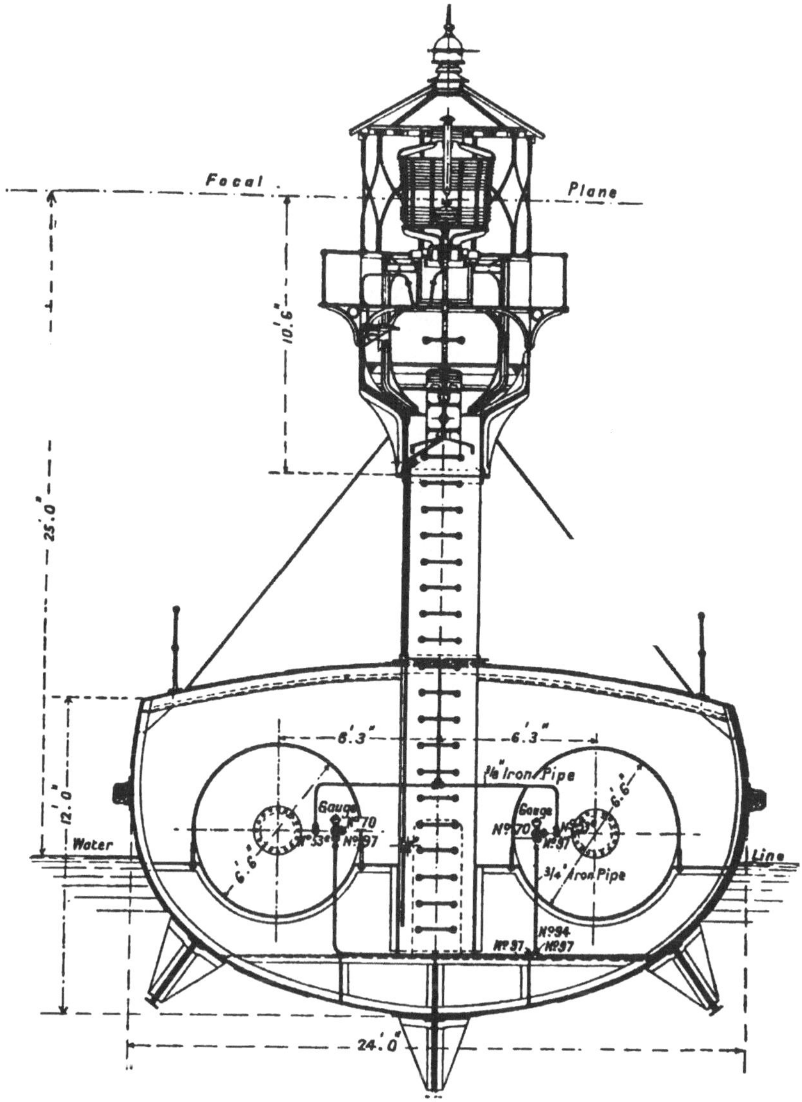


FIG. 228.—Section of Otter Rock Lightship.

structures, welded throughout, afford the most satisfactory method of inclosure, and they are, at the same time, less liable to admit water in case of concussion. In certain instances, closely rivetted joints seem, nevertheless, to have answered all requirements.

As regards the method of suspending the illuminating apparatus of a buoy, it is evident that in order to ensure its remaining upright amid tidal currents and storm swells, it should be attached in such a manner as to swing freely. The principle of attachments of this kind will be explained more fully in connection with lightships. Except for Wigham lamps and in boat-shaped structures, ordinary attachments may, however, be made to serve.

**Lightships.**—Buoys are not invulnerable, and it is quite within the bounds of possibility that a luminous buoy may be extinguished from one cause or another, though the occurrence is by no means common. Where there are heavy seas and strong currents, however, lighted buoys are exposed to undue risks, and signals of a more reliable character are desirable. Lightships are much steadier under these conditions: their oscillation is less, and they are not so liable to be put out of action. Moreover, they have the additional advantages of presenting a more conspicuous bulk and a more striking individuality.

Steadiness is one of the most essential qualities of a lightship, and the attainment of it, so far as such a thing can be realised amid unstable surroundings, involves the suppression of synchronism in the periods of oscillation, respectively of the vessel and of the waves. Synchronism is more likely to occur transversely under the action of rolling, than longitudinally under the action of pitching. In order to avoid it, ballast should be located at some distance from the centre of gravity of the vessel, so that the moment of inertia of the latter about its longitudinal axis may be as great as possible, and the metacentric arm reduced to the minimum consistent with stability. Furthermore, the steadiness of lightships is promoted by deep central and markedly protruding bilge keels, to all of which ballast may advantageously be affixed. In some of the latest examples of lightships, the keels have a depth exceeding 3 feet.

DIMENSIONS OF REPRESENTATIVE LIGHTSHIPS OF VARIOUS NATIONALITIES.

| Nationality. | Name or Locality. | Length. | Beam. | Depth. | Draught. | Height of Focal Plane above Water. |
|---|---|---|---|---|---|---|
| | | ft. | ft. | ft. | ft. | ft. |
| British . | Longsand . . | 60 | 24 | 12 | 6¾ | 30 |
| French . | Talais . . . | 61 | 20 | 9 | 6¼ | 33 |
| ,, . | Snouw . . . | 65½ | 20 | 13 | 11¾ | 33 |
| British . | Gaspar Point, R. Hooghly . . | 75 | 28 | 12 | 6¾ | 35 |
| ,, . | R. Mersey . . | 103¼ | 21¼ | 11 | 9½ | 30 |
| French . | Sandettié . . | 115 | 20½ | 16¾ | 15 | 39¼ |
| British . | R. Mersey . . | 118¾ | 21 | 11¼ | 9½ | 30 |
| German . | Fehrmarbelt . . | 134½ | 24¼ | 17 | ... | ... |

The dimensions of lightships have materially increased of recent years. A length of 60 or 70 feet used to be considered a maximum, but now several boats in the English and French services have lengths of over 100 feet, and one boat in the German service has a length of 134½ feet. The depth and draught of these vessels manifests a proportionate increase, but the beam has, if anything, tended to diminish, or, at least, to remain stationary, as will be evident from an inspection of the accompanying table, which classifies the leading dimensions of certain representative vessels of all three nationalities.

**Suspension of Floating-lights.**—In order to maintain verticality, the illuminating apparatus of a lightship is supported on gimbals. In catoptric[1] lights, the mirror and lamp are suspended in this manner from above. The dioptric[1] apparatus is generally hung in the form of a pendulum swinging about a horizontal axis located immediately beneath the lamp. The pendulum, or rod, is weighted and counterweighted above and below, the weights being adjusted in such a way that the period of oscillation of the lamp is considerably longer than that of the vessel, so that the maximum inclination of the former may not exceed 5 or 6 degrees. Manifestly, the apparatus must not only be sufficiently sensitive to maintain its verticality, but it must also admit of free and ready response to change of direction, and this is secured by attaching the gimbals to a horizontal circle rotating on steel balls. In the event of exceptionally heavy rolling on the part of the vessel, the possibility of collision between the pendulum and the lower part of the lantern may be guarded against by the provision of a thick annular pad of india-rubber on the weighted portion of the latter, or by restricting the swing of the pendulum with the aid of check chains and flexible guys.

**Lightship Attendance.**—The reliability and automatic continuity of the compressed oil gas illuminating apparatus has very largely done away with the necessity for crews on board lightships. In many cases now these vessels are unattended, except at long intervals for the purpose of supplying fresh gas. This has effected considerable economy in maintenance expenses, and extended the scope of utility.

The liability, however, of all floating objects to displacement, is the inherent weakness of the lightship, as also of the light-buoy. A displaced signal is much worse than none at all. Beacons and lighthouses, therefore, from their very fixity, possess uncontrovertible merits as regards accuracy of alignment, and it is usual to rely mainly upon them in so far as they happen to be available for this purpose.

**Lighthouses and Luminous Beacons.**—The earliest type of the lighthouse was the lighted beacon, usually situated upon a natural eminence or upon a tower. It was an iron-barred grate, or receptacle, for wood and coal, which was ignited at night-time. These signals were, therefore, most crude and primitive, and often gave out more smoke than light. Moreover, they lent themselves to easy reproduction and imitation for illicit ends. In this form they have long since disappeared into the lumber of the past. Their present development is the harbour light, a lantern attached to the top of an upright mast, which is used at the entrances of minor ports.

[1] See p. 268.

FIG. 229.—Lightship.

The lighthouse is a tall structure, occasionally of wood, but much more commonly of stone or iron, rising oftentimes to a considerable height above the water level. When, however, a natural headland or cliff lends itself to the purpose, the structure is not necessarily lofty, and, indeed, for channel lighting, no great height is essential. The building is usually planned in a series of stages or floors, the lantern containing the illuminating apparatus being located at the summit.

The practice of channel lighting, developing through luminous buoys to lightships, attains its highest degree of utility and perfection in the lighthouse. Sources of illumination for lighthouse use are not only numerically and potentially greater than those available for buoyage service, but they are also of a much more diverse nature, including electricity, coal gas, mineral, vegetable and animal oils, oil gas, and acetylene. The feeble and ineffective candle, which maintained its footing for at least thirty years of the last century, has now entirely disappeared. Its great modern prototype is the electric arc, the crater of which possesses an intrinsic radiance of over 55,000 candle-power per square inch of illuminating surface. Great and remarkable, indeed, have been the strides of late years in the development of lighthouses. Beams of light can now be projected far beyond the limits of their geographical range. The mariner sees their reflection in the sky before he comes within direct visual contact with them. This, of course, applies to landfall lights, and not to the class of lights which form the subjective basis of this chapter. Channel lighting is achieved perfectly satisfactorily with the aid of lights of a far lower calibre. The electric light is rarely, if ever, employed for this purpose. For general use, the incandescent petroleum vapour burner is more convenient and much less costly, and its light is sufficiently powerful for all stations other than those of primary and special importance. The system has, indeed, only been introduced into this country since the beginning of the present century, but it was adopted in the French lighthouse service several years previously, and it may now be said to have attained general recognition. Prior to this, the wick burner was so prevalent as to be practically universal, either flat, as in the earlier instances, or cylindrical on the Argand principle, with as many as six, eight, or ten wicks arranged in concentric rings. Despite the inferiority of the wick apparatus to the electric arc, its illuminative power was a very long way ahead of the feeble glimmer emitted by the cluster of tallow candles which lit up the summit of Eddystone a century ago.

RED SAND LIGHTHOUSE.

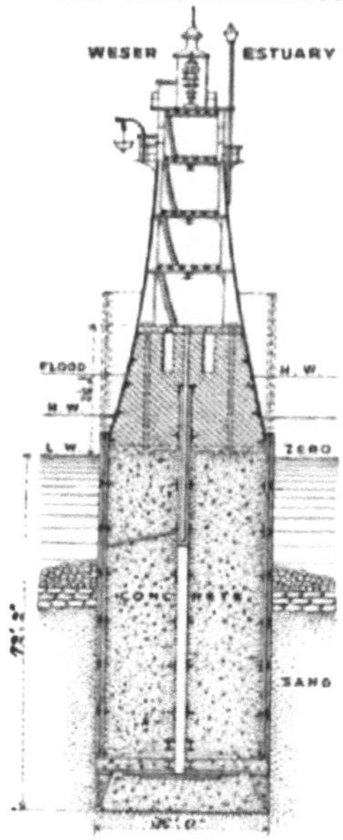


FIG. 230.

Twenty-four of these candles unitedly gave a light equivalent to sixty-seven standard candles.[1] In the later Eddystone of 1882, Douglass burners, with six concentric wicks, attained an aggregate of nearly 80,000 candle-power.

But the true standard of comparison is not so much the gross illumination as the intensity per unit of area. It is this intrinsic intensity which confers upon a beam its penetrative power. The brightness of the flame section of the old wick burners ranged up to 70, or, at most, 80 candle-power per square inch, according to the number of the wicks. With incandescent mantles there is, generally speaking, an increase in intensity of 300 per cent., equivalent to about 200 to 250 candle-power, while at the same time the consumption of oil is reduced by nearly one-half.

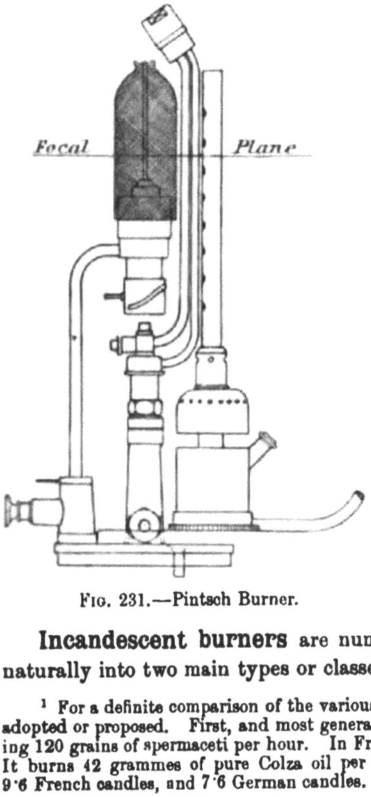


FIG. 231.—Pintsch Burner.

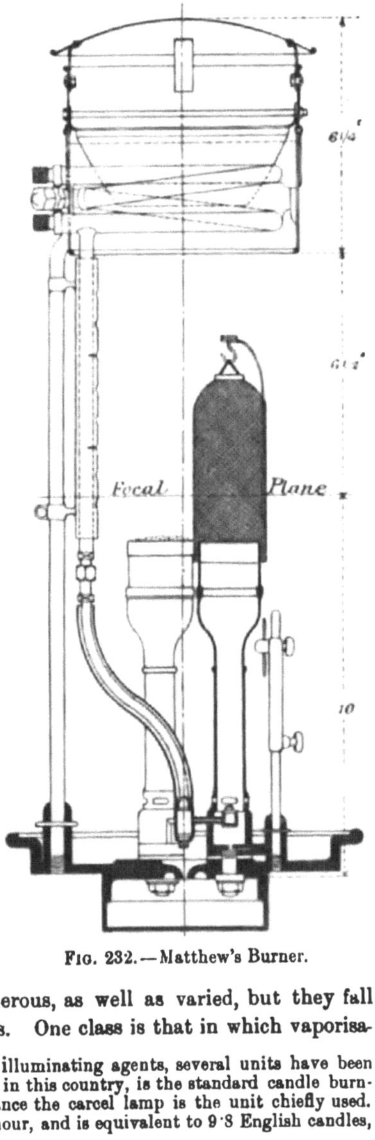


FIG. 232.—Matthew's Burner.

**Incandescent burners** are numerous, as well as varied, but they fall naturally into two main types or classes. One class is that in which vaporisa-

[1] For a definite comparison of the various illuminating agents, several units have been adopted or proposed. First, and most general in this country, is the standard candle burning 120 grains of spermaceti per hour. In France the carcel lamp is the unit chiefly used. It burns 42 grammes of pure Colza oil per hour, and is equivalent to 9·8 English candles, 9·6 French candles, and 7·6 German candles.

tion is effected by the direct heat of the mantle, and for this purpose the vaporising chamber, or tube, is placed close to and above the mantle. In the other class, a separate heater is used for vaporisation. Obviously, this entails additional apparatus; but, on the other hand, there is less interference with the luminous range of the mantle. In all cases, some temporary source of

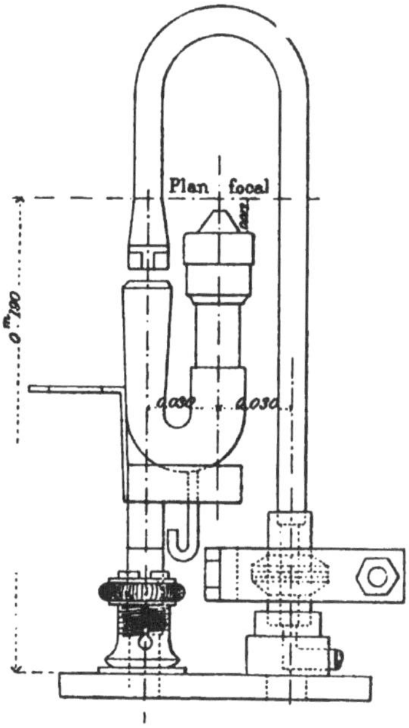


FIG. 233.—Luchaire Burner.

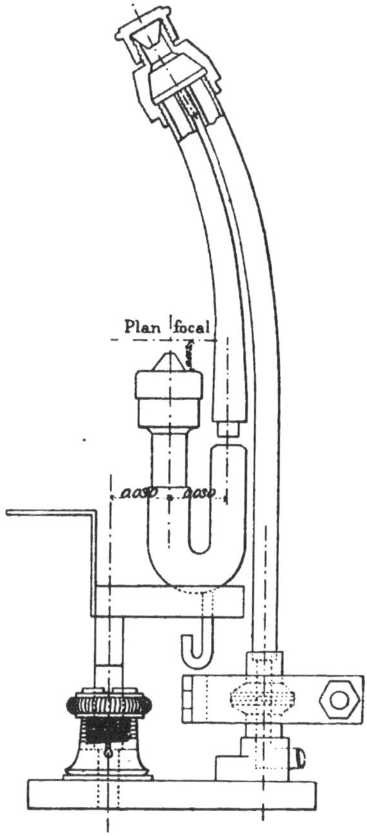


FIG. 234.—French Burner.

heat is required for a preliminary five or ten minutes, until the action of the burner becomes automatic.

Burners of the first type include the **Pintsch** burner, the **Matthews** burner, and the **Luchaire** burner, amongst those of German, English, and French manufacture respectively.

In the first-named (fig. 231), the oil is forced under compressed air pressure of 45 lbs. per square inch into a metal chamber located immediately above the mantle, whence, after vaporisation, it is conducted downwards to a mixing chamber, where it is combined with air before passing into the burner.

In the **Matthews** burner (fig. 232), vaporisation is effected in a brass tube coiled above the mantle and inclosed in a metal hood.

In the **Luchaire** burner (fig. 233), the vaporising tube takes the form of the letter U inverted, and arches the mantle.

Manifestly, the piping necessary to convey the oil to and from the source of vaporisation obstructs the rays emanating from certain sections of the mantle.

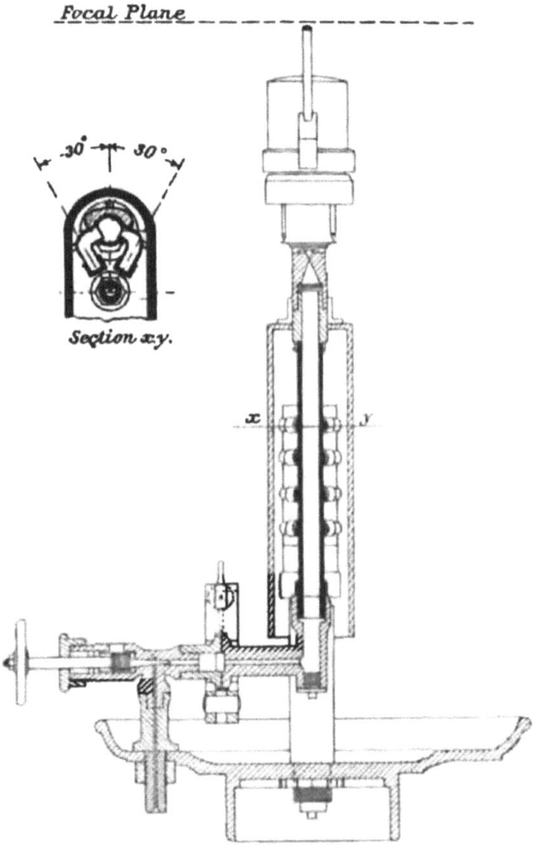


FIG. 235.—Scott Burner.

Burners of the second type include the **Chance** burner and the **Scott** burner.

In the **Chance** burner the dual horizontal vaporising tubes are contained in a metal chamber below the burner proper, and they are heated by a subsidiary burner, deriving its gaseous supply from the product of vaporisation.

In the **Scott** burner (fig. 235), a series of subsidiary burners, inclosed in a metal cover, heat the conducting tube in a vertical position. These subsidiary burners likewise draw their alimentation by means of a bye-pass from the mixing chamber.

The difficulty attaching to the use of incandescent burners is the fragility of the mantles. The average life of an ordinary mantle is perhaps six or seven days. Yet mantles have been known to last for over thirty days with care. The vaporising coils and tubes last from four to six months, and require frequent and constant cleaning.

**Acetylene** has also been used as an illuminant, but only to a very limited extent, although the dangers formerly attending its production have now been largely overcome by the improved nature of the generating cylinders. It is still undesirable to use this illuminant in the highly compressed liquid condition, from which explosions have resulted.

Dissolved acetylene, however, is largely used in motor cars, and is now being tested by Trinity House. Acetylene in this form certainly gives indications of being the coming illuminant for buoys, etc.

**Light Concentration.**—The source of light being but one of the factors in the determination of lighthouse efficiency, we now turn our attention to the methods adopted for the concentration and intensification of the issuing rays.

The simple, undirected flame is wasteful of light; that is, much of the light is lost to useful purposes. Most lighthouses stand upon the coast-line, and the area of radiation, therefore, frequently includes a large sector of land over which illumination is entirely unnecessary. Also, apart from this cause, a good deal of light is lost by diffusion and dispersion.

To remedy these defects, reflectors were introduced as far back as the latter half of the eighteenth century. At first spherical in form, the mirror ultimately became parabolic, concentrating the emergent rays of the light from the focus along a path parallel to the horizontal axis. This constitutes the **catoptric** principle. Catoptric reflectors are of two types: first, the paraboloid, formed by the generation of a parabola about its own axis, and sending the light rays in a single direction only; and secondly, the dual (upper and lower) surfaces formed by the horizontal rotation of a parabola round a vertical axis through the focus. This system, while confining the light within vertical limits, distributes it equally throughout a horizontal plane.

The **dioptric**, or **lenticular**, principle of ray concentration, based on the refractive properties of lenses, is due to Augustin Fresnel, who initiated it, or rather, applied it in an elementary form in 1822. As then exemplified, it consisted of a plano-convex lens set vertically in front of the light, so that all rays passing through the lens were transmitted horizontally. To the central lens were then added a number of parallel lenses of triangular form, which served to refract a certain proportion of the rays passing above and below it. The amount of non-utilised light was still considerable.

Fresnel disposed his lenses so as to form a cylinder completely inclosing the light, which thus illuminated the entire circumference uniformly. Stevenson devised a variation known as the **holophotal** system, by which the light was surrounded by a series of panels, each containing a circular central lens with annular adjuncts, the result being a concentration of the rays in a corresponding series of pencils, with intervening sectors of darkness. Such an arrangement lends itself to the production of flash-lights by the

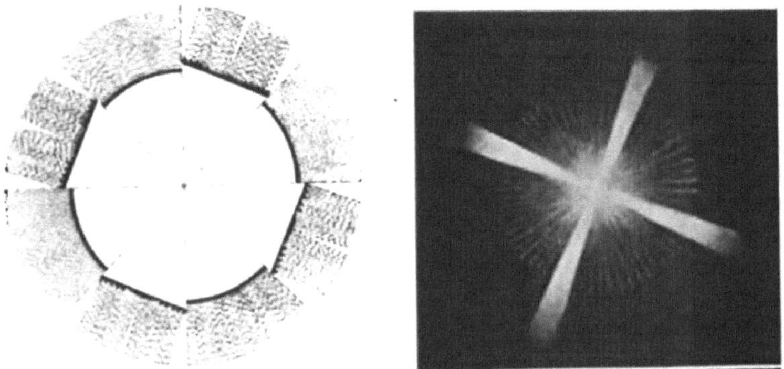

FIGS. 238.—Fixed and Flashing Light. 4 flashing panels of 45° each; 4 fixed panels of 45° each.

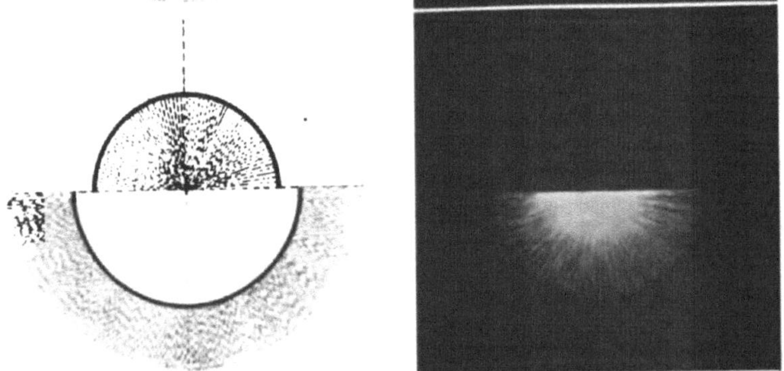

FIGS. 237.—Fixed Light of 180° and Dioptric Mirror of 180°.

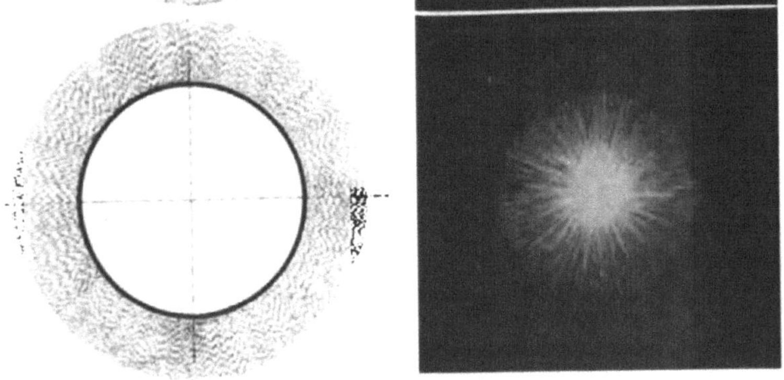

FIGS. 236.—Fixed Light of 360°.

revolution of the lenticular apparatus around the light, the illuminations being alternated with brief periods of obscurity, either total or partial, according as the apparatus is holophotal throughout, or combined with a fixed light.

With the object of still further strengthening the serviceable illumination, mirrors were placed above and below the lenses so as to reflect many of the

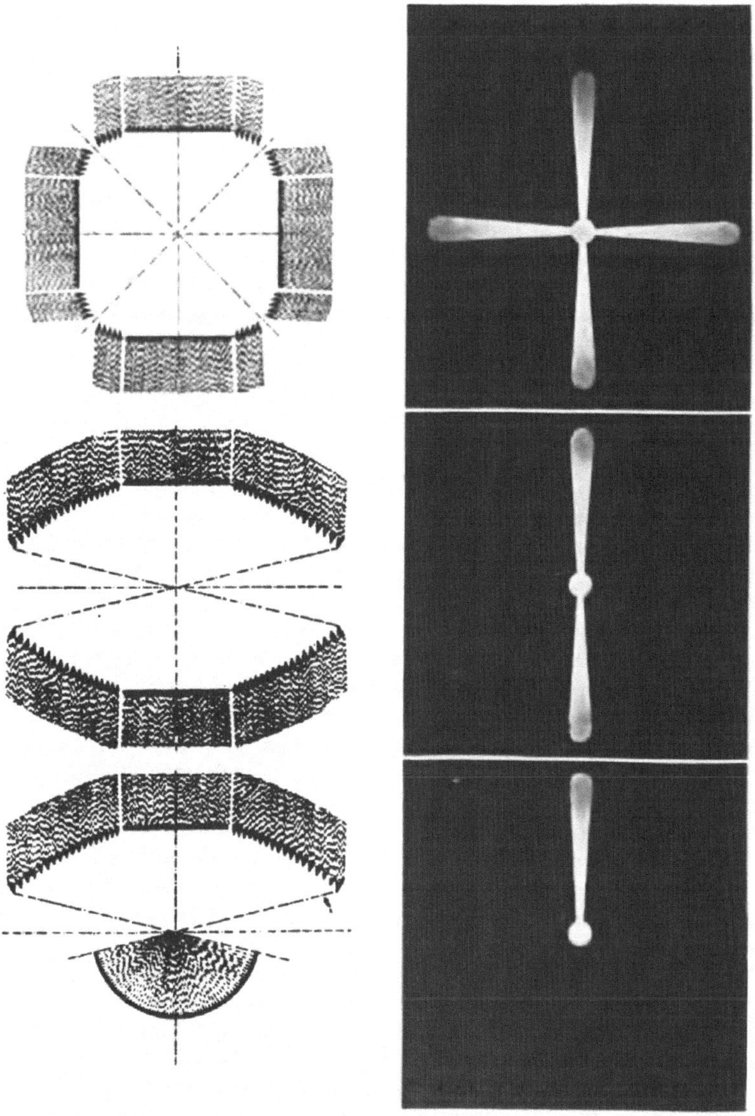

Figs. 239.—Single Flashing Light. 1 panel of 157°; Dioptric mirror 170°.

Figs. 240.—Single Flashing Light. 2 panels of 157°.

Figs. 241.—Single Flashing Light. 4 panels of 90°.

rays which escaped the latter. This conjunction of the reflective and refractive principles led to the adoption of the term **catadioptric** to distinguish it. The mirrors were eventually replaced by lenticular prisms capable of effecting the same object, the rays entering the prisms being entirely reflected at one of the surfaces. Mirrors behind the light were similarly replaced. Altogether, a very high percentage of the total illumination is utilised. In consequence of improvements recently effected, the vertical angle of Fresnel lenses now reaches 80 degrees, and the upper and lower prisms have, in many instances, been suppressed.

Yet another application of lens concentration is to be found in the **Azimuthal Condensing System** of Thomas Stevenson, in which holophotes and vertical prisms are employed to concentrate the light in special horizontal directions. This is exemplified in the apparatus constructed in connection with the Oronsay Lighthouse, where the light, as a leading light, is required to be seen along two intersecting axes of unequal length, in the one case for a distance of 15 miles, and in the other case for a distance of 7 miles. The dark or landward sector embraces an arc of rather more than 180 degrees.

A curious and interesting application of ray deflection is to be found at Armish Rock in the Hebrides. The lighthouse is situated on a rock separated from the Island of Lewis by a channel 500 feet in width. It contains no source of illumination itself, but it receives on a mirror a pencil of light rays from a lighthouse on Lewis. This ray is then deflected by prisms to pass onward in the directions required for the purposes of navigation.

As at present constructed, a modern lenticular panel consists of a central, circular, plano-convex lens with annular adjuncts, and upper and lower catadioptrical elements. If used for regular flashes, the optical apparatus will not uncommonly be divided into four panels, each comprising a luminous angle of 90 degrees; but the number of panels may be decreased or augmented at will. Thus there may be six panels of 60 degrees or eight of 45 degrees, and so on. Yet it must be borne in mind that with an increase in the number of panels, there is a corresponding decrease in the intensity of the light. The beam of maximum power is attained by a single panel of about 160 degrees, with a lenticular mirror behind the light capable of reflecting to the focus all rays impinging upon it. Apparatus of this concentration calls for very rapid rotation, such as would be incompatible with the old system of revolution on rollers. The required rotation is actually achieved by supporting the column of the lenticular apparatus in a bath of mercury, which very materially reduces the friction of movement.

The biform, triform, and quadriform arrangements of superimposed lights, depending as they do upon the increase in total illumination instead of upon the unit intensity, have more or less ceased to be generally utilised. They are mainly serviceable in misty weather, the separate burners of which they are composed being individually ignitable and extinguishable at will.

**Range of Light.**—The distance penetrable by rays of light varies obviously with the transparency and opacity of the atmosphere. If the medium were a vacuum, the range would be proportional to the square root of the luminous intensity. This relationship, however, cannot be realised

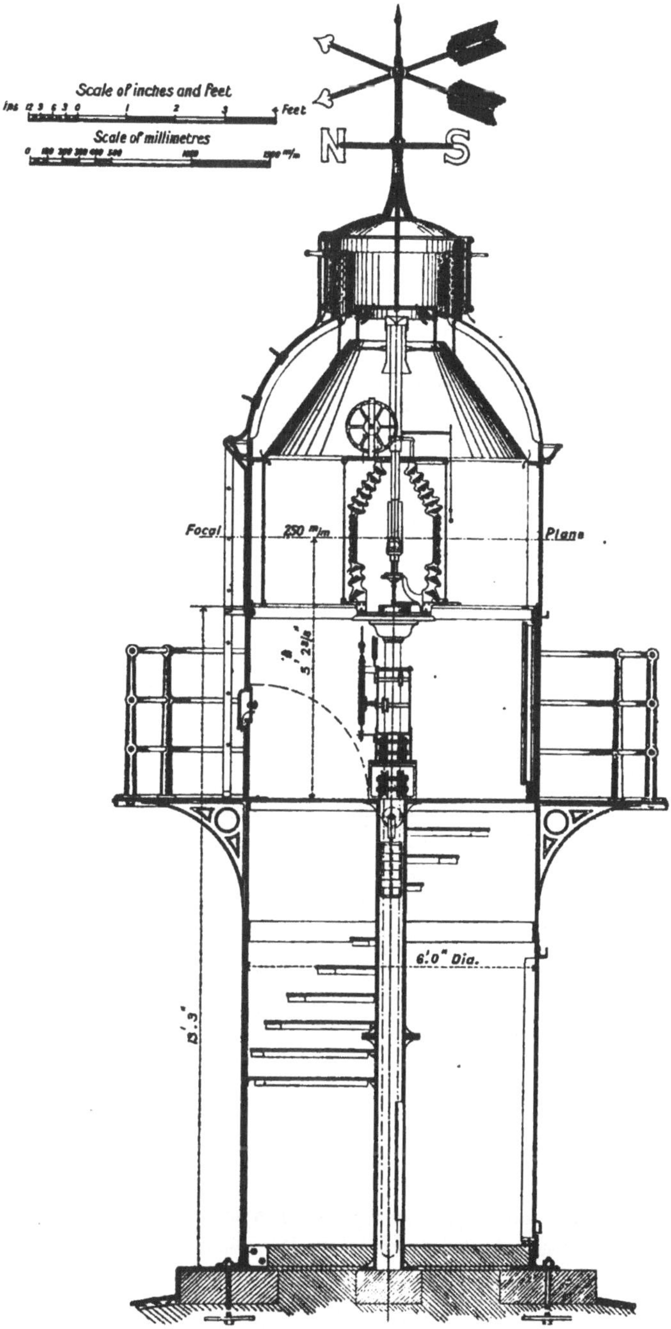


FIG. 242.—Tower Lantern and Fourth Order Occulting Light.

in a medium such as that which envelops the British coasts: often foggy, only occasionally very clear. And as localities vary in their meteorological experiences, no serviceable standard can be devised. The French have a system of denoting the visible range of their lights during two percentages

FIGS 243, 244.—Single Flashing Apparatus of 250 mm. focal distance, consisting of 1 panel of 161½° horizontal angle and dioptric mirror.

of the whole year, viz., 50 per cent. corresponding to clear weather only, and 90 per cent. corresponding to the inclusion of moderately misty and variable weather. It is difficult to see how such a system could be successfully applied to British pharology, where climatic changes are more frequent and much more adverse.

Powerful lights in clear weather may easily exceed their geographical range, *i.e.* the distance at which, owing to the earth's rotundity, they cease to reach the eye of the observer. This distance varies with the respective heights of the light and of the observer, and also with the degree of latitude. The latter affects the radius of curvature (R); but, assuming it to be known as also the levels (H and $h$) of the source of light and of the station of observation, the limiting distance D is given by the formula

$$D = k \{ \sqrt{RH} + \sqrt{Rh} \}$$

where $k$ is a coefficient representing the effect of atmospheric refraction.

**Identification of Light Signals.**—For the purpose of identification, various characteristics are conferred upon lights. Formerly colours were largely relied upon, but the great difference in range of the three chief varieties of light, viz., white, red, and green, militates very much against the efficiency of the method. Red cannot be seen at half the distance penetrable by white light, and green is even less powerful. At a distance of two miles, a white light of 3 candle-power is readily discernible, while from 30 to 40 candle-power would be requisite to bring a red or green light into equal prominence. Moreover, there was not much scope for variation with merely three alternatives.

With the introduction of the **group flashing system**, devised by Dr Hopkinson in 1875, a new and preferable means of identification came into vogue, and its utility has been still further extended by the introduction of **lightning flash-lights** under the inspiration of M. Bourdelles. The first consists of a definite series of illuminations and eclipses, variable in extent and sequence. The latter derives its title from the extreme rapidity of its appearance and disappearance, the period of visibility being the minimum required for imprinting a distinct visual impression. Further exposure is now found to be unnecessary, as it is covered by the persistence of the image on the retina. Light rays, accordingly, instead of being uselessly expended in emphasising their effect on one point, may be deflected to another, with much more serviceable results. The principle, in fact, is that of a highly concentrated beam rotating rapidly and reappearing at intervals of a few seconds. The duration of the flash, though short, is ample for recognition, and its frequent appearance, besides affording greater scope for characterisation, enables the mariner to verify his position with greater assurance than was feasible with an arrangement of slowly moving lights It was not impossible, of course, to provide a considerable number of flashes under the system of rotation on rollers, but it could only be done by increasing the number of lens panels at the expense of luminous power; whereas, by the introduction of the mercury seating, a much greater rotary velocity may be imparted to the apparatus, with less effort and friction. In some cases of older lenticular apparatus, the number of panels was as high as twenty-four. The diffusion of light, therefore, was very great, and the beams suffered correspondingly in intensity and penetration.

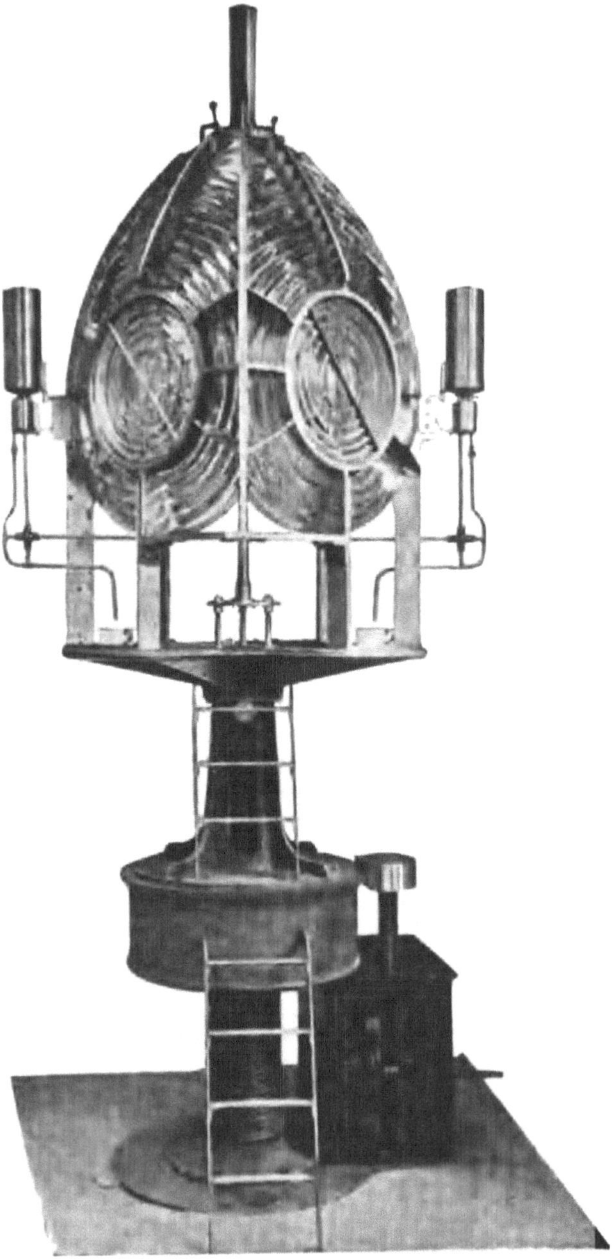

FIG. 245.—Single Flashing Apparatus of 500 mm. focal distance, consisting of 4 panels, each of 90° horizontal angle.

The minimum duration of a visible flash has been determined by laboratory experiments as one-tenth of a second; but under conditions obtaining in marine navigation, the period will be increased to, at least, one-third of a second for white lights and very much more for coloured lights. If the duration were restricted to its exact experimental limit, it is quite conceivable, and even likely, that the occurrence of a flash might escape detection by those on board a vessel rolling in a heavy sea. The flash might also be eclipsed by any intervening vessel.

Flashes are either uniformly regular or arranged in groups of two, three, or more. The interval between successive flashes need not exceed—and, indeed, will be preferably limited to—five seconds, so as to afford frequent opportunities for observation. In grouped flashes, eclipses of 2½ to 4 seconds are common. Certain lighthouses signal a definite number, as, for example, that at Minot's Ledge, U.S.A., which constantly repeats the figures 143 in a series of flashes separated from each other in the same group by an interval of 2 seconds, the groups being separated by an interval of 3 seconds, and the entire signal followed by an eclipse of 15 seconds. The whole period covers 30 seconds, which is the time of a single revolution of the apparatus.

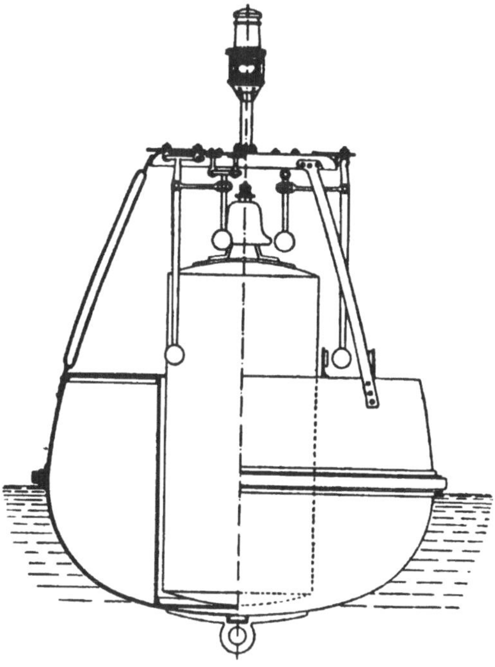

FIG. 246.—Bell-buoy.

**Sound Signals.**—Lighting, while effective enough as a guiding agency in darkness, is practically useless in fog, and reliance has then to be placed upon sound as a warning medium. This is accomplished in various ways, principally by foghorns and syrens on the shore, and by bell-buoys and whistling buoys on the water. Of foghorns, it need simply be said that they consist of a brass trumpet through which a strong blast of compressed air or steam is expelled at definite intervals. They are raucous and unmusical in the extreme. The syren gives out a note of high frequency due to the impulsion of air or steam through a series of holes in a rapidly revolving disc.

Of **bell-buoys** there are several different arrangements. A fixed bell may be struck by pendant clappers, or by a set of balls rolling freely in horizontal grooves or cylinders. When the water is smooth, as is commonly the case in foggy weather, neither of these appliances can be counted upon to emit signals, depending as they do upon the swaying action of waves. In that event, an automatic lever apparatus, worked by the agency of gas, as in the Pintsch system, has been found useful. The gas forces up a diaphragm until it works a lever which closes the inlet valve and opens the outlet, simultaneously actuating

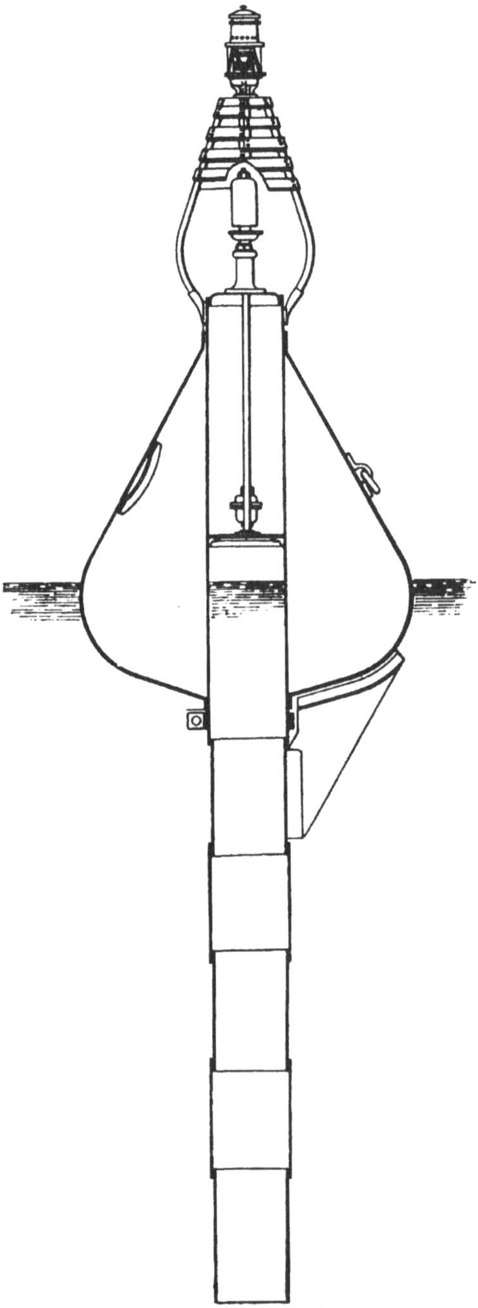

FIG. 247.—Courtenay Whistling Buoy.

another lever, which, in conjunction with a strong spring, impels the hammer against the rim of the bell.

**Whistling buoys** are actuated by the rise and fall of the buoy in a swell. In the Courtenay type, air is drawn into a long central tube during the period of rising. The entrance is controlled by a valve, and when the buoy descends the imprisoned air is expelled through the whistle, emitting a penetrating sound of no particular musical value. Whistling buoys are best adapted to fairly deep water where there is an almost constant swell.

Sound, however, in air is but an imperfect medium for the notification of danger. It gives no reliable indication, and indeed often conveys a very misleading impression as to locality. Zones of silence are found to lie within the sonorous area.[1] Yet, until fogs are dissipable by human agency, it is difficult to see what other means could be universally substituted; and, certainly, it is reliable in so far as it signifies the imminence of danger, though in many cases the exact location of the warning is a matter of conjecture.

**Subaqueous Signals.**—In water, the sense of direction is more determinable, though, even then, with approximation merely. A system of submarine signalling through the agency of a bell struck by a clapper at depths varying from 10 feet to 30 feet below the surface of the water, has recently been promoted by the Boston (U.S.A.) Submarine Signalling Company. The sound is transmitted through the water to a receiver fixed in the ship's bottom, and thence to a megaphone, with results which have been considered very satisfactory so long as the instrument is immersed to depths of not less than 10 feet and preferably of about 25 feet. The distance traversed has reached 8, 10, and even 15 miles. By turning the ship in various directions, the quarter from which the sound emanates can be easily determined as the sound waves only affect the receiver when it faces the direction from which they come. There is manifestly much scope for the development of this principle of sound transmission. The chief difficulty hitherto has been that of ensuring a constant and regular striking of the warning bell without the necessity for human attendance. The discovery of some convenient automatic action which is at once simple and reliable should lead to the general adoption of a system which is much more effective than that of the transmission of sound in air.[2]

[1] In some interesting experiments carried out off the Isle of Wight, fog sound-signals could not be heard at all at 2 miles from the coast, although they were distinctly audible 10 miles out and were heard again at half a mile from land.

[2] At present there are three systems in vogue: viz., (1) Bells suspended from lightships and struck by the agency of compressed air controlled by a code-ringing device in the engine room; (2) Bells supported by buoys and worked by the aid of discs, acting on the principle of a sea anchor, so that the difference in movement between disc and buoy operates delicate mechanism; and (3) Bells supported on tripods resting on the sea floor, and having electrical communication with a shore station.

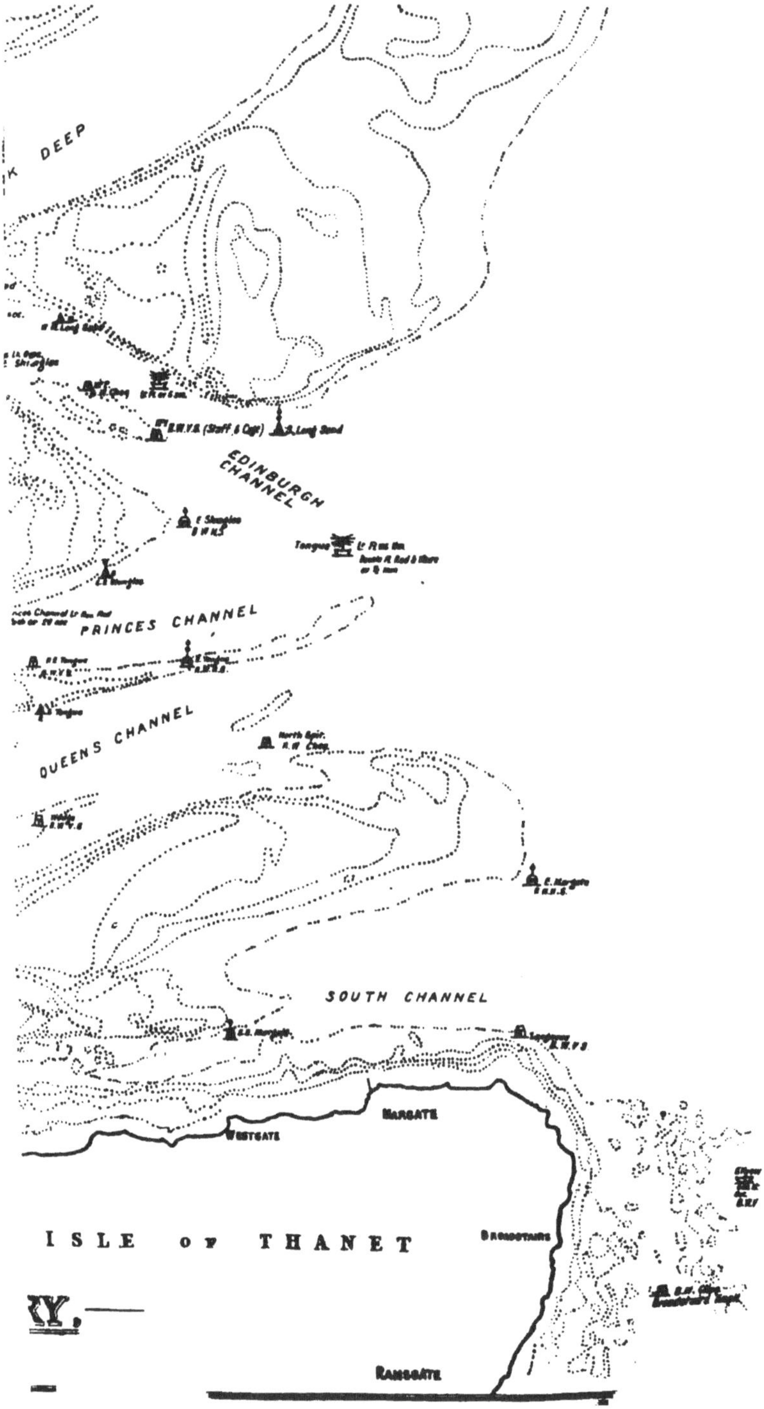
K DEEP
S. Long Sand
EDINBURGH CHANNEL
PRINCES CHANNEL
QUEENS CHANNEL
SOUTH CHANNEL
MARGATE
WESTGATE
ISLE OF THANET
BROADSTAIRS
RAMSGATE

# INDEX.